AutoCAD® 2023 Tutorial
First Level 2D Fundamentals

Randy H. Shih
Oregon Institute of Technology

SDC PUBLICATIONS

SDC Publications
P.O. Box 1334
Mission, KS 66222
913-262-2664
www.SDCpublications.com
Publisher: Stephen Schroff

ISBN-13: 978-1-63057-501-4
ISBN-10: 1-63057-501-1

Printed and bound in the United States of America.

Preface

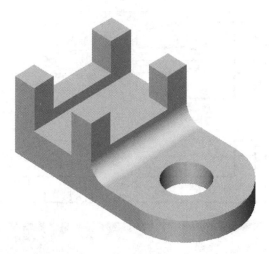

The primary goal of *AutoCAD® 2023 Tutorial First Level 2D Fundamentals* is to introduce the aspects of **Computer Aided Design and Drafting (CADD)**. This text is intended to be used as a training guide for students and professionals. This text covers AutoCAD 2023, and the lessons proceed in a pedagogical fashion to guide you from constructing basic shapes to making multiview drawings. This text takes a hands-on, exercise-intensive approach to all the important 2D CAD techniques and concepts. This textbook contains a series of ten tutorial style lessons designed to introduce beginning CAD users to AutoCAD 2023. This text is also helpful to AutoCAD users upgrading from a previous release of the software. The new improvements and key enhancements of the software are incorporated into the lessons. The 2D-CAD techniques and concepts discussed in this text are also designed to serve as the foundation to the more advanced parametric feature-based CAD packages such as Autodesk® Inventor.

The basic premise of this book is that the more designs you create using AutoCAD 2023, the better you learn the software. With this in mind, each lesson introduces a new set of commands and concepts, building on previous lessons. This book does not attempt to cover all of AutoCAD 2023's features, only to provide an introduction to the software. It is intended to help you establish a good basis for exploring and growing in the exciting field of Computer Aided Engineering.

Acknowledgments

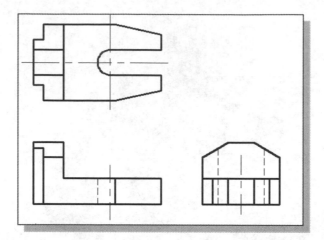

This book would not have been possible without a great deal of support. First, special thanks to two great teachers, Prof. George R. Schade of University of Nebraska-Lincoln and Mr. Denwu Lee from Taiwan, who taught me the fundamentals, the intrigue, and the sheer fun of Computer Aided Engineering.

The effort and support of the editorial and production staff of SDC Publications is gratefully acknowledged. I would especially like to thank Mr. Stephen Schroff for his support and helpful suggestions during this project.

I am grateful that the Mechanical and Manufacturing Engineering Technology Department of Oregon Institute of Technology has provided me with an excellent environment in which to pursue my interests in teaching and research. I would especially like to thank Emeritus Professor Charles Hermach for helpful comments and encouragement.

Finally, truly unbounded thanks are due to my wife Hsiu-Ling and our daughter Casandra for their understanding and encouragement throughout this project.

Randy H. Shih
Klamath Falls, Oregon
Spring, 2022

Notes on using this book to prepare for the AutoCAD Certified User Examination

This edition of the text can also be used as a preparation guide for the AutoCAD Certified User examination. The book's content has been expanded to include a majority of the topics covered in the examination. The reference tables beginning on the next page and located at the beginning of each chapter show the textbook's coverage of Certified User Examination performance tasks. Students taking the certification exam can use the reference tables as both a check list for topics that they need to understand and as a way of finding particular topics. The tables are provided as a reference only. It is important for the user to work through the book chapters in sequential order, as each chapter builds on the skills learned in previous chapters.

PLEASE NOTE:

Every effort has been made to cover the exam objectives included in the AutoCAD Certified User Examination. However, because the format and topics covered by the examination are constantly changing, students planning to take the Certified User Examination are advised to visit the Autodesk website and obtain information regarding the format and details about the AutoCAD Certified User Examination.

Certified User Reference Guide

AutoCAD Certified User Examination Reference Guide

The AutoCAD Certified User examination includes 12 sections. The following tables show where the performance tasks for each section are covered in this book.

This Reference Guide is provided to give you a checklist of performance tasks covered on the Certified User examination and to show you on which page(s) specific tasks are covered.

Certified User Reference Guide

Section 1: Introduction to AutoCAD

Objectives: Describe and set the *Workspace*.

Section 2: Creating Drawings

Objectives: Create and edit geometry using the *Dynamic Input* interface.
Use running *Object Snaps* and object snap overrides to select *Snap* points in the drawing.
Use *Polar Tracking* and *PolarSnap* efficiently and effectively.
Use *Object Tracking* to position geometry.
Use the *Units* command to set drawing units.

Certified User Reference Guide

Section 3: Manipulating Objects

Objectives: Use several different *Selecting Objects* methods to select objects.
Select Objects for grip editing and identify the type of editing that can
be done using grips.
Move objects in the drawing using *Object Snaps*, *Coordinate* entry,
and *Object Snap Tracking* for precise placement.
Use the *Copy* command to copy objects in the drawing.
Use the *Rotate* command to rotate objects in the drawing.
Use the *Mirror* command to mirror objects in the drawing.
Use the *Array* command to pattern objects in the drawing.

Certified User Reference Guide

Section 4: Drawing Organization and Inquiry Commands

Objectives: Use *Layer tools*.

Describe the use and effect of the *ByLayer* property.

Use the Match Properties command to apply the *Properties* from a source object to destination objects.

Use the inquiry commands (Distance, Radius, Angle, Area, List, and ID) to obtain geometric information from the drawing.

Section 5: Altering Objects

Objectives: Use the *Offset* command to create parallel and offset geometry.

Use the *Fillet* command to create radius geometry connecting two objects.

Use the *Join* command to combine multiple objects into a single object.

Certified User Reference Guide

Section 6: Working With Layouts

Objectives: Create a new layout.
           Create and manipulate *Viewports*.

Certification Examination Performance Task	Covered in this book on Chapter – Page

Section 7: Annotating the Drawing

Objectives: Use the *Multiline Text* command to create and format paragraphs of text.
           Create and use *Text Styles*.

Certification Examination Performance Task	Covered in this book on Chapter – Page

Section 8: Dimensioning

Objectives: Create different types of *Dimensions* on linear objects.
           Create and modify *Dimension Styles* to control the appearance of dimensions.
           Create and edit *Multileaders*.

Certification Examination Performance Task	Covered in this book on Chapter – Page

Certified User Reference Guide

Section 9: Hatching Objects

Objectives: Create *Hatch* patterns and fills.

Section 10: Working With Reusable Content

Objectives: Use the *Block* command to create a block definition.
Use the *Insert* command to insert a block reference in a drawing.

Section 11: Creating Additional Drawing Objects

Objectives: Work with *polylines*.
Edit *polylines*.

Section 12: Plotting Your Drawing

Objectives: Create and modify *Page Setup*.

The AutoCAD Certified User Examination is a performance-based exam. The examination is comprised of approximately 30 questions to be completed in 50 minutes. The test items will require you to use the AutoCAD software to perform specific tasks and then answer questions about the tasks. Performance-based testing is defined as **testing by doing**. This means you actually perform the given task then answer the questions regarding the task. Performance-based testing is widely accepted as a better way of ensuring the users have the skills needed, rather than just recalling information.

For detailed information, visit http://www.autodesk.com/certification.

Certified User Reference Guide

Tips on Taking the AutoCAD Certified User Examination

1. **Study**: The first step to maximize your potential on an exam is to sufficiently prepare for it. You need to be familiar with the AutoCAD package, and this can only be achieved by doing drawings and exploring the different commands available. The AutoCAD Certified User exam is designed to measure your familiarity with the AutoCAD software. You must be able to perform the given task and answer the exam questions correctly and quickly.

2. **Make Notes**: Take notes on what you learn either while attending classroom sessions or going through study material. Use these notes as a review guide before taking the actual test.

3. **Time Management**: Manage the time you spend on each question. Always remember you do not need to score 100% to pass the exam. Also, keep in mind that some questions are weighed more heavily and may take more time to answer.

4. **Be Cautious**: Devote some time to ponder and think of the correct answer. Ensure that you interpret all the options correctly before selecting from available choices.

5. **Use Common Sense**: If you are unable to get the correct answer and unable to eliminate all distracters, then you need to select the best answer from the remaining selections. This may be a task of selecting the best answer from amongst several correct answers, or it may be selecting the least incorrect answer from amongst several poor answers.

6. **Take Your Time**: The examination has a time limit. If you encounter a question you cannot answer in a reasonable amount of time, use the Save As feature to save a copy of the data file, and mark the question for review. When you review the question, open your copy of the data file and complete the performance task. After you verify that you have entered the answer correctly, unmark the question so it no longer appears as marked for review.

7. **Don't Act in Haste**: Don't go into panic mode while taking a test. Always read the question carefully before you look out for choices in hand. Use the *Review* screen to ensure you have reviewed all the questions you may have marked for review. When you are confident that you have answered all questions, end the examination to submit your answers for scoring. You will receive a score report once you have submitted your answers.

8. **Relax before exam**: In order to avoid last minute stress, make sure that you arrive 10 to 15 minutes early and relax before taking the exam.

Every effort has been made to cover the exam objectives included in the AutoCAD Certified User Examination. However, the format and topics covered by the examination are constantly changing; students planning to take the Certified User Examination are advised to visit the Autodesk website and obtain information regarding the format and details about the examination.

Table of Contents

Introduction
Getting Started

Chapter 1
AutoCAD Fundamentals

Chapter 2
Basic Object Construction Tools

Chapter 3
Geometric Construction and Editing Tools

Chapter 4
Object Properties and Organization

Chapter 5
Orthographic Views in Multiview Drawings

Chapter 6
AutoCAD 2D Isometric Drawings

Chapter 7
Basic Dimensioning and Notes

Chapter 8
Templates and Plotting

Chapter 9
Parametric Drawing Tools

Chapter 10
Auxiliary Views and Editing with GRIPS

Chapter 11
Section Views

Chapter 12
Assembly Drawings and Blocks

Index

Notes:

Introduction
Getting Started

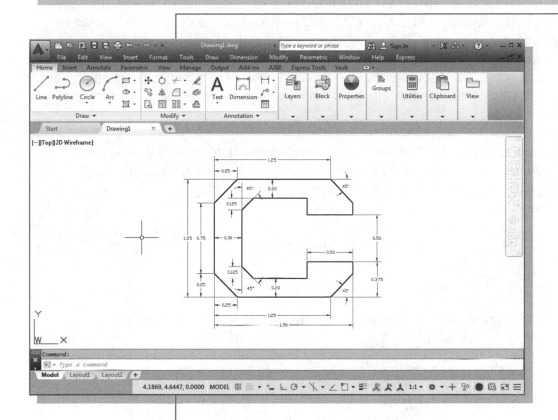

Learning Objectives

♦ **Development of Computer Aided Design**
♦ **Why use AutoCAD 2023?**
♦ **Getting Started with AutoCAD 2023**
♦ **The AutoCAD Startup Dialog Box and Units Setup**
♦ **AutoCAD 2023 Screen Layout**
♦ **Mouse Buttons**

Introduction

Computer Aided Design (CAD) is the process of doing designs with the aid of computers. This includes the generation of computer models, analysis of design data, and the creation of drawings. **AutoCAD 2023** is a computer-aided-design (CAD) software developed by *Autodesk Inc*. The **AutoCAD 2023** software is a tool that can be used for design and drafting activities. The two-dimensional and three-dimensional models created in **AutoCAD 2023** can be transferred to other computer programs for further analysis and testing. The computer models can also be used in manufacturing equipment such as machining centers, lathes, mills, or rapid prototyping machines to manufacture the product.

The rapid changes in the field of **computer aided engineering** (CAE) have brought exciting advances in industry. Recent advances have made the long-sought goal of reducing design time, producing prototypes faster, and achieving higher product quality closer to a reality.

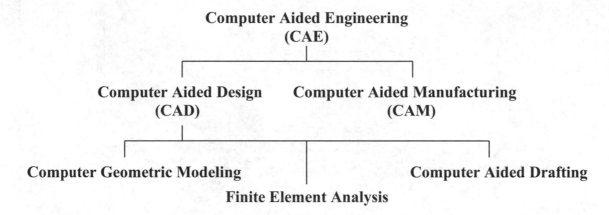

Development of Computer Geometric Modeling

Computer Aided Design is a relatively new technology, and its rapid expansion in the last fifty years is truly amazing. Computer modeling technology advanced along with the development of computer hardware. The first-generation CAD programs, developed in the 1950s, were mostly non-interactive; CAD users were required to create program codes to generate the desired two-dimensional (2D) geometric shapes. Initially, the development of CAD technology occurred mostly in academic research facilities. The Massachusetts Institute of Technology, Carnegie-Mellon University, and Cambridge University were the lead pioneers at that time. The interest in CAD technology spread quickly and several major industry companies, such as General Motors, Lockheed, McDonnell, IBM, and Ford Motor Co., participated in the development of interactive CAD programs in the 1960s. Usage of CAD systems was primarily in the automotive industry, aerospace industry, and government agencies that developed their own programs for their specific needs. The 1960s also marked the beginning of the development of finite element analysis methods for computer stress analysis and computer aided manufacturing for generating machine tool paths.

The 1970s are generally viewed as the years of the most significant progress in the development of computer hardware, namely the invention and development of **microprocessors**. With the improvement in computing power, new types of 3D CAD programs that were user-friendly and interactive became reality. CAD technology quickly expanded from very simple **computer aided drafting** to very complex **computer aided design**. The use of 2D and 3D wireframe modelers was accepted as the leading-edge technology that could increase productivity in industry. The developments of surface modeling and solid modeling technology were taking shape by the late 1970s, but the high cost of computer hardware and programming slowed the development of such technology. During this time period, the available CAD systems all required extremely expensive room-sized mainframe computers.

In the 1980s, improvements in computer hardware brought the power of mainframes to the desktop at less cost and with more accessibility to the general public. By the mid-1980s, CAD technology had become the main focus of a variety of manufacturing industries and was very competitive with traditional design/drafting methods. It was during this period of time that 3D solid modeling technology had major advancements, which boosted the usage of CAE technology in industry.

In the 1990s, CAD programs evolved into powerful design/manufacturing/management tools. CAD technology has come a long way, and during these years of development, modeling schemes progressed from two-dimensional (2D) wireframe to three-dimensional (3D) wireframe, to surface modeling, to solid modeling and, finally, to feature-based parametric solid modeling.

The first-generation CAD packages were simply 2D **Computer Aided Drafting** programs, basically the electronic equivalents of the drafting board. For typical models, the use of this type of program would require that several to many views of the objects be created individually as they would be on the drafting board. The 3D designs remained in the designer's mind, not in the computer database. The mental translation of 3D objects to 2D views is required throughout the use of the packages. Although such systems have some advantages over traditional board drafting, they are still tedious and labor intensive. The need for the development of 3D modelers came quite naturally, given the limitations of the 2D drafting packages.

The development of the 3D wireframe modeler was a major leap in the area of computer modeling. The computer database in the 3D wireframe modeler contains the locations of all the points in space coordinates, and it is sufficient to create just one model rather than multiple models. This single 3D model can then be viewed from any direction as needed. The 3D wireframe modelers require the least computer power and achieve reasonably good representation of 3D models. But because surface definition is not part of a wireframe model, all wireframe images have the inherent problem of ambiguity.

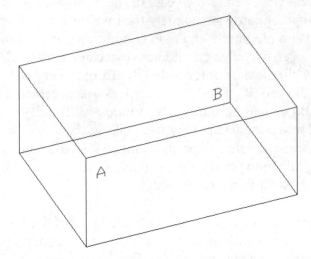

Wireframe Ambiguity: Which corner is in front, A or B?

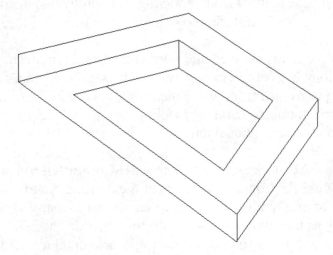

A non-realizable object: Wireframe models contain no surface definitions.

Surface modeling is the logical development in computer geometry modeling to follow the 3D wireframe modeling scheme by organizing and grouping edges that define polygonal surfaces. Surface modeling describes the part's surfaces but not its interiors. Designers are still required to interactively examine surface models to ensure that the various surfaces on a model are contiguous throughout. Many of the concepts used in 3D wireframe and surface modelers are incorporated in the solid modeling scheme, but it is solid modeling that offers the most advantages as a design tool.

In the solid modeling presentation scheme, the solid definitions include nodes, edges, and surfaces, and it is a complete and unambiguous mathematical representation of a precisely enclosed and filled volume. Unlike the surface modeling method, solid modelers start with a solid or use topology rules to guarantee that all of the surfaces are stitched together properly. Two predominant methods for representing solid models are **constructive solid geometry** (CSG) representation and **boundary representation** (B-rep).

The CSG representation method can be defined as the combination of 3D solid primitives. What constitutes a "primitive" varies somewhat with the software but typically includes a rectangular prism, a cylinder, a cone, a wedge, and a sphere. Most solid modelers allow the user to define additional primitives, which can be very complex.

In the B-rep representation method, objects are represented in terms of their spatial boundaries. This method defines the points, edges, and surfaces of a volume, and/or issues commands that sweep or rotate a defined face into a third dimension to form a solid. The object is then made up of the unions of these surfaces that completely and precisely enclose a volume.

By the 1990s, a new paradigm called *concurrent engineering* had emerged. With concurrent engineering, designers, design engineers, analysts, manufacturing engineers, and management engineers all work closely right from the initial stages of the design. In this way, all aspects of the design can be evaluated, and any potential problems can be identified right from the start and throughout the design process. Using the principles of concurrent engineering, a new type of computer modeling technique appeared. The technique is known as the *feature-based parametric modeling technique.* The key advantage of the *feature-based parametric modeling technique* is its capability to produce very flexible designs. Changes can be made easily, and design alternatives can be evaluated with minimum effort. Various software packages offer different approaches to feature-based parametric modeling, yet the end result is a flexible design defined by its design variables and parametric features.

In this text, we will concentrate on creating designs using two-dimensional geometric construction techniques. The fundamental concepts and use of different **AutoCAD 2023** commands are presented using step-by-step tutorials. We will begin with creating simple geometric entities and then move toward creating detailed working drawings and assembly drawings. The techniques presented in this text will also serve as the foundation for entering the world of three-dimensional solid modeling using packages such as **AutoCAD Mechanical Desktop, AutoCAD Architecture** and **Autodesk Inventor**.

Why use AutoCAD 2023?

AutoCAD was first introduced to the public in late 1982 and was one of the first CAD software products that were available for personal computers. Since 1984, **AutoCAD** has established a reputation for being the most widely used PC-based CAD software around the world. By 2017, it was estimated that there were over 6 million **AutoCAD** users in more than 150 countries worldwide. **AutoCAD 2023** is the thirty-seventh release, with many added features and enhancements, of the original **AutoCAD** software produced by *Autodesk Inc.*

CAD provides us with a wide range of benefits; in most cases, the result of using CAD is increased accuracy and productivity. First of all, the computer offers much higher accuracy than the traditional methods of drafting and design. Traditionally, drafting and detailing are the most expensive cost elements in a project and the biggest bottleneck.

With CAD systems, such as **AutoCAD 2023,** the tedious drafting and detailing tasks are simplified through the use of many of the CAD geometric construction tools, such as *grids*, *snap*, *trim,* and *auto-dimensioning*. Dimensions and notes are always legible in CAD drawings, and in most cases, CAD systems can produce higher quality prints compared to traditional hand drawings.

CAD also offers much-needed flexibility in design and drafting. A CAD model generated on a computer consists of numeric data that describe the geometry of the object. This allows the designers and clients to see something tangible and to interpret the ramifications of the design. In many cases, it is also possible to simulate operating conditions on the computer and observe the results. Any kind of geometric shape stored in the database can be easily duplicated. For large and complex designs and drawings, particularly those involving similar shapes and repetitive operations, CAD approaches are very efficient and effective. Because computer designs and models can be altered easily, a multitude of design options can be examined and presented to a client before any construction or manufacturing actually takes place. Making changes to a CAD model is generally much faster than making changes to a traditional hand drawing. Only the affected components of the design need to be modified and the drawings can be plotted again. In addition, the greatest benefit is that, once the CAD model is created, it can be used over and over again. The CAD models can also be transferred into manufacturing equipment such as machining centers, lathes, mills, or rapid prototyping machines to manufacture the product directly.

CAD, however, does not replace every design activity. CAD may help, but it does not replace the designer's experience with geometry and graphical conventions and standards for the specific field. CAD is a powerful tool, but the use of this tool does not guarantee correct results; the designer is still responsible for using good design practice and applying good judgment. CAD will supplement these skills to ensure that the best design is obtained.

CAD designs and drawings are stored in binary form, usually as CAD files, to magnetic devices such as diskettes and hard disks. The information stored in CAD files usually requires much less physical space in comparison to traditional hand drawings. However, the information stored inside the computer is not indestructible. On the contrary, the electronic format of information is very fragile and sensitive to the environment. Heat or cold can damage the information stored on magnetic storage devices. A power failure while you are creating a design could wipe out the many hours you spent working in front of the computer monitor. It is a good habit to save your work periodically, just in case something might go wrong while you are working on your design. In general, one should save one's work onto a storage device at an interval of every 15 to 20 minutes. You should also save your work before you make any major modifications to the design. It is also a good habit to periodically make backup copies of your work and put them in a safe place.

Getting Started with AutoCAD 2023

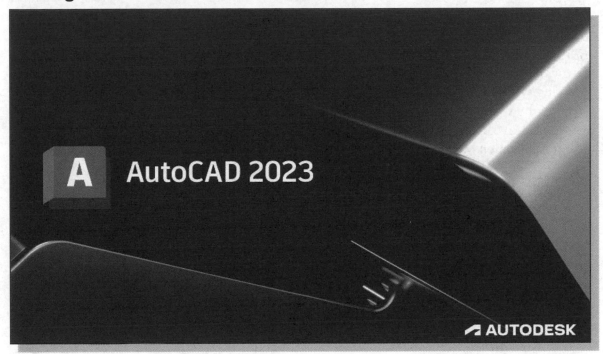

How to start **AutoCAD 2023** depends on the type of workstation and the particular software configuration you are using. With most *Windows* systems, you may select the **AutoCAD 2023** option on the *Start* menu or select the **AutoCAD 2023** icon on the *Desktop*. Consult with your instructor or technical support personnel if you have difficulty starting the software.

The program takes a while to load, so be patient. Eventually the **AutoCAD 2023** main *drawing screen* will appear on the screen. Click **New Drawing** as shown in the below figure. The tutorials in this text are based on the assumption that you are using **AutoCAD 2023**'s default settings. If your system has been customized, some of the settings may not work with the step-by-step instructions in the tutorials. Contact your instructor and technical support to restore the default software configuration.

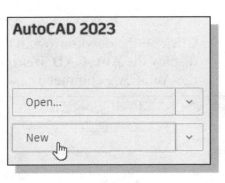

AutoCAD 2023 Screen Layout

The default **AutoCAD 2023** *drawing screen* contains the *pull-down* menus, the *Standard* toolbar, the *InfoCenter Help system,* the *scrollbars,* the *command prompt area*, the *Status Bar*, and the *Ribbon Tabs* and *Panels* that contain several *control panels* such as the *Draw and Modify* panel and the *Annotation* panel. You may resize the **AutoCAD 2023** drawing window by clicking and dragging at the edges of the window, or relocate the window by clicking and dragging at the window title area.

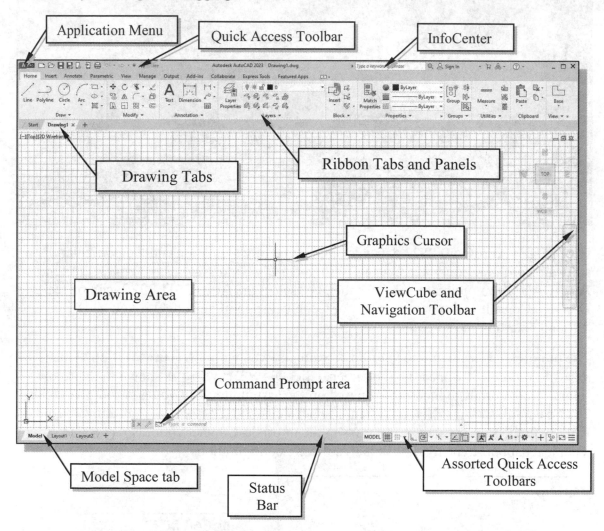

❖ Click on the down-arrow in the *Quick Access* bar and select **Show Menu Bar** to display the **AutoCAD** *Menu* bar. Note that the menu bar provides access to all of the AutoCAD commands.

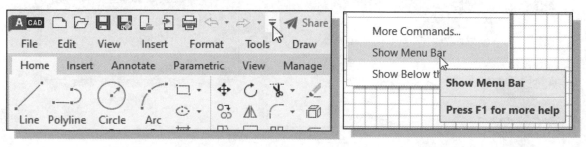

- **Application Menu**

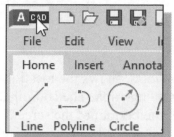

  The *Application Menu* at the top of the main window contains commonly used file operations.

- **Quick Access Toolbar**

  The *Quick Access* toolbar at the top of the *AutoCAD* window allows us quick access to frequently used commands, such as Qnew, Open, Save and also the Undo command. Note that we can customize the quick access toolbar by adding and removing sets of options or individual commands.

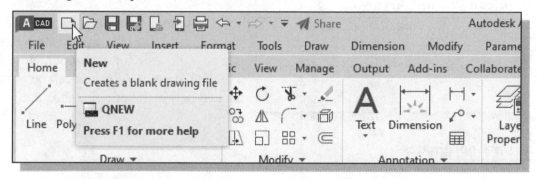

- **AutoCAD Menu Bar**

  The *Menu* bar is the pull-down menu where all operations of AutoCAD can be accessed.

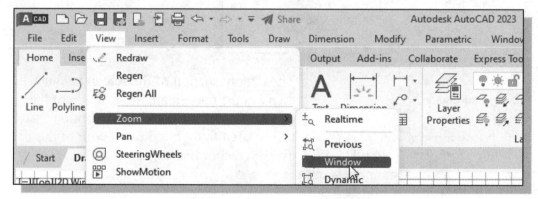

- **Layout tabs**

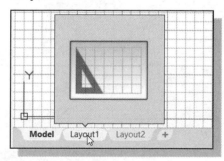

  The Model/Layout tabs allow us to switch/create between different **model space** and **paper space**.

- **Drawing Area**
  The *Drawing Area* is the area where models and drawings are displayed.

- **Graphics Cursor or Crosshairs**

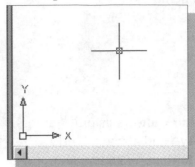

The *graphics cursor*, or *crosshairs*, shows the location of the pointing device in the Drawing Area. The coordinates of the cursor are displayed at the bottom of the screen layout. The cursor's appearance depends on the selected command or option.

- **Command Prompt Area**

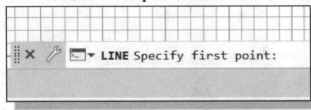

The *Command Prompt Area* provides status information for an operation and it is also the area for data input. Note that the *Command Prompt* can be docked below the drawing area as shown.

- **Cursor Coordinates**

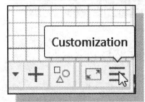

To switch on the **AutoCAD Coordinates Display**, use the *Customization option* at the bottom right corner.

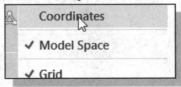

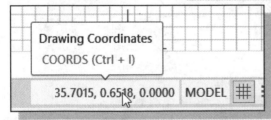

The bottom section of the screen layout displays the coordinate information of the cursor. Note that the quick-key option, [Ctrl+I], can be used to toggle the behavior of the displayed coordinates.

- **Status Toolbar**
  Next to the cursor coordinate display is the *Status* toolbar, showing the status of many commonly used display and construction options.

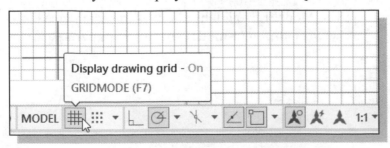

- **Ribbon Tabs and Panels**
  The top section of the screen layout contains customizable icon panels, which contain groups of buttons that allow us to pick commands quickly, without searching through a menu structure. These panels allow us to quickly access the commonly used commands available in AutoCAD.

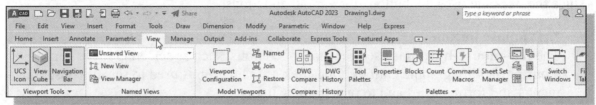

- **Draw and Modify Toolbar Panels**
  The *Draw* and *Modify* toolbar panels are the two main panels for creating drawings; the toolbars contain icons for basic draw and modify commands.

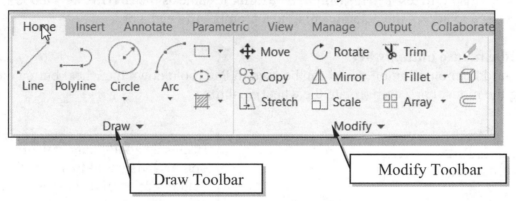

Draw Toolbar

Modify Toolbar

- **Layers Control Toolbar Panel**

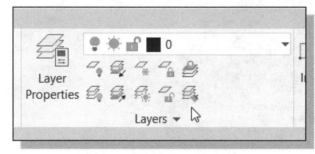

The *Layers Control* toolbar panel contains tools to help manipulate the properties of graphical objects.

- **Viewport/View/Display Controls**

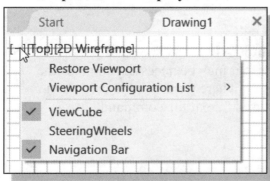

The *Viewport/View/Display controls panel* is located at the upper left corner of the graphics area and it can be used to quickly access viewing related commands, such as Viewport and Display style.

Mouse Buttons

AutoCAD 2023 utilizes the mouse buttons extensively. In learning **AutoCAD 2023**'s interactive environment, it is important to understand the basic functions of the mouse buttons. It is highly recommended that you use a mouse or a tablet with **AutoCAD 2023** since the package uses the buttons for various functions.

- **Left mouse button**
  The **left-mouse-button** is used for most operations, such as selecting menus and icons or picking graphic entities. One click of the button is used to select icons, menus and form entries and to pick graphic items.

- **Right mouse button**
  The **right-mouse-button** is used to bring up additional available options. The software also utilizes the **right-mouse-button** the same as the **ENTER** key and is often used to accept the default setting to a prompt or to end a process.

- **Middle mouse button/wheel**
  The middle mouse button/wheel can be used to Pan (hold down the wheel button and drag the mouse) or Zoom (rotate the wheel) real time.

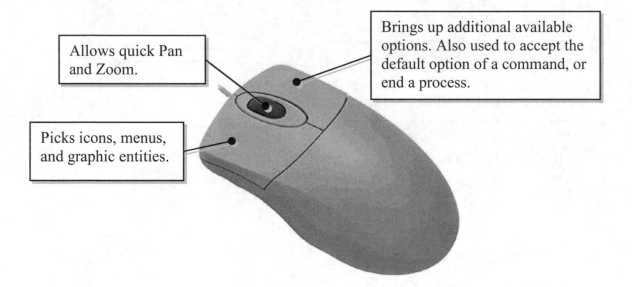

Allows quick Pan and Zoom.

Brings up additional available options. Also used to accept the default option of a command, or end a process.

Picks icons, menus, and graphic entities.

[Esc] – Canceling commands

The [**Esc**] key is used to cancel a command in **AutoCAD 2023**. The [**Esc**] key is located near the top-left corner of the keyboard. Sometimes, it may be necessary to press the [**Esc**] key twice to cancel a command; it depends on where we are in the command sequence. For some commands, the [**Esc**] key is used to exit the command.

Online Help

Several types of online help are available at any time during an **AutoCAD 2023** session. The **AutoCAD 2023** software provides many online help options:

* **Autodesk Exchange**:
  Autodesk Exchange is a central portal in AutoCAD 2023; AutoCAD Exchange provides a user interface for Help, learning aids, tips and tricks, videos, and downloadable apps. By default, *Autodesk Exchange* is displayed at **startup**. This allows access to a dynamic selection of tools from the *Autodesk community*; note that an internet connection is required to use this option.

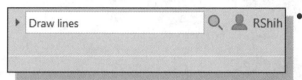

* To use *Autodesk Exchange*, simply type a question in the *input box* to search through the Autodesk's *Help* system as shown.

* A list of the search results appears in the *Autodesk Help* window, and we can also determine the level and type of searches of the associated information.

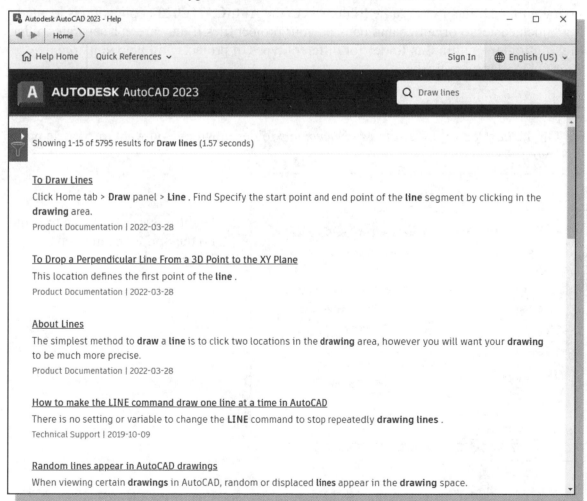

Leaving AutoCAD 2023

To leave **AutoCAD 2023**, use the left-mouse-button and click the **Application Menu** button at the top left corner of the **AutoCAD 2023** screen window, then choose **Exit AutoCAD** from the pull-down menu or type *QUIT* in the command prompt area.

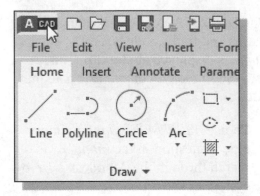
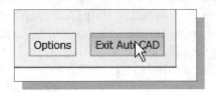

Creating a CAD File Folder

It is a good practice to create a separate folder to store your CAD files. You should not save your CAD files in the same folder where the **AutoCAD 2023** application is located. It is much easier to organize and back up your project files if they are in a separate folder. Making folders within this folder for different types of projects will help you organize your CAD files even further.

➢ To create a new folder in the *Microsoft Windows* environment:

1. On the *desktop* or under the *My Documents* folder in which you want to create a new folder.

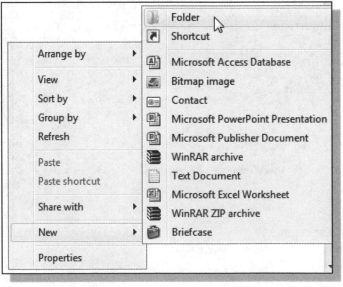

2. Right-mouse-click once to bring up the option menu, then select New→ Folder.

3. Type a name for the new folder, and then press **ENTER**.

Chapter 1
AutoCAD 2D Fundamentals

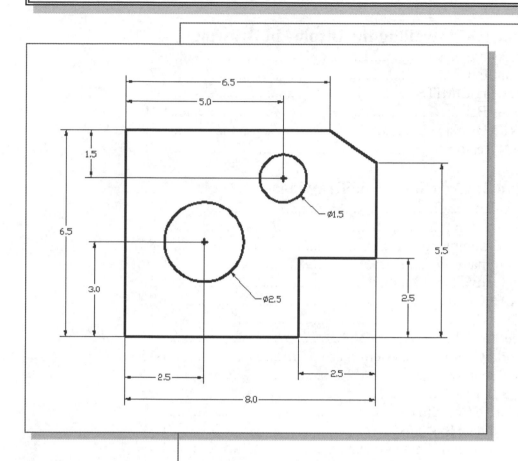

Learning Objectives

- ♦ **Create and Save AutoCAD drawing files**
- ♦ **Use the AutoCAD visual reference commands**
- ♦ **Draw, using the LINE and CIRCLE commands**
- ♦ **Use the ERASE command**
- ♦ **Define Positions using the Basic Entry methods**
- ♦ **Use the AutoCAD Pan Realtime option**

AutoCAD Certified User Examination Objectives Coverage

This table shows the pages on which the objectives of the Certified User Examination are covered in Chapter 1.

Introduction

Learning to use a CAD system is similar to learning a new language. It is necessary to begin with the basic alphabet and learn how to use it correctly and effectively through practice. This will require learning some new concepts and skills as well as learning a different vocabulary. Today, the majority of the Mechanical CAD systems are capable of creating three-dimensional solid models. Nonetheless, all CAD systems create designs using basic geometric entities, and many of the constructions used in technical designs are based upon two-dimensional planar geometry. The method and number of operations that are required to accomplish the basic planar constructions are different from one system to another.

In order to become effective and efficient in using a CAD system, we must learn to create geometric entities quickly and accurately. In learning to use a CAD system, **lines** and **circles** are the first two, and perhaps the most important two, geometric entities that one should master the skills of creating and modifying. Straight lines and circles are used in almost all technical designs. In examining the different types of planar geometric entities, the importance of lines and circles becomes obvious. Triangles and polygons are planar figures bounded by straight lines. Ellipses and splines can be constructed by connecting arcs with different radii. As one gains some experience in creating lines and circles, similar procedures can be applied to create other geometric entities. In this chapter, the different ways of creating lines and circles in **AutoCAD 2023** are examined.

Starting Up AutoCAD 2023

1. Select the **AutoCAD 2023** option on the *Program* menu or select the **AutoCAD 2023** icon on the *Desktop*. Click **New** to start a new drawing.

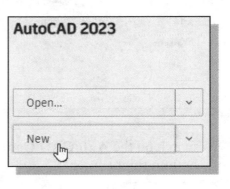

❖ Once the program is loaded into memory, the **AutoCAD 2023** main drawing screen will appear on the screen.

➢ Note that AutoCAD automatically assigns generic names, *Drawing 1*, as new drawings are created. In our example, AutoCAD opened the graphics window using the default system units and assigned the drawing name *Drawing1*.

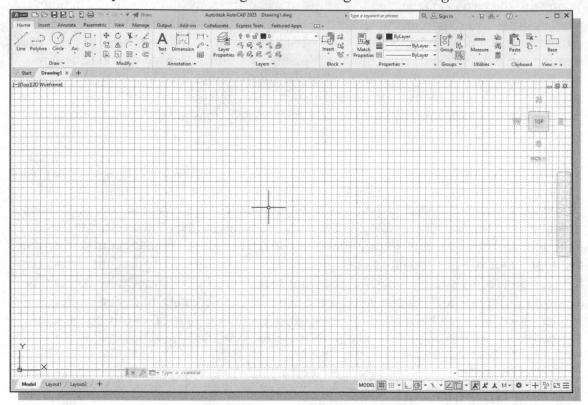

2. If necessary, click on the down-arrow in the *Quick Access* bar and select **Show Menu Bar** to display the **AutoCAD *Menu Bar***. The *Menu Bar* provides access to all AutoCAD commands.

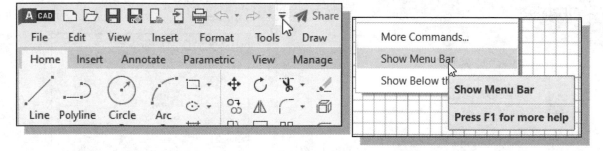

3. To switch on the **AutoCAD Coordinates Display**, use the *Customization option* at the bottom right corner.

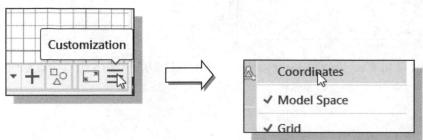

Drawing Units Setup

Every object we construct in a CAD system is measured in **units**. We should determine the system of units within the CAD system before creating the first geometric entities.

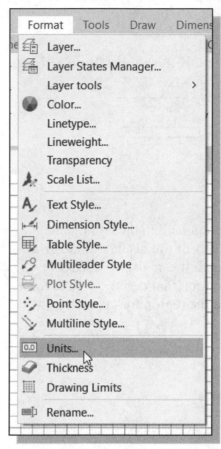

1. In the *Menu Bar* select:
[Format] → [Units]

- The AutoCAD *Menu Bar* contains multiple pull-down menus where all of the AutoCAD commands can be accessed. Note that many of the menu items listed in the pull-down menus can also be accessed through the *Quick Access* toolbar and/or *Ribbon* panels.

2. Click on the *Length Type* option to display the different types of length units available. Confirm the *Length Type* is set to **Decimal**.

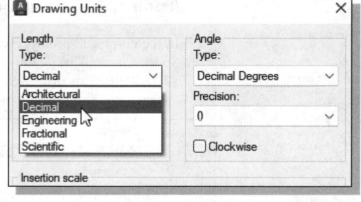

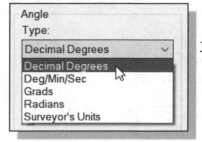

3. On your own, examine the other settings that are available.

- Also note the Insertion Scale section will show the default measurement system, such as the *English* units, inches.

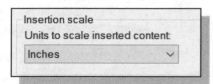

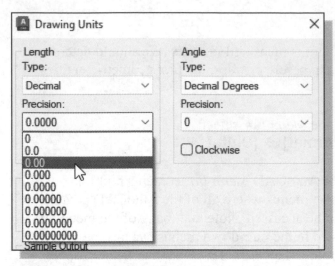

4. Set the *Precision* to **two digits** after the decimal point as shown in the above figure.

5. Pick **OK** to exit the *Drawing Units* dialog box.

Drawing Area Setup

Next, we will set up the **Drawing Limits** by entering a command in the command prompt area. Setting the Drawing Limits controls the extents of the display of the *grid*. It also serves as a visual reference that marks the working area. It can also be used to prevent construction outside the grid limits and as a plot option that defines an area to be plotted and/or printed. Note that this setting does not limit the region for geometry construction.

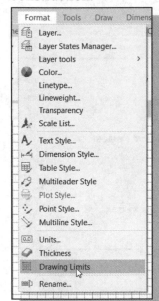

1. In the *Menu Bar* select:
    [Format] → [Drawing Limits]

2. In the command prompt area, the message "*Reset Model Space Limits: Specify lower left corner or [On/Off] <0.00,0.00>:*" is displayed. Press the **ENTER** key once to accept the default coordinates <**0.00,0.00**>.

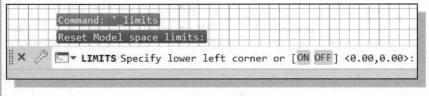

3. In the command prompt area, the message "*Specify upper right corner <12.00,9.00>:*" is displayed. Press the **ENTER** key again to accept the default coordinates <**12.00,9.00**>.

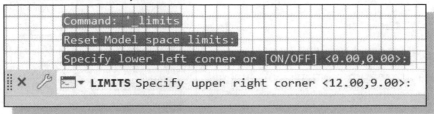

4. On your own, move the graphics cursor near the upper-right corner inside the drawing area and note that the drawing area is unchanged. (The Drawing Limits command is used to set the drawing area, but the display will not be adjusted until a display command is used.)

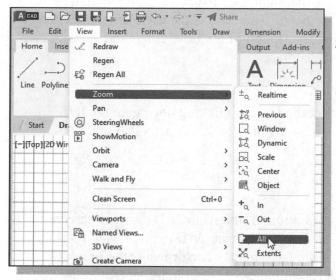

5. Inside the *Menu Bar* area select:
[View] → [Zoom] → [All]

❖ The **Zoom All** command will adjust the display so that all objects in the drawing are displayed to be as large as possible. If no objects are constructed, the Drawing Limits are used to adjust the current viewport.

6. Move the graphics cursor near the upper-right corner inside the drawing area and note that the display area is updated.

7. Hit the function key [**F7**] once to turn **off** the display of the *Grid* lines.

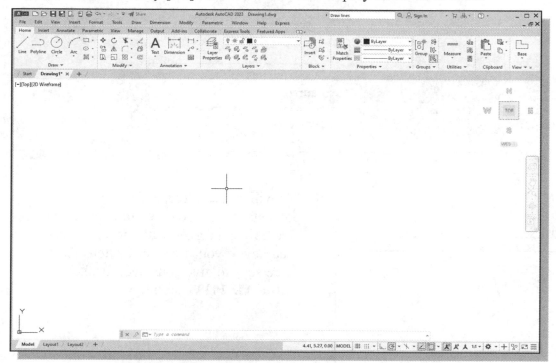

• Note that function key [**F7**] is a quick key, which can be used to quickly toggle on/off the grid display. Also, note the *command prompt area* can be positioned to dock below the drawing area or float inside the drawing area as shown.

Drawing Lines with the Line Command

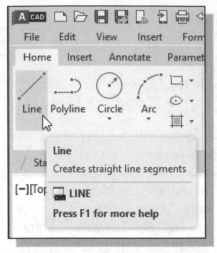

1. Move the graphics cursor to the first icon in the *Draw* panel. This icon is the **Line** icon. Note that a brief description of the Line command appears next to the cursor.

2. Select the icon by clicking once with the **left-mouse-button**, which will activate the Line command.

3. In the command prompt area, near the bottom of the AutoCAD drawing screen, the message "**Line** *Specify first point:*" is displayed. AutoCAD expects us to identify the starting location of a straight line. Move the graphics cursor inside the graphics window and watch the display of the coordinates of the graphics cursor at the bottom of the AutoCAD drawing screen. The three numbers represent the location of the cursor in the X, Y, and Z directions. We can treat the graphics window as if it was a piece of paper, and we are using the graphics cursor as if it was a pencil with which to draw.

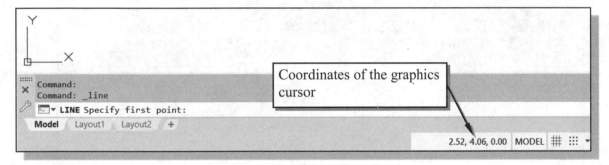

Coordinates of the graphics cursor

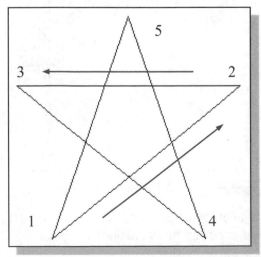

❖ We will create a freehand sketch of a five-point star using the Line command. Do not be overly concerned with the actual size or accuracy of your freehand sketch. This exercise is to give you a feel for the **AutoCAD 2023** user interface.

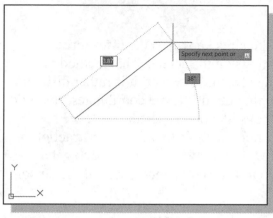

4. We will start at a location about one-third from the bottom of the graphics window. Left-click once to position the starting point of our first line. This will be *point 1* of our sketch. Next, move the cursor upward and toward the right side of *point 1*. Notice the rubber-band line that follows the graphics cursor in the graphics window. Left-click again (*point 2*) and we have created the first line of our sketch.

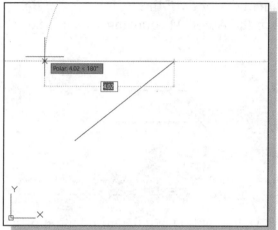

5. Move the cursor to the left of *point 2* and create a horizontal line about the same length as the first line on the screen.

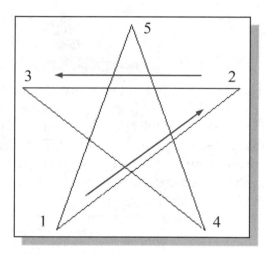

6. Repeat the above steps and complete the freehand sketch by adding three more lines (from *point 3* to *point 4*, *point 4* to *point 5*, and then connect to *point 5* back to *point 1*).

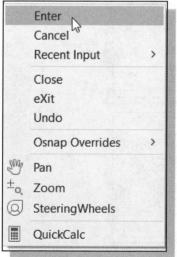

7. Notice that the Line command remains activated even after we connected the last segment of the line to the starting point *(point 1)* of our sketch. Inside the graphics window, **click once** with the **right-mouse-button** and a pop-up menu appears on the screen.

8. Select **Enter** with the left-mouse-button to end the Line command. (This is equivalent to hitting the [**ENTER**] key on the keyboard.)

9. Move the cursor near *point 2* and *point 3*, and estimate the length of the horizontal line by watching the displayed coordinates for each point.

Visual Reference

The method we just used to create the freehand sketch is known as the **interactive method**, where we use the cursor to specify locations on the screen. This method is perhaps the fastest way to specify locations on the screen. However, it is rather difficult to try to create a line of a specific length by watching the displayed coordinates. It would be helpful to know what one inch or one meter looks like on the screen while we are creating entities. **AutoCAD 2023** provides us with many tools to aid the construction of our designs. For example, the *GRID* and *SNAP MODE* options can be used to get a visual reference as to the size of objects and learn to restrict the movement of the cursor to a set increment on the screen.

The *GRID* and *SNAP MODE* options can be turned *ON* or *OFF* through the *Status Bar*. The *Status Bar* area is located at the bottom left of the AutoCAD drawing screen, next to the cursor coordinates.

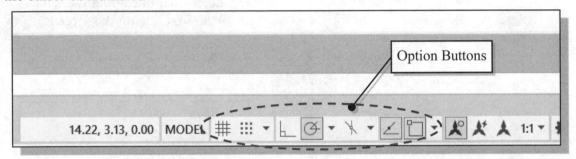

The second button in the *Status Bar* is the *SNAP MODE* option and the third button is the *GRID DISPLAY* option. Note that the buttons in the *Status Bar* area serve two functions: (1) the status of the specific option, and (2) as toggle switches that can be used to turn these special options *ON* and *OFF*. When the corresponding button is *highlighted*, the specific option is turned *ON*. Using the buttons is a quick and easy way to make changes to these *drawing aid* options. The buttons in the *Status Bar* can also be switched on and off in the middle of another command.

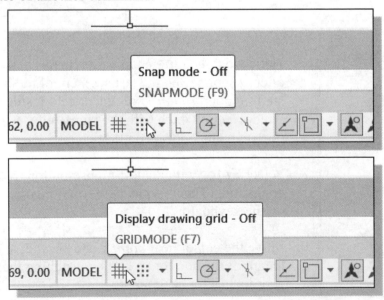

GRID ON

1. Left-click the **GRID** button in the *Status Bar* to turn **ON** the *GRID DISPLAY* option. (Notice in the command prompt area, the message *"<Grid on>"* is also displayed.)

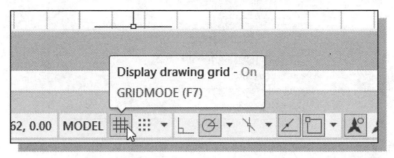

2. Move the cursor inside the graphics window, and estimate the distance in between the grid lines by watching the coordinates displayed at the bottom of the screen.

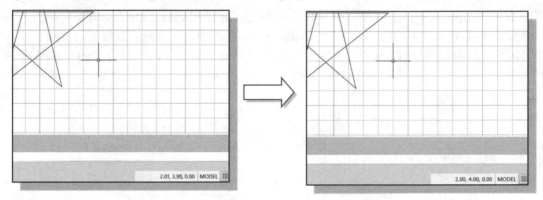

➢ The *GRID* option creates a pattern of lines that extends over an area on the screen. Using the grid is similar to placing a sheet of grid paper under a drawing. The grid helps you align objects and visualize the distance between them. The grid is not displayed in the plotted drawing. The default grid spacing, which means the distance in between two lines on the screen, is 0.5 inches. We can see that the sketched horizontal line in the sketch is about 4 inches long.

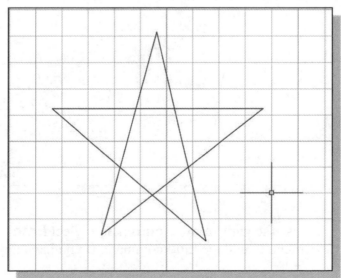

SNAP MODE ON

1. Left-click the **SNAP MODE** button in the *Status Bar* to turn **ON** the *SNAP* option.

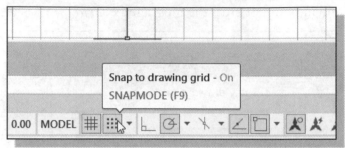

2. Move the cursor inside the graphics window, and move the cursor diagonally on the screen. Observe the movement of the cursor and watch the *coordinates display* at the bottom of the screen.

➤ The *SNAP* option controls an invisible rectangular grid that restricts cursor movement to specified intervals. When *SNAP* mode is on, the screen cursor and all input coordinates are snapped to the nearest point on the grid. The default snap interval is 0.5 inches and aligned to the grid points on the screen.

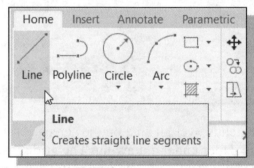

3. Click on the **Line** icon in the *Draw* toolbar. In the command prompt area, the message "*Line Specify first point:*" is displayed.

4. On your own, create another sketch of the five-point star with the *GRID* and *SNAP* options switched *ON*.

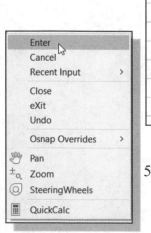

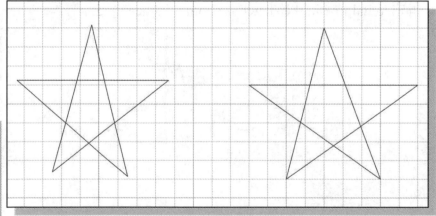

5. Use the **right-mouse-button** and select **Enter** in the pop-up menu to end the Line command if you have not done so.

Using the Erase Command

One of the advantages of using a CAD system is the ability to remove entities without leaving any marks. We will erase two of the lines using the **Erase** command.

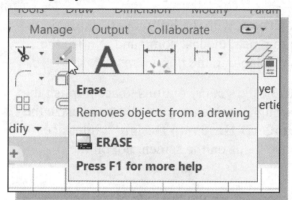

1. Pick **Erase** in the *Modify* toolbar. (The icon is a picture of an eraser at the end of a pencil.) The message "*Select objects*" is displayed in the command prompt area and AutoCAD awaits us to select the objects to erase.

2. Left-click the **SNAP MODE** button on the *Status Bar* to turn **OFF** the *SNAP MODE* option. We can toggle the *Status Bar* options *ON* or *OFF* in the middle of another command.

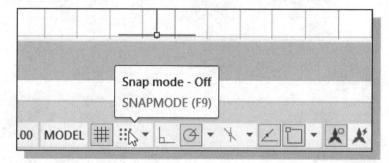

3. Select any two lines on the screen; the selected lines are highlighted as shown in the figure below.

➤ To **deselect** an object from the selection set, hold down the [**SHIFT**] key and select the object again.

4. **Right-mouse-click** once to accept the selections. The selected two lines are erased.

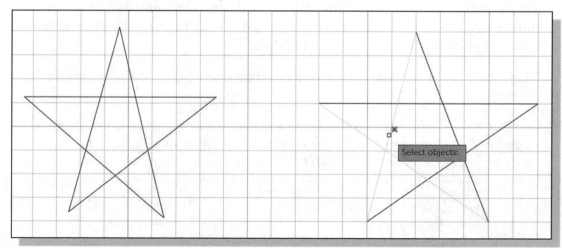

Repeat the Last Command

1. Inside the graphics window, click once with the **right-mouse-button** to bring up the pop-up option menu.

2. Pick **Repeat Erase**, with the left-mouse-button, in the pop-up menu to repeat the last command. Notice the other options available in the pop-up menu.

➤ **AutoCAD 2023** offers many options to accomplish the same task. Throughout this text, we will emphasize the use of the **AutoCAD Heads-up Design™** interface, which means we focus on the screen, not on the keyboard.

3. Move the cursor to a location that is above and toward the left side of the entities on the screen. Left-mouse-click **once** to start a corner of a rubber-band window.

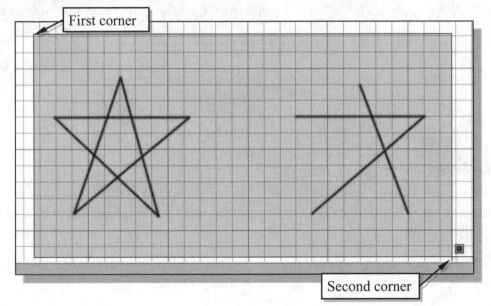

First corner

Second corner

4. Move the cursor toward the right and below the entities, and then click once with the left-mouse-click to enclose all the entities inside the **selection window**. Notice all entities that are inside the window are selected. (Note the *enclosed window selection* direction is from **top left** to **bottom right**.)

5. Inside the graphics window, **right-mouse-click** once to proceed with erasing the selected entities.

➤ On your own, create a free-hand sketch of your choice using the **Line** command. Experiment with using the different commands we have discussed so far. Reset the status buttons so that **only** the *GRID DISPLAY* option is turned *ON* as shown.

The CAD Database and the User Coordinate System

Designs and drawings created in a CAD system are usually defined and stored using sets of points in what is called **world space**. In most CAD systems, the world space is defined using a three-dimensional *Cartesian coordinate system*. Three mutually perpendicular axes, usually referred to as the X-, Y-, and Z-axes, define this system. The intersection of the three coordinate axes forms a point called the **origin**. Any point in world space can then be defined as the distance from the origin in the X-, Y- and Z-directions. In most CAD systems, the directions of the arrows shown on the axes identify the positive sides of the coordinates.

A CAD file, which is the electronic version of the design, contains data that describes the entities created in the CAD system. Information such as the coordinate values in world space for all endpoints, center points, etc., along with the descriptions of the types of entities, are all stored in the file. Knowing that AutoCAD stores designs by keeping coordinate data helps us understand the inputs required to create entities.

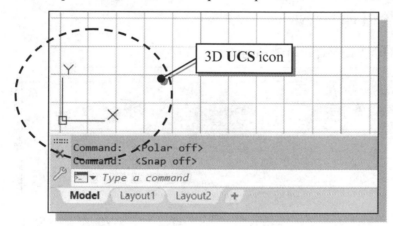

The icon near the bottom left corner of the default AutoCAD graphics window shows the positive X-direction and positive Y-direction of the coordinate system that is active. In AutoCAD, the coordinate system that is used to create entities is called the **user coordinate system** (UCS). By default, the **user coordinate system** is aligned to the **world coordinate system** (WCS). The **world coordinate system** is a coordinate system used by AutoCAD as the basis for defining all objects and other coordinate systems defined by the users. We can think of the **origin** of the **world coordinate system** as a fixed point being used as a reference for all measurements. The default orientation of the Z-axis can be considered as positive values in front of the monitor and negative values inside the monitor.

Changing to the 2D UCS Icon Display

In **AutoCAD 2023**, the **UCS** icon is displayed in various ways to help us visualize the orientation of the drawing plane.

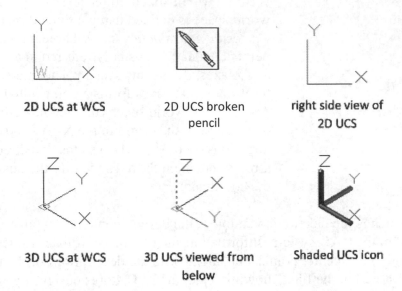

2D UCS at WCS   **2D UCS broken pencil**   **right side view of 2D UCS**

3D UCS at WCS   **3D UCS viewed from below**   **Shaded UCS icon**

1. Click on the **[View]** pull-down menu and select

    [Display] → [UCS Icon] → [Properties…]

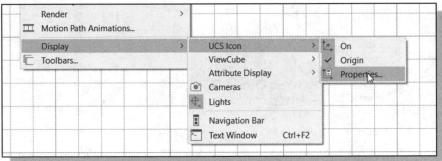

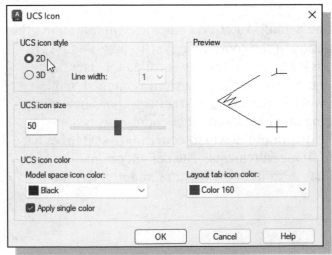

2. In the *UCS icon style* section, switch to the **2D** option as shown.

3. Click **OK** to accept the settings.

❖   Note the W symbol in the UCS icon indicates that the UCS is aligned to the **world coordinate system**.

Cartesian and Polar Coordinate Systems

In a two-dimensional space, a point can be represented using different coordinate systems. The point can be located, using a *Cartesian coordinate system*, as X and Y units away from the origin. The same point can also be located using the *polar coordinate system*, as r and θ units away from the origin.

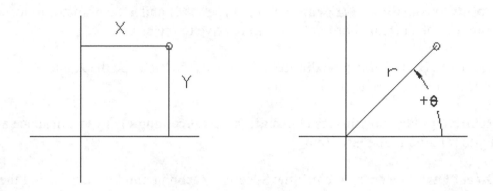

For planar geometry, the polar coordinate system is very useful for certain applications. In the polar coordinate system, points are defined in terms of a radial distance, r, from the origin and an angle θ between the direction of r and the positive X axis. The default system for measuring angles in **AutoCAD 2023** defines positive angular values as counterclockwise from the positive X-axis.

Absolute and Relative Coordinates

AutoCAD 2023 also allows us to use *absolute* and *relative coordinates* to quickly construct objects. **Absolute coordinate values** are measured from the current coordinate system's origin point. **Relative coordinate values** are specified in relation to previous coordinates.

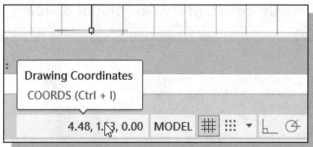

➤ Note that the *coordinate display area* can also be used as a toggle switch; each left-mouse-click will toggle the coordinate display *on* or *off*.

In **AutoCAD 2023**, the *absolute* coordinates and the *relative* coordinates can be used in conjunction with the *Cartesian* and *polar* coordinate systems. By default, AutoCAD expects us to enter values in *absolute Cartesian coordinates*, distances measured from the current coordinate system's origin point. We can switch to using the *relative coordinates* by using the @ symbol. The @ symbol is used as the *relative coordinates specifier*, which means that we can specify the position of a point in relation to the previous point.

Defining Positions

In AutoCAD, there are five methods for specifying the locations of points when we create planar geometric entities.

➢ **Interactive method**: Use the cursor to select on the screen.

➢ **Absolute coordinates (Format: X,Y)**: Type the X and Y coordinates to locate the point on the current coordinate system relative to the origin.

➢ **Relative rectangular coordinates (Format: @X,Y)**: Type the X and Y coordinates relative to the last point.

➢ **Relative polar coordinates (Format: @Distance<angle)**: Type a distance and angle relative to the last point.

➢ **Direct Distance entry technique**: Specify a second point by first moving the cursor to indicate direction and then entering a distance.

GRID Style Setup

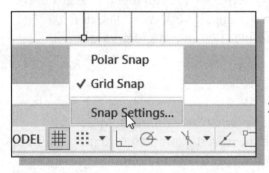

1. In the *Status Bar* area, **right-mouse-click** on *Snap Mode* and choose **[Snap settings]**.

2. In the *Drafting Settings* dialog box, select the **Snap and Grid** tab if it is not the page on top.

3. Change *Grid Style* to **Display dotted grid in 2D model Space** as shown in the below figure.

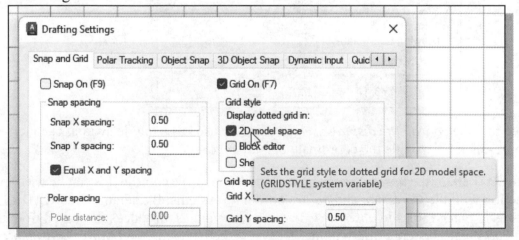

4. Pick **OK** to exit the *Drafting Settings* dialog box.

The Guide Plate

We will next create a mechanical design using the different coordinate entry methods.

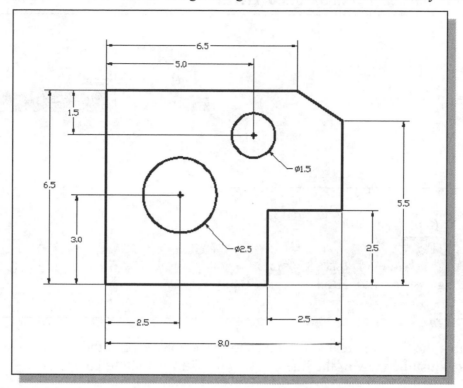

❖ The rule for creating CAD designs and drawings is that they should be created at **full size** using real-world units. The CAD database contains all the definitions of the geometric entities and the design is considered as a virtual, full-sized object. Only when a printer or plotter transfers the CAD design to paper is the design scaled to fit on a sheet. The tedious task of determining a scale factor so that the design will fit on a sheet of paper is taken care of by the CAD system. This allows the designers and CAD operators to concentrate their attention on the more important issues – the design.

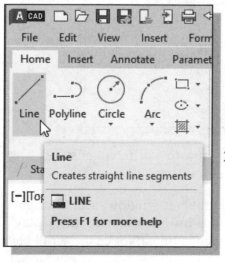

1. Select the **Line** command icon in the *Draw* toolbar. In the command prompt area, near the bottom of the AutoCAD graphics window, the message "*Line Specify first point:*" is displayed. AutoCAD expects us to identify the starting location of a straight line.

2. We will locate the starting point of our design at the origin of the *world coordinate system*.

   Command: Line Specify first point: **0,0**
   (Type **0,0** and press the [**ENTER**] key once.)

3. We will create a horizontal line by entering the absolute coordinates of the second point.
    Specify next point or [Undo]: **5.5,0 [ENTER]**

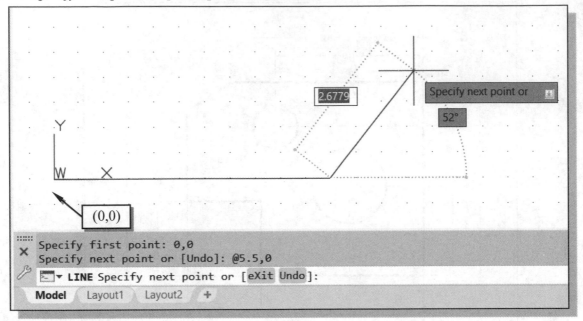

*   Note that the line we created is aligned to the bottom edge of the drawing window. Let us adjust the view of the line by using the **Pan Realtime** command.

4. In the *Menu Bar* area select: **[View] → [Pan] → [Realtime]**

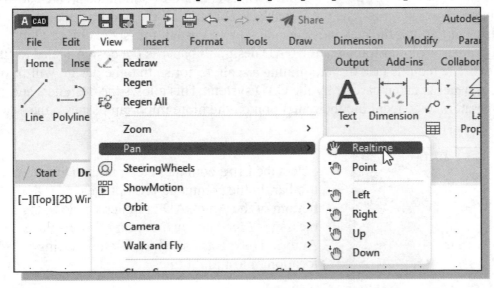

❖   The available **Pan** commands enable us to move the view to a different position. The *Pan-Realtime* function acts as if you are using a video camera.

5. Move the cursor, which appears as a hand inside the graphics window, near the center of the drawing window, then push down the **left-mouse-button** and drag the display toward the right and top side until we can see the sketched line. (Notice the scroll bars can also be used to adjust viewing of the display.)

6. Press the **[Esc]** key to exit the *Pan-Realtime* command. Notice that AutoCAD goes back to the **Line** command.

7. We will create a vertical line by using the *relative rectangular coordinates entry method*, relative to the last point we specified:
    Specify next point or [Close/Undo]: **@0,2.5** **[ENTER]**

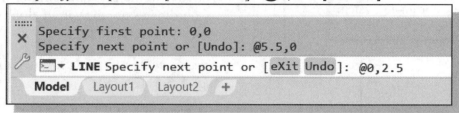

8. Left-click once on the *coordinates display area* to switch to a different coordinate display option. Click again to see the other option. Note the coordinates display has changed to show the length of the new line and its angle. Each click will change the display format of the cursor coordinates.

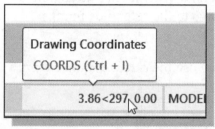

9. On your own, left-click on the coordinates display area to observe the switching of the coordinate display; set the display back to using the world coordinate system.

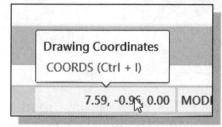

10. We can mix any of the entry methods in positioning the locations of the endpoints. Move the cursor to the *Status Bar* area, and turn **ON** the *SNAP MODE* option.

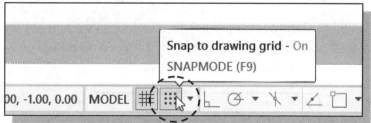

❖ Note that the **Line** command is resumed as the settings are adjusted.

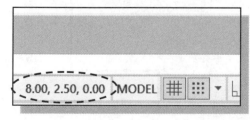

11. Create the next line by picking the location, world coordinates **(8,2.5)**, on the screen.

12. We will next use the *relative polar coordinates entry method*; distance is **3** inches with an angle of **90** degrees, relative to the last point we specified:
    Specify next point or [Close/Undo]: **@3<90** **[ENTER]**

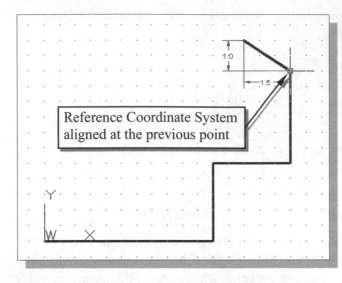

Reference Coordinate System aligned at the previous point

13. Using the *relative rectangular coordinates entry method* to create the next line, we can imagine a *reference coordinate system* aligned at the previous point. Coordinates are measured along the two reference axes.

Specify next point or [Close/Undo]:
@-1.5,1 [ENTER]

(**-1.5** and **1** inches are measured relative to the reference point.)

14. Move the cursor directly to the left of the last point and use the *direct distance entry technique* by entering **6.5 [ENTER]**.

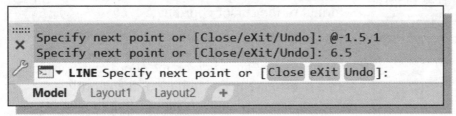

```
Specify next point or [Close/eXit/Undo]: @-1.5,1
Specify next point or [Close/eXit/Undo]: 6.5
LINE Specify next point or [Close eXit Undo]:
Model    Layout1    Layout2    +
```

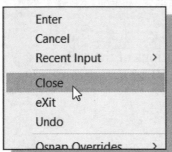

Enter
Cancel
Recent Input   >
Close
eXit
Undo
Osnap Overrides   >

15. For the last segment of the sketch, we can use the **Close** option to connect back to the starting point. Inside the graphics window, **right-mouse-click** and a *pop-up menu* appears on the screen.

16. Select **Close** with the left-mouse-button to connect back to the starting point and end the Line command.

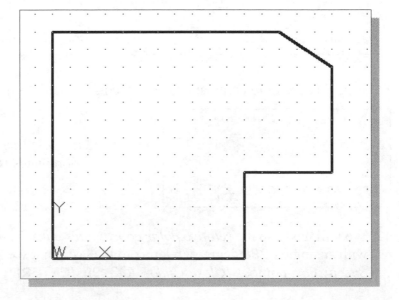

Creating Circles

The menus and toolbars in **AutoCAD 2023** are designed to allow the CAD operator to quickly activate the desired commands.

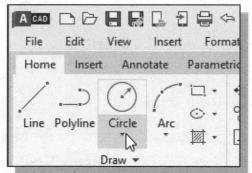

1. In the *Draw* toolbar, click on the little triangle below the circle icon. Note that the little triangle indicates additional options are available.

2. In the option list, select: **[Center, Diameter]**

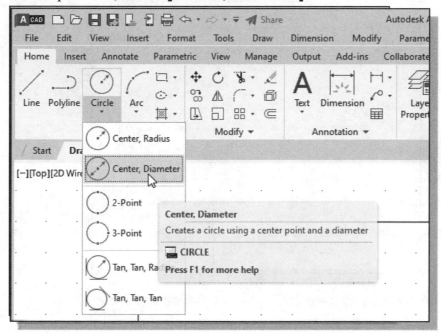

Notice the different options available under the circle submenu:

- **Center, Radius**: Draws a circle based on a center point and a radius.

- **Center, Diameter**: Draws a circle based on a center point and a diameter.

- **2 Points**: Draws a circle based on two endpoints of the diameter.

- **3 Points**: Draws a circle based on three points on the circumference.

- **TTR–Tangent, Tangent**, **Radius**: Draws a circle with a specified radius tangent to two objects.

- **TTT–Tangent, Tangent, Tangent**: Draws a circle tangent to three objects.

3. In the command prompt area, the message "*Specify center point for circle or [3P/2P/Ttr (tan tan radius)]:*" is displayed. AutoCAD expects us to identify the location of a point or enter an option. We can use any of the four coordinate entry methods to identify the desired location. We will enter the **world coordinates (2.5,3)** as the center point for the first circle.

    Specify center point for circle or [3P/2P/Ttr (tan tan radius)]: **2.5,3 [ENTER]**

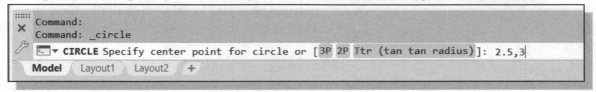

4. In the command prompt area, the message "*Specify diameter of circle:*" is displayed.
    Specify diameter of circle: **2.5 [ENTER]**

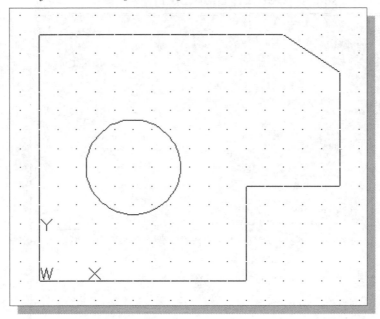

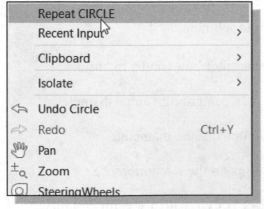

5. Inside the graphics window, right-mouse-click to bring up the pop-up option menu.

6. Pick **Repeat CIRCLE** with the left-mouse-button in the pop-up menu to repeat the last command.

7. Using the *relative rectangular coordinates entry method*, relative to the center-point coordinates of the first circle, we specify the relative location as (**@2.5,2**).

    Specify center point for circle or [3P/2P/Ttr (tan tan radius)]: **@2.5,2 [ENTER]**

8. In the command prompt area, the message "*Specify Radius of circle: <2.50>*" is displayed. The default option for the Circle command in AutoCAD is to specify the *radius* and the last radius used is also displayed in brackets.

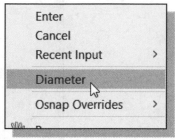

9. Inside the graphics window, **right-mouse-click** to bring up the pop-up option menu and select **Diameter** as shown.

10. In the command prompt area, enter **1.5** as the diameter.

   Specify Diameter of circle<2.50>: **1.5 [ENTER]**

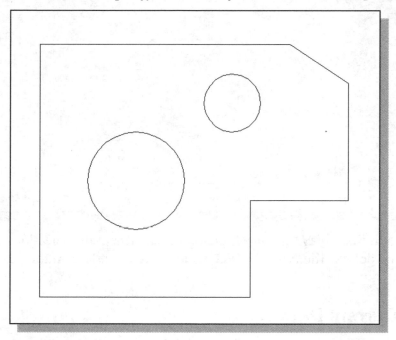

Save the CAD Design

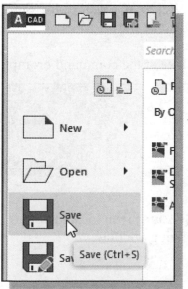

1. In the *Application Menu*, select:

   [Application] → [Save]

❖ Note the command can also be activated with the quick-key combination of **[Ctrl]+[S]**.

2. In the *Save Drawing As* dialog box, select the folder in which you want to store the CAD file and enter **GuidePlate** in the *File name* box.

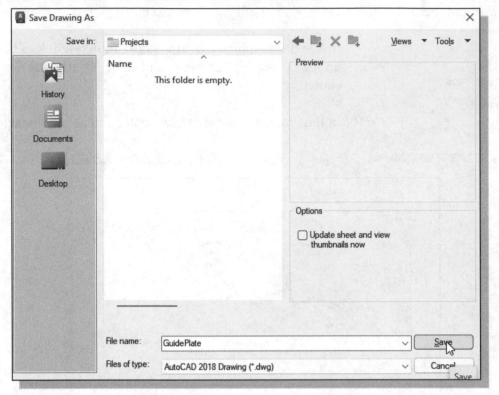

3. Click **Save** in the *Save Drawing As* dialog box to accept the selections and save the file. Note the default file type is DWG, which is the standard AutoCAD drawing format.

Close the Current Drawing

Several options are available to close the current drawing:

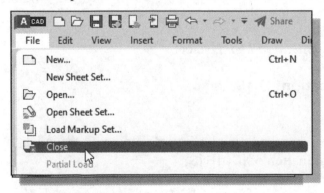

➢ Select **[File]** ➔ **[Close]** in the *Pull-down Menu* as shown.

➢ Enter **Close** at the command prompt.

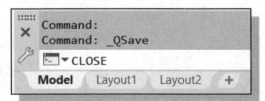

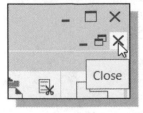

➢ Another option is to click on the **[Close]** icon located at the upper-right-hand corner of the drawing window.

The Spacer Design

We will next create the spacer design using more of AutoCAD's drawing tools.

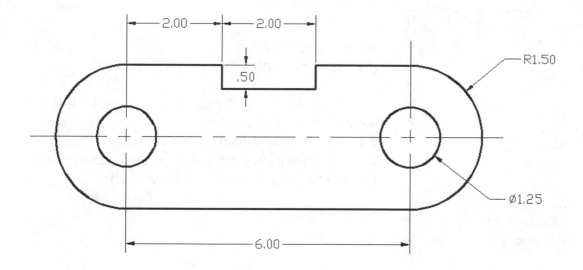

Start a New Drawing

1. In the *Application Menu*, select **[New]** to start a new drawing.

2. The **Select Template** dialog box appears on the screen. Click **Open** to accept the default **acad.dwt** as the template to open.

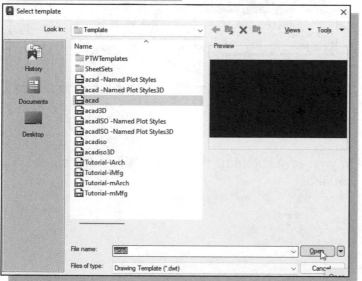

> The dwt file type is the AutoCAD template file format. An AutoCAD template file contains pre-defined settings to reduce the amount of tedious repetitions.

Drawing Units Setup

Every object we construct in a CAD system is measured in **units**. We should determine the system of units within the CAD system before creating the first geometric entities.

1. In the *Menu Bar* select:
   [Format] → [Units]

- The AutoCAD *Menu Bar* contains multiple pull-down menus where all of the AutoCAD commands can be accessed. Note that many of the menu items listed in the pull-down menus can also be accessed through the *Quick Access* toolbar and/or *Ribbon* panels.

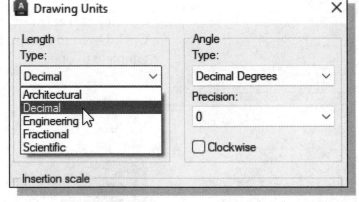

2. Click on the *Length Type* option to display the different types of length units available. Confirm the *Length Type* is set to **Decimal**.

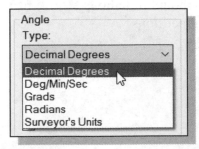

3. On your own, examine the other settings that are available.

4. In the *Drawing Units* dialog box, set the *Length Type* to **Decimal**. This will set the measurement to the default *English* units, inches.

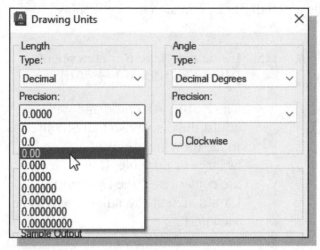

5. Set the *Precision* to **two digits** after the decimal point as shown in the above figure.

6. Pick **OK** to exit the *Drawing Units* dialog box.

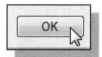

Drawing Area Setup

Next, we will set up the **Drawing Limits** by entering a command in the command prompt area. Setting the Drawing Limits controls the extents of the display of the *grid*. It also serves as a visual reference that marks the working area. It can also be used to prevent construction outside the grid limits and as a plot option that defines an area to be plotted/printed. Note that this setting does not limit the region for geometry construction.

1. In the *Menu Bar* select:
    [Format] → [Drawing Limits]

2. In the command prompt area, the message "*Reset Model Space Limits: Specify lower left corner or [On/Off] <0.00,0.00>:*" is displayed. Press the **ENTER** key once to accept the default coordinates **<0.00,0.00>**.

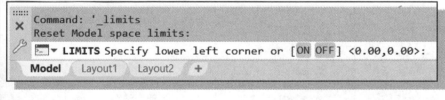

3. In the command prompt area, the message "*Specify upper right corner <12.00,9.00>:*" is displayed. Press the **ENTER** key again to accept the default coordinates **<12.00,9.00>**.

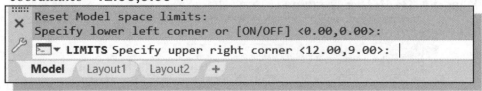

4. On your own, move the graphics cursor near the upper-right corner inside the drawing area and note that the drawing area is unchanged. (The Drawing Limits command is used to set the drawing area, but the display will not be adjusted until a display command is used.)

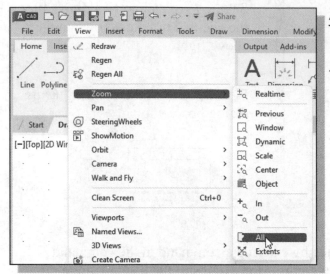

5. Inside the *Menu Bar* area select:
    [View] → [Zoom] → [All]

❖ The **Zoom All** command will adjust the display so that all objects in the drawing are displayed to be as large as possible. If no objects are constructed, the Drawing Limits are used to adjust the current viewport.

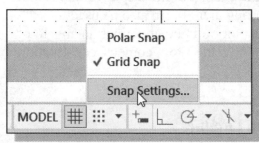

6. Move the graphics cursor near the upper-right corner inside the drawing area, and note that the display area is updated.

7. In the *Status Bar* area, **right-mouse-click** on *SnapMode* and choose **[Snap Settings]**.

8. In the *Drafting Settings dialog box*, switch **on** the **Snap** and **Grid** options as shown.

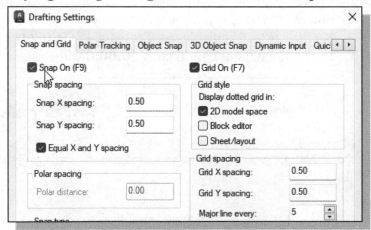

➤ On your own, exit the *Drafting Settings dialog box* and confirm that only *GRID DISPLAY* and *SNAP MODE* are turned *ON* as shown.

Use the Line Command

1. Select the **Line** command icon in the *Draw* toolbar. In the command prompt area, near the bottom of the AutoCAD graphics window, the message "*Line Specify first point:*" is displayed. AutoCAD expects us to identify the starting location of a straight line.

2. To further illustrate the usage of the different input methods and tools available in AutoCAD, we will **start the line segments at an arbitrary location**. Start at a location that is somewhere in the lower left side of the graphics window.

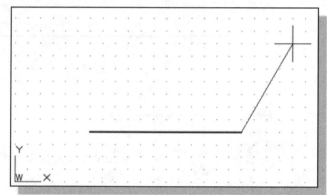

3. We will create a horizontal line by using the *relative rectangular coordinates entry method*, relative to the last point we specified:
    @6,0 [ENTER]

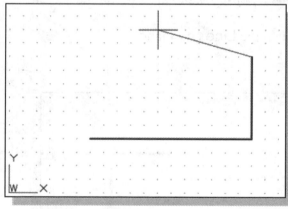

4. Next, create a vertical line by using the *relative polar coordinates entry method*, relative to the last point we specified: **@3<90 [ENTER]**

5. Next, we will use the direct input method. First, move the cursor directly to the left of the last endpoint of the line segments.

6. On your own, turn the mouse wheel to zoom in and drag with the middle mouse to reposition the display.

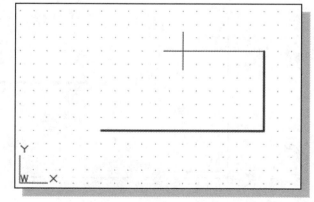

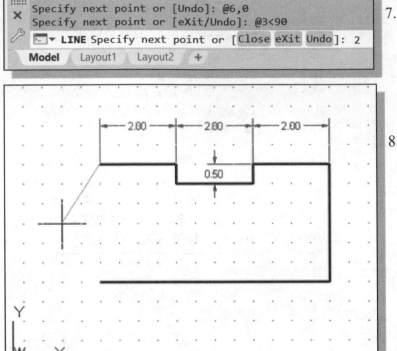

```
Specify next point or [Undo]: @6,0
Specify next point or [eXit/Undo]: @3<90
⌐▼ LINE Specify next point or [Close eXit Undo]: 2
Model   Layout1   Layout2   +
```

7. Use the *direct distance entry technique* by entering **2** [ENTER].

8. On your own, repeat the above steps and create the four additional line segments, using the dimensions as shown.

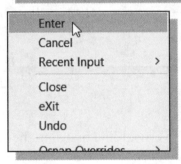

```
Enter
Cancel
Recent Input        >
Close
eXit
Undo
Osnap Overrides     >
```

-   To end the line command, we can either hit the [Enter] key on the keyboard or use the **Enter** option, **right-mouse-click** and a *pop-up menu* appears on the screen.

9. On your own, end the Line command.

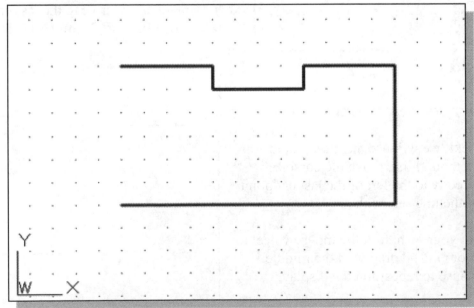

Use the Erase Command

The vertical line on the right was created as a construction line to aid the construction of the rest of the lines for the design. We will use the Erase command to remove it.

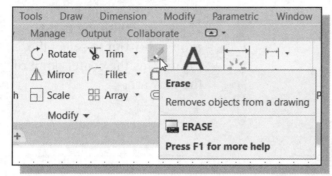

1. Pick **Erase** in the *Modify* toolbar. The message "*Select objects*" is displayed in the command prompt area and AutoCAD awaits us to select the objects to erase.

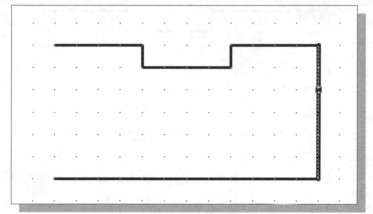

2. Select the vertical line as shown.

3. Click once with the **right-mouse-button** to accept the selection and delete the line.

Using the Arc Command

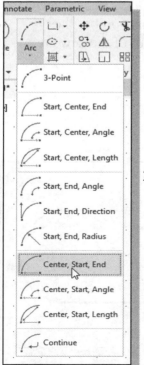

1. Click the down-arrow icon of the **Arc** command in the *Draw* toolbar to display the different Arc construction options.

➢ AutoCAD provides eleven different ways to create arcs. Note that the different options are used based on the geometry conditions of the design. The more commonly used options are the **3-Points** option and the **Center-Start-End** option.

2. Select the **Center-Start-End** option as shown. This option requires the selection of the center point, start point and end point location, in that order, of the arc.

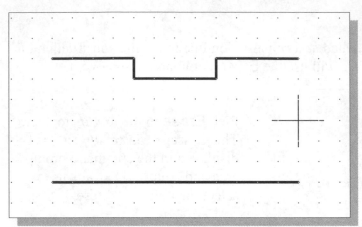

3. Move the cursor to the middle of the two horizontal lines and align the cursor to the two endpoints as shown. Click once with the **right-mouse-button** to select the location as the center point of the new arc.

4. Move the cursor downward and select the right endpoint of the bottom horizontal line as the start point of the arc.

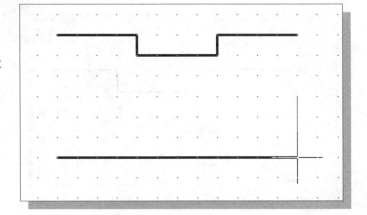

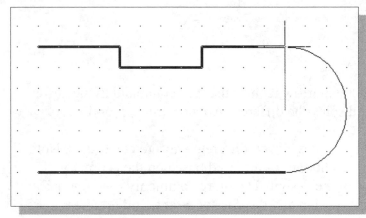

5. Move the cursor to the right endpoint of the top horizontal line as shown. Pick this point as the endpoint of the new arc.

6. On your own, repeat the above steps and create the other arc as shown. Note that in most CAD packages, positive angles are defined as going counterclockwise; therefore, the starting point of the second arc should be at the endpoint on top.

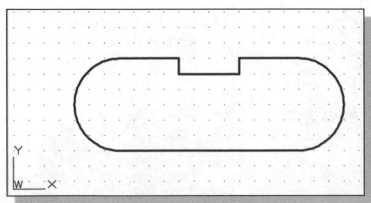

Using the Circle Command

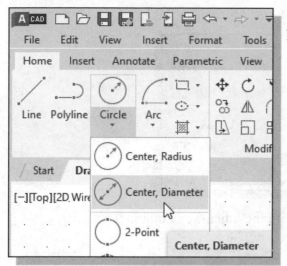

1. Select the **[Circle]** → **[Center, Diameter]** option as shown.

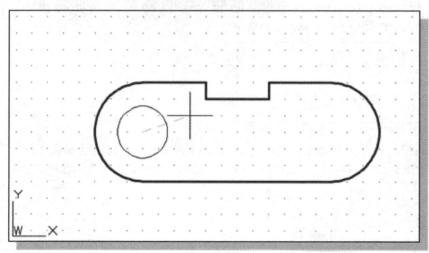

2. Select the same location for the arc center as the center point for the new circle.

3. In the command prompt area, the message "*Specify diameter of circle:*" is displayed. *Specify diameter of circle:* **1.25 [ENTER]**

4. On your own, create the other circle and complete the drawing as shown.

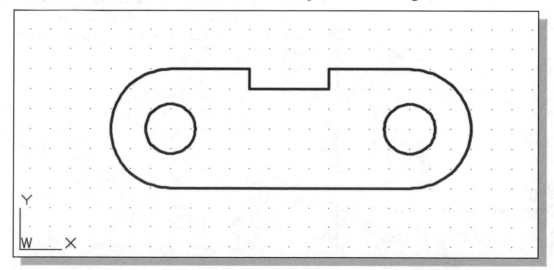

Saving the CAD Design

1. In the *Quick Access Toolbar*, select **[Save]**.

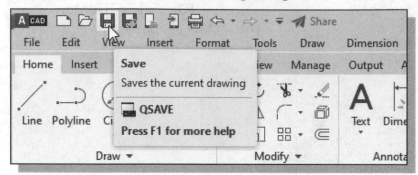

❖ Note the command can also be activated with the quick-key combination of **[Ctrl]+[S]**.

2. In the *Save Drawing As* dialog box, select the folder in which you want to store the CAD file and enter **Spacer** in the *File name* box.

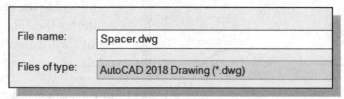

3. Click **Save** in the *Save Drawing As* dialog box to accept the selections and save the file. Note the default file type is DWG, which is the standard AutoCAD drawing format.

Exit AutoCAD 2023

❖ To exit **AutoCAD 2023**, select **Exit AutoCAD** in the *Menu Bar* or type **QUIT** at the command prompt. Note the command can also be activated with the quick-key combination of **[Ctrl]+[Q]**.

Review Questions: (Time: 20 minutes)

1. What are the advantages and disadvantages of using CAD systems to create engineering drawings?

2. What is the default AutoCAD filename extension?

3. How do the **GRID** and **SNAP** options assist us in sketching?

4. List and describe the different **coordinate entry methods** available in AutoCAD.

5. When using the Line command, which option allows us to quickly create a line-segment connecting back to the starting point?

6. List and describe the two types of coordinate systems commonly used for planar geometry.

7. Which key do you use to quickly cancel a command?

8. When you use the Pan command, do the coordinates of objects get changed?

9. Find information on how to draw ellipses in AutoCAD through the *Autodesk Exchange*, and create the following arc. If it is desired to position the center of the ellipse to a specific location, which ellipse command is more suitable?

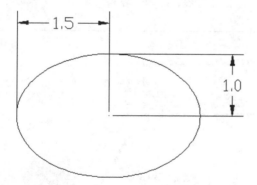

10. Find information on how to draw arcs in AutoCAD through the *Autodesk Exchange* and create the following arc. List and describe two methods to create arcs in AutoCAD.

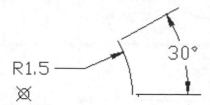

Exercises:

(All dimensions are in inches.) (Time: 90 minutes)

1. Angle Spacer

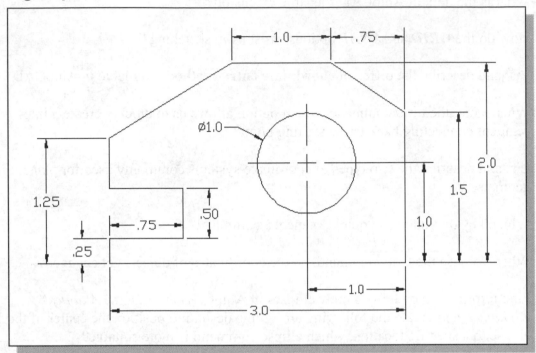

2. Base Plate

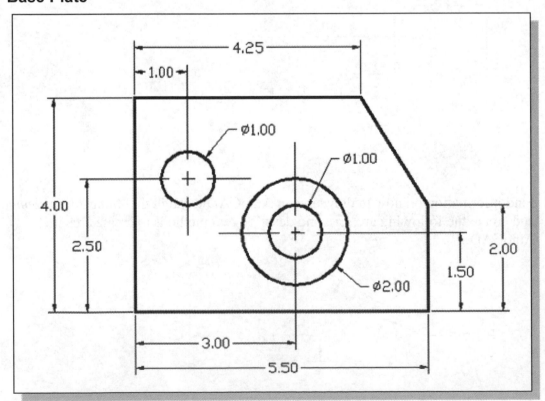

3. T-Clip

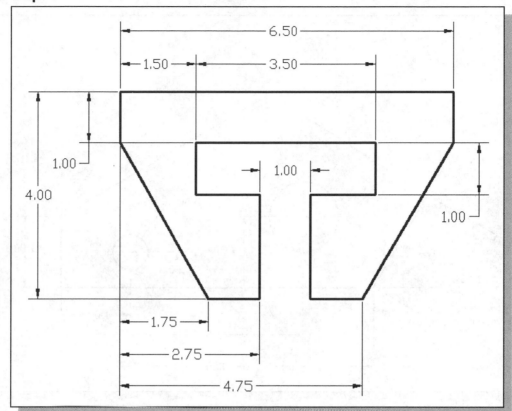

4. Channel Plate

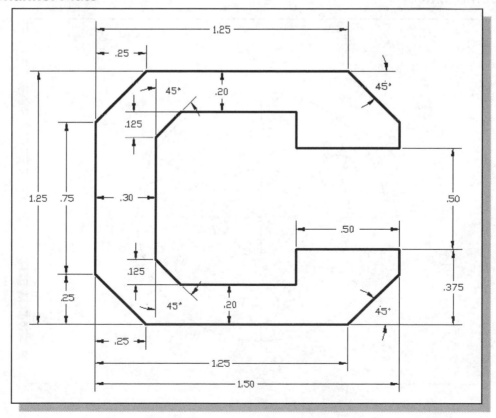

5. Slider Block

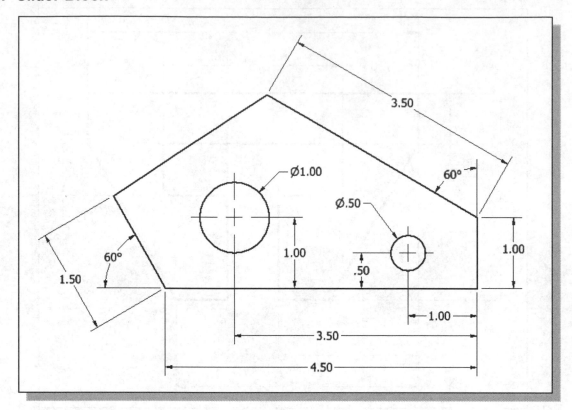

6. Circular Spacer

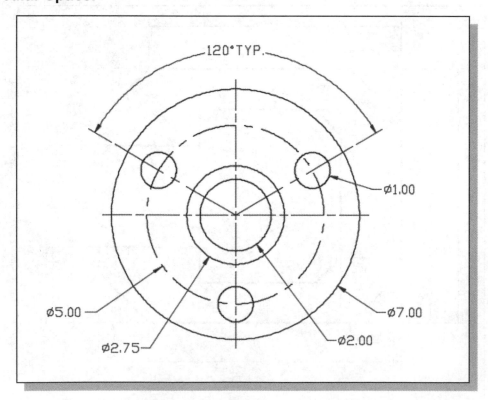

7. Angle Base

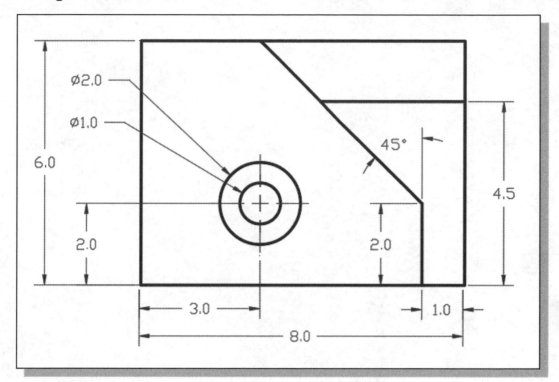

8. Index Key

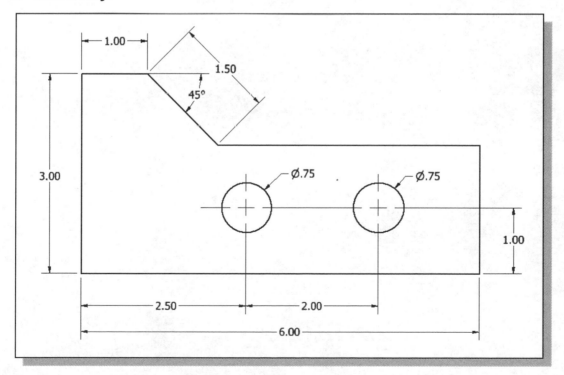

Notes:

Chapter 2
Basic Object Construction Tools

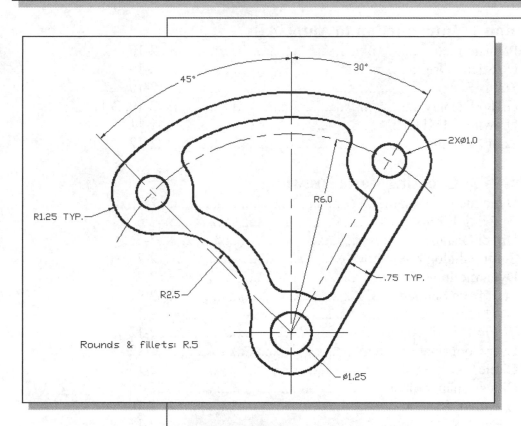

Learning Objectives

♦ **Referencing the WCS**
♦ **Use the Startup dialog box**
♦ **Set up GRID & SNAP intervals**
♦ **Display AutoCAD's toolbars**
♦ **Set up and use OBJECT SNAPS**
♦ **Edit using the TRIM command**
♦ **Use the POLYGON command**
♦ **Create TTR circles**
♦ **Create Tangent lines**

AutoCAD Certified User Examination Objectives Coverage

This table shows the pages on which the objectives of the Certified User Examination are covered in Chapter 2.

Certified User Reference Guide

Introduction

The main characteristic of any CAD system is its ability to create and modify 2D/3D geometric entities quickly and accurately. Most CAD systems provide a variety of object construction and editing tools to relieve the designer of the tedious drudgery of this task, so that the designer can concentrate more on design content. It is important to note that CAD systems can be used to replace traditional drafting with pencil and paper, but the CAD user must have a good understanding of the basic geometric construction techniques to fully utilize the capability of the CAD systems.

One of the major enhancements of AutoCAD 2006 was the introduction of the *Dynamic Input* feature. This addition, which is also available in AutoCAD 2023, greatly enhanced the **AutoCAD Heads-up Design™** interface.

The use of the **User Coordinate System (UCS)** and the **World Coordinate System (WCS)** is further discussed in this chapter. In working CAD, one simple approach to creating designs in CAD systems is to create geometry by referencing the **World Coordinate System**. The general procedure of this approach is illustrated in this chapter.

In this chapter, we will examine the *Dynamic Input* options, the basic geometric construction and editing tools provided by **AutoCAD 2023**. We will first look at the *Dynamic Input* options, and tools such as *UNITS*, *GRID*, *SNAP MODE* intervals set up and the *OSNAP* option, followed by construction tools such as circles and polygons; we will also look at the basic Trim command.

Starting Up AutoCAD 2023

1. Start **AutoCAD 2023** by selecting the *Autodesk* folder in the **Start** menu as shown. Once the program is loaded into the memory, click **New Drawing** to start a new drawing.

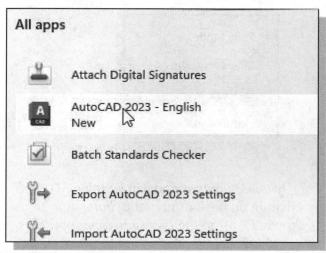

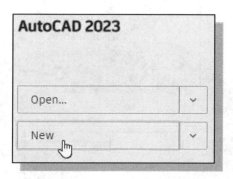

Dynamic Input

In AutoCAD 2023, the **Dynamic Input** feature provides the user with **visual tooltips** and **entry options** right on the screen.

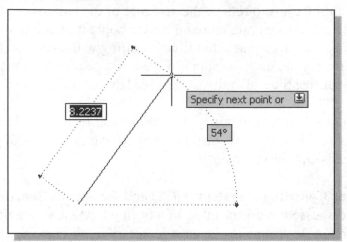

Dynamic Input provides a convenient command interface near the cursor to help the user focus in the graphics area. When *Dynamic Input* is *ON*, tooltips display information near the cursor that is dynamically updated as the cursor moves.

The tooltips also provide a place for user entry when a command is activated. The actions required to complete a command remain the same as those for the command line. Note that **Dynamic Input** is **not** designed to replace the *command line*. The main advantage of using the *Dynamic Input* options is to keep our attention near the cursor.

The *Dynamic Input* features can be used to enhance the **five methods** for specifying the locations of points as described in the *Defining Positions* section of Chapter 1.

1. To switch on the **AutoCAD Dynamic Input option**, use the *Customization option* at the bottom right corner.

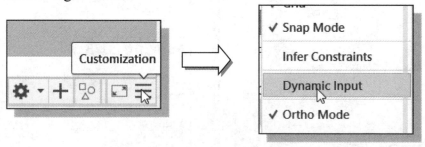

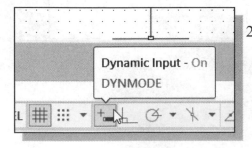

2. The *Dynamic Input* option can be toggled on/off by clicking on the button in the *Status Bar* area as shown. Confirm the *Dynamic Input* option is switched **ON** before proceeding to the next section.

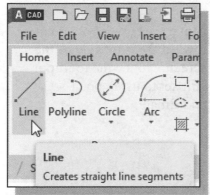

3. Click on the **Line** icon in the *Draw* toolbar. In the command prompt area, the message "*Line Specify first point:*" is displayed.

4. Move the cursor inside the *Drawing Area* and notice the displayed tooltip, which shows the coordinates of the cursor position.

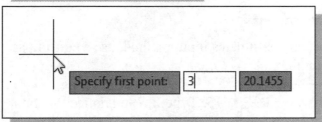

5. Type **3** and notice the input is entered in the first entry box.

6. Hit the **TAB** key once to move the input focus to the second entry box.

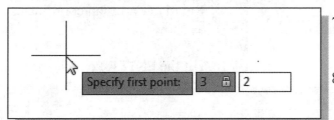

7. Type **2** and notice the input is displayed in the second entry box.

8. Hit the **ENTER** key once to accept the inputs. Note that the point is placed using the world coordinates.

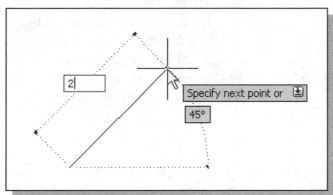

9. Move the cursor upward and toward the right side of the screen. Notice the tooltip is set to use polar coordinates by default.

10. Type **2** and notice the input is displayed in the entry box as shown.

11. Hit the **TAB** key once to move the input focus to the second entry box.

12. Type **30** and notice the input is displayed in the angle entry box.

13. Hit the **ENTER** key once to create the line that is 2 units long and at an angle of 30 degrees.

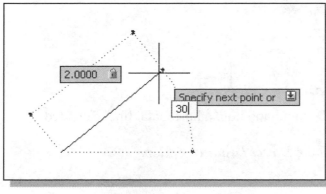
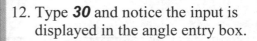
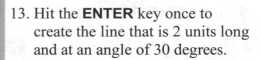

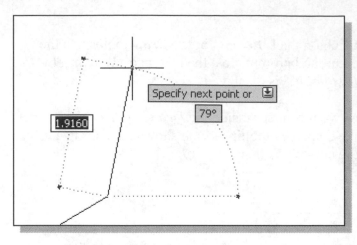

14. Move the cursor upward and toward the right side of the screen. Notice the tooltip is still set to using polar coordinates.

➢ To switch to using the Relative Cartesian coordinates input method, use a **comma** as the **specifier** after entering the first number.

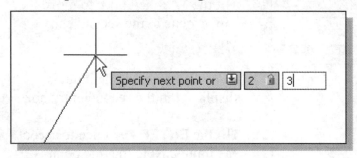

15. Type **2,3** and notice the input option is now set to using Relative Cartesian coordinates as shown.

16. Hit the **ENTER** key once to accept the inputs.

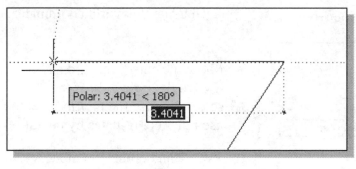

17. Move the cursor toward the left side of the last position until the angle is near 180 degrees as shown.

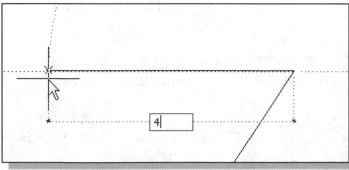

18. Type **4** and notice the input is displayed on the screen.

19. Hit the **ENTER** key once to accept the input and note a horizontal line is created.

➢ In effect, we just created a line using the **Direct Distance** option.

20. Hit the [**Esc**] ley once to exit the Line command.

21. In the *Status Bar* area, **right-click** on *Dynamic Input* and choose **Settings**.

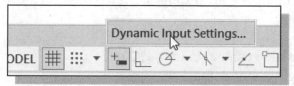

❖ The *Settings* dialog box provides different controls to what is displayed when *Dynamic Input* is on.

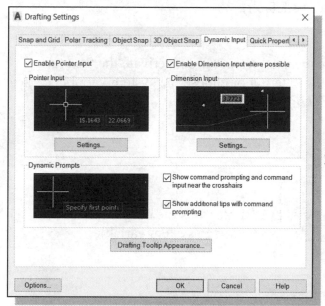

➢ Note that the *Dynamic Input* feature has three components: ***Pointer Input***, ***Dimensional Input***, and ***Dynamic Prompts***.

22. On your own, toggle *ON/OFF* the three options and create additional line segments to see the different effects of the settings.

The Rocker-Arm Design

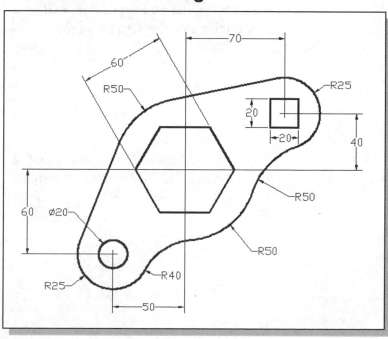

❖ Before continuing to the next page, on your own make a rough freehand sketch on a piece of paper to show the steps that you plan to use in creating the design. Be aware that there are many different approaches to accomplishing the same task.

Activate the Startup Option

In **AutoCAD 2023**, we can use the *Startup* dialog box to establish different types of drawing settings. The startup dialog box can be activated through the use of the **STARTUP** system variable.

The STARTUP system variable can be set to 0, 1, 2 or 3:
- 0: Starts a drawing without defined settings.
- 1: Displays the *Create New Drawing* dialog box.
- 2: Displays a *New Tab* with options; a custom dialog box can be used.
- 3: Displays a *New Tab* with options (default).

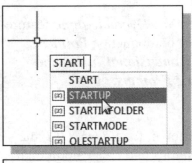

1. In the *command prompt area*, activate the **Startup** option by entering or choose the system variable name: ***STARTUP* [ENTER]**

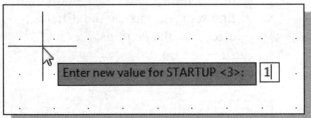

2. Enter **1** as the new value for the *Startup* system variable.

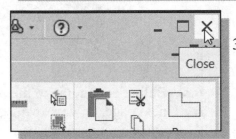

3. To show the effect of the *Startup* option, **exit** AutoCAD by clicking on the **Close** icon as shown.

4. Restart AutoCAD by selecting the **AutoCAD 2023** option through the *Start* menu.

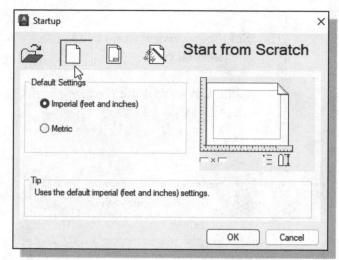

5. The *Startup* dialog box appears on the screen with different options to assist in the creation of drawings. Move the cursor on top of the four icons and notice the different options available:
   (1) **Open a drawing**
   (2) **Start from Scratch**
   (3) **Use a Template**
   (4) **Use a Setup Wizard**

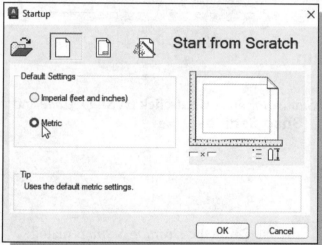

6. In the *Startup* dialog box, select the **Start from Scratch** option as shown in the figure.

7. Choose **Metric** to use the metric settings.

8. Click **OK** to accept the setting.

Drawing Units Display Setup

1. On your own, activate the display of the AutoCAD *Menu Bar* if it is not displayed (Refer to page 1-4 for the procedure.)

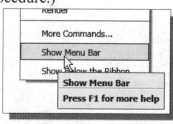

2. Click the *Menu Bar* area and select **[Format] → [Units]**.

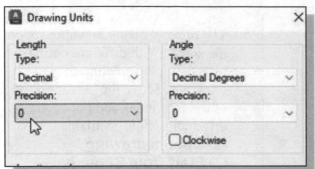

3. Set the *Precision* to **no digits** after the decimal point.

4. Click **OK** to exit the *Drawing Units* dialog box.

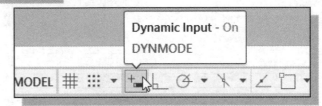

5. On your own, adjust the option settings so that only the **Dynamic Input** option is turned *ON* in the *Status Bar* area.

GRID and SNAP Intervals Setup

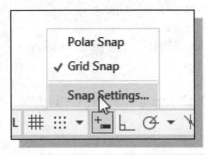

1. In the *Status Bar* area, **right-click** on *Snap Mode* and choose **[Snap Settings]**.

2. In the *Drafting Settings* dialog box, select the **Snap and Grid** tab if it is not the page on top.

3. Change *Grid Spacing* and *Snap Spacing* to **10** for both X and Y directions.

4. Switch *ON* the *Display dotted grid in 2D model Space* option as shown.

5. Switch *ON* the *Grid On* and *Snap On* options as shown.

6. Click **OK** to exit the *Drawing Units* dialog box.

Drawing Area Setup

Next, we will set up the **Drawing Limits**. Setting the Drawing Limits controls the extents of the display of the *grid*. It also serves as a visual reference that marks the working area. Note that this setting can also be adjusted through the use of the command prompt area.

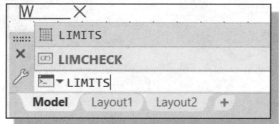

1. Click inside the *command prompt* area.

2. Inside the *command prompt area*, enter **Limits** and press the **[Enter]** key.

3. In the command prompt area, near the bottom of the AutoCAD drawing screen, the message "*Reset Model Space Limits: Specify lower left corner or [On/Off] <0,0>:*" is displayed. Enter **-200,-150** through the *Dynamic Input* entry boxes.

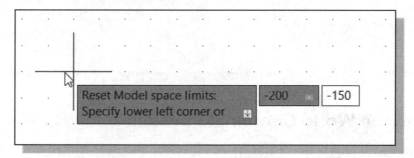

4. In the command prompt area, the message "*Specify upper right corner <420,297>:*" is displayed. Enter **200,150** as the new upper right coordinates as shown.

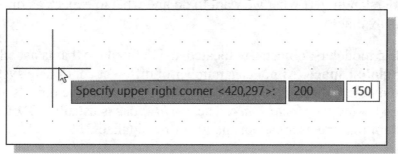

5. On your own, use the *Menu Bar* and confirm the **[View] → [Display] → [UCS Icon] → [Origin]** option is switched **ON** as shown. (The little checked icon next to the option indicates it is switched **ON**.)

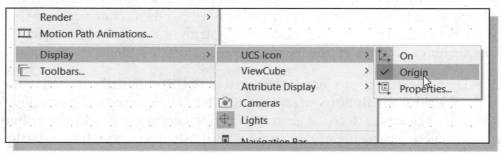

6. On your own, use the **Zoom Extents** command, under the **View** pull-down menu, to reset the display.

❖   Notice the *UCS Icon*, which is aligned to the origin, is displayed near the center of the Drawing Area.

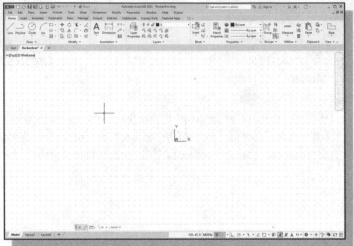

Referencing the World Coordinate System

Design modeling software is becoming more powerful and user friendly, yet the system still does only what the user tells it to do. When using a geometric modeler, we therefore need to have a good understanding of what the inherent limitations are. We should also have a good understanding of what we want to do and what to expect, as the results are based on what is available.

In most geometric modelers, objects are located and defined in what is usually called **world space** or **global space**. Although a number of different coordinate systems can be used to create and manipulate objects in a 3D modeling system, the objects are typically defined and stored using the *world space*. The *world space* is usually a **3D Cartesian coordinate system** that the user cannot change or manipulate.

In most engineering designs, models can be very complex, and it would be tedious and confusing if only one coordinate system were available in CAD systems. Practical CAD systems provide the user with definable **Local Coordinate Systems (LCS)** or **User Coordinate Systems (UCS)**, which are measured relative to the world coordinate system. Once a local coordinate system is defined, we can then create geometry in terms of this more convenient system. For most CAD systems, the default construction coordinate system is initially aligned to the world coordinate system.

In AutoCAD, the default **User Coordinate System (UCS)** is initially aligned to the XY plane of the **World Coordinate System (WCS)**. One simple approach to creating designs in CAD systems is to create geometry by referencing the **World Coordinate System**. The general procedure of this approach is illustrated in the following sections.

Creating Circles

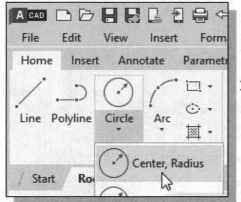

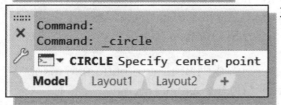

1. Select the **Circle – Center, Radius** command icon in the *Draw* toolbar.

2. On your own, dock the command prompt dialog box to the lower left of the graphics window. Note that you can adjust the size of the box by dragging the edge of the box.

3. In the *command prompt area*, the message "*Circle Specify center point for the circle or [3P/2P/Ttr (tan tan radius)]:*" is displayed. Select the **origin** of the world coordinate system as the center point.

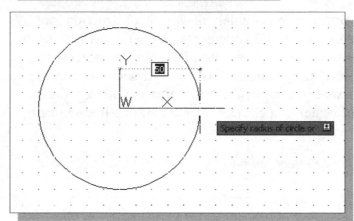

4. In the command prompt area, the message "*Specify radius of circle or [Diameter]:*" is displayed. AutoCAD expects us to identify the radius of the circle. Set the radius to **50** by observing the tooltips as shown.

5. Hit the [**SPACE BAR**] once to repeat the circle command.

6. On your own, select **70,40** as the absolute coordinate values of the center point coordinates of the second circle.

7. On your own, set the radius of the circle to **25**.

8. On your own, repeat the above procedure and create another circle (radius **25**) at absolute coordinates of **-50,-60** as shown in the figure.

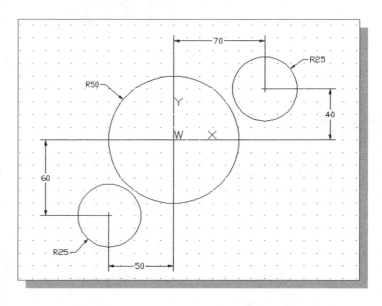

Object Snap Toolbar

1. Move the cursor to the *Menu Bar* area and choose **[Tools]** → **[Toolbars]** → **[AutoCAD]**.

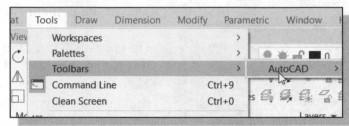

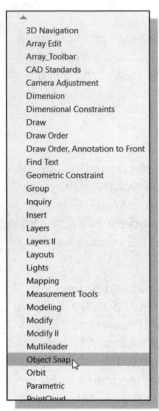

❖ AutoCAD provides 50+ predefined toolbars for access to frequently used commands, settings, and modes. A *checkmark* (next to the item) in the list identifies the toolbars that are currently displayed on the screen.

2. Select **Object Snap**, with the left-mouse-button, to display the *Object Snap* toolbar on the screen.

❖ **Object Snap** is an extremely powerful construction tool available on most CAD systems. During an entity's creation operations, we can snap the cursor to points on objects such as endpoints, midpoints, centers, and intersections. For example, we can turn on **Object Snap** and quickly draw a line to the center of a circle, the midpoint of a line segment, or the intersection of two lines.

3. Move the cursor over the icons in the *Object Snap* toolbar and read the description of each icon.

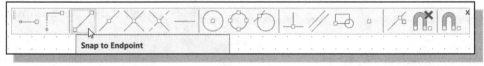

Snap to Endpoint

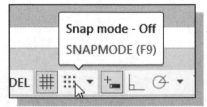

Snap mode - Off
SNAPMODE (F9)

4. We will next turn **OFF** the *GRID SNAP* option by toggling off the **SNAP Mode** button in the *Status Bar* area.

5. On your own, reset the option buttons in the *Status Bar* area, so that only the *GRID DISPLAY* option is switched **ON**.

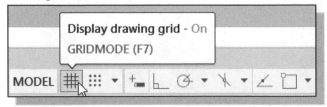

Display drawing grid - On
GRIDMODE (F7)

Using the Line Command

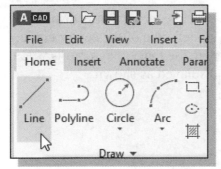

1. Select the **Line** command icon in the *Draw* toolbar. In the command prompt area, near the bottom of the AutoCAD drawing screen, the message "*Line Specify first point:*" is displayed.

2. Pick **Snap to Tangent** in the *Object Snap* toolbar. In the command prompt area, the message "*_tan to*" is displayed. AutoCAD now expects us to select a circle or an arc on the screen.

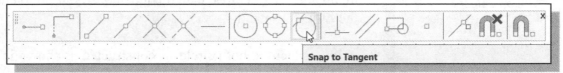

❖ The **Snap to Tangent** option allows us to snap to the point on a circle or arc that, when connected to the last point, forms a line tangent to that object.

3. Pick a location that is near the top left side of the smaller circle on the right; note the tangent symbol is displayed as shown.

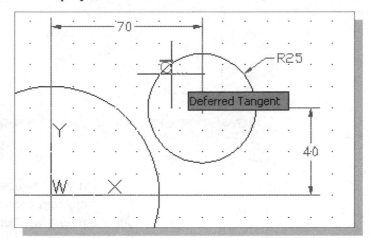

❖ Note that the note "Deferred Tangent" indicates that AutoCAD will calculate the tangent location when the other endpoint of the line is defined.

4. Pick **Snap to Tangent** in the *Object Snap* toolbar. In the command prompt area, the message "*_tan to*" is displayed. AutoCAD now expects us to select a circle or an arc on the screen.

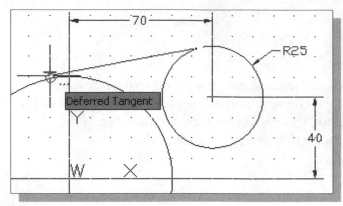

5. Pick a location that is near the top left side of the center circle; note the tangent symbol is displayed as shown.

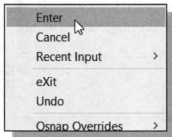

6. Inside the *Drawing Area*, **right-click** to activate the option menu and select **Enter** with the left-mouse-button to end the **Line** command.

❖ A line tangent to both circles is constructed as shown in the figure.

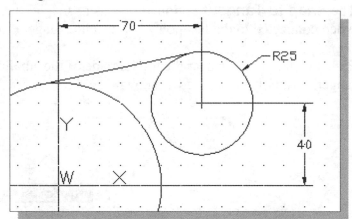

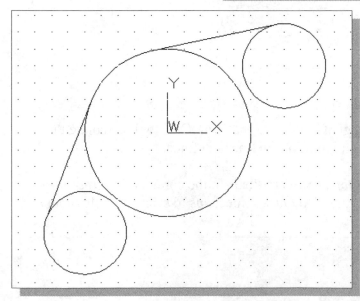

7. On your own, repeat the above steps and create the other tangent line between the center circle and the circle on the left. Your drawing should appear as the figure.

Creating TTR Circles

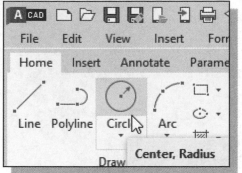

1. Select the **Circle** command icon in the *Draw* toolbar. In the command prompt area, the message "*Specify center point for circle or [3P/2P/Ttr (tan tan radius)]:*" is displayed.

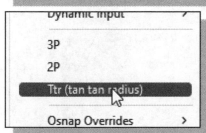

2. Inside the Drawing Area, right-click to activate the option menu and select the **Ttr (tan tan radius)** option. This option allows us to create a circle that is tangent to two objects.

3. Pick a location near the **bottom of the smaller circle** on the right. We will create a circle that is tangent to this circle and the center circle.

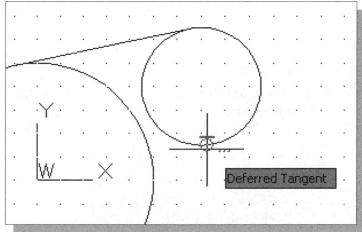

4. Pick the **center circle** by selecting a location that is near the right side of the circle. AutoCAD interprets the locations we selected as being near the tangency.

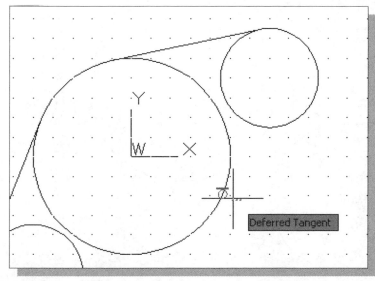

5. In the command prompt area, the message "*Specify radius of circle*" is displayed. Enter **50** as the radius of the circle.

    Specify radius of circle: **50 [ENTER]**

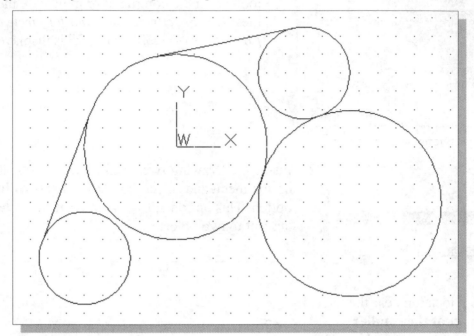

6. On your own, repeat the above steps and create the other TTR circle (radius **40**). Your drawing should appear as the figure below.

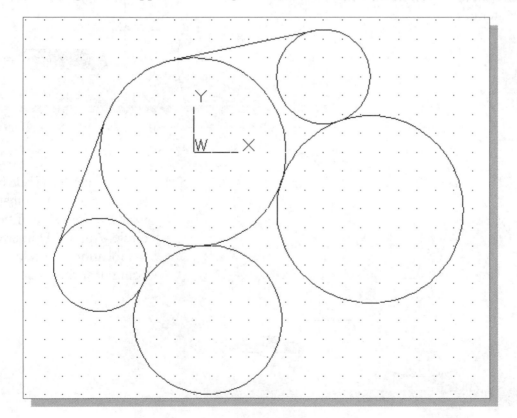

Using the Trim Command

The **Trim** command shortens an object so that it ends precisely at a selected boundary.

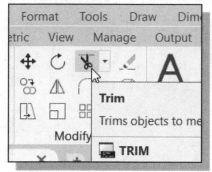

1. Select the **Trim** command icon in the *Modify* toolbar, and click on the down-triangle to display additional icons as shown.

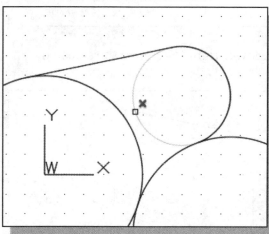

2. The message *"Select object to trim or shift-select object to extend or [Project/Edge/Undo]:"* is displayed in the command prompt area. Move the cursor on the **left** section of the upper right circle and note the selected portion is trimmed as shown.

❖ In AutoCAD, several options are available with the Trim command: (1) We can specify cutting edges at which an object is to stop. Valid cutting edge objects include most 2D geometry such as lines, arcs, circles, ellipses, polylines, splines, and text. For 3D objects, a 2D projection method can also be used, where objects are projected onto the XY plane of the current user coordinate system (UCS). (2) We can select objects to trim by defining a two-point rectangle. (3) We can erase an entire object while in the trim command.

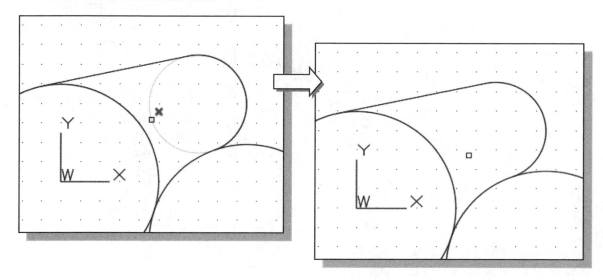

3. Select the upper right side of the center circle to remove the selected portion.

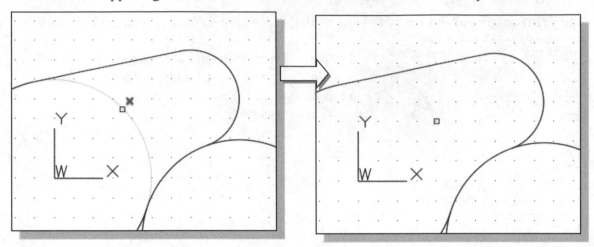

4. Select the upper right side of the circle on the lower right to remove the selected portion.

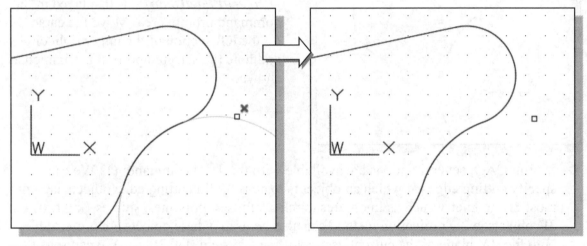

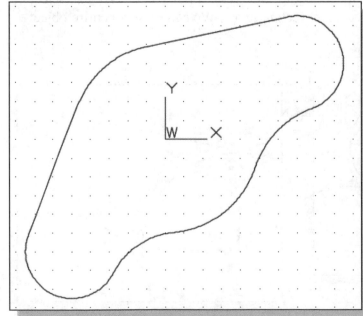

5. On your own, trim the other geometry so that the drawing appears as shown.

6. Inside the *Drawing Area*, **right-click** to activate the option menu and select **Enter** with the left-mouse-button to end the **Trim** command.

Using the Polygon Command

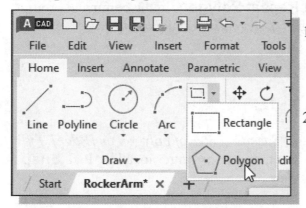

1. Select the **Polygon** command icon in the *Draw* toolbar. Click on the triangle icon next to the rectangle icon to display the additional icon list.

2. Enter **6** to create a six-sided hexagon.
    polygon Enter number of sides <4>: **6** **[ENTER]**

3. The message "*Specify center of polygon or [Edge]:*" is displayed. Since the center of the large circle is aligned to the origin of the WCS, the center of the polygon can be positioned using several methods. Set the center point to the origin by entering the absolute coordinates. *Specify center of polygon or [Edge]:* **0,0 [ENTER]**

4. In the *command prompt area*, the message "*Enter an option [Inscribed in circle/ Circumscribed about circle] <I>:*" is displayed. Click **Circumscribed about circle** to select the *Circumscribed about circle* option.

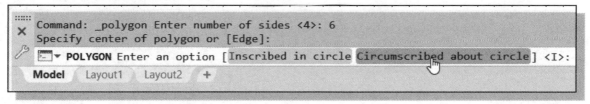

5. In the command prompt area, the message *Specify radius of circle:*" is displayed. Enter **30** as the radius.
    Specify radius of circle: **30 [ENTER]**

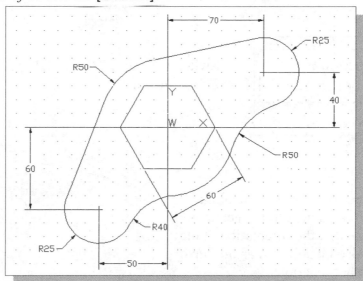

❖ Note that the polygon option [Inscribed in circle/Circumscribed about circle] allows us to create either **flat to flat** or **corner to corner** distance.

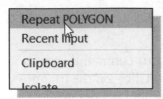

6. Inside the Drawing Area, **right-click** to activate the option menu and select **Repeat Polygon**. In the command prompt area, the message "_polygon Enter number of sides <6>:" is displayed.

7. Enter **4** to create a four-sided polygon.
    _polygon Enter number of sides <6>: **4 [ENTER]**

8. In the command prompt area, the message "*Specify center of polygon or [Edge]:*" is displayed. Let's use the *Object Snap* options to locate its center location. Pick **Snap to Center** in the *Object Snap* toolbar as shown.

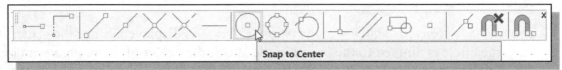

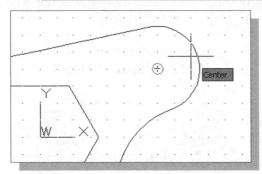

9. Move the cursor on top of the arc on the right and notice the center point is automatically highlighted. Select the arc to accept the highlighted location.

10. Inside the *Drawing Area*, **right-click** to activate the option menu and select **Circumscribed about circle**.

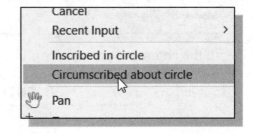

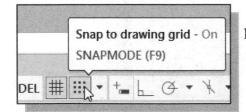

11. Switch *ON* the *GRID SNAP* option in the *Status Bar* as shown.

12. Create a square by selecting one of the adjacent grid points next to the center point as shown. Note that the orientation of the polygon can also be adjusted as the cursor is moved to other locations.

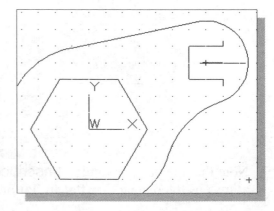

Creating a Concentric Circle

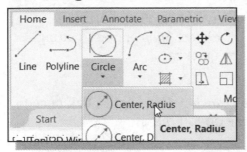

1. Select the **Circle, Radius** command icon in the *Draw* toolbar. In the command prompt area, the message *"Specify center point for circle or [3P/2P/Ttr (tan tan radius)]:"* is displayed.

2. Let's use the *Object Snap* options to assure the center location is aligned properly. Pick **Snap to Center** in the *Object Snap* toolbar as shown.

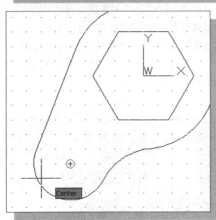

3. Move the cursor on top of the lower arc on the left and notice the center point is automatically highlighted. Select the arc to accept the highlighted location.

4. In the command prompt area, the message *"Specify radius of circle <25>"* is displayed. Enter **10** to complete the Circle command.
   Specify radius of circle <25>: **10** **[ENTER]**

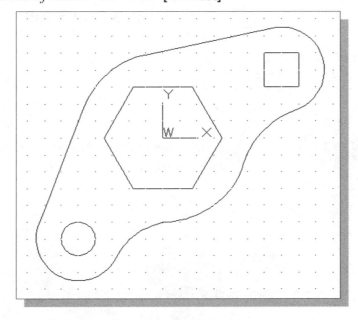

Using the QuickCalc Calculator to Measure Distance and Angle

AutoCAD also provides several tools that will allow us to measure distance, area, perimeter, and even mass properties. With the use of the *Object Snap* options, getting measurements of the completed design can be done very quickly.

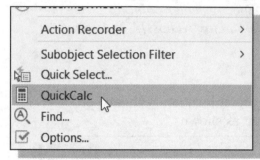

1. Inside the *Drawing Area*, **right-click** once to bring up the option menu.

2. Select **QuickCalc** in the *option menu* as shown.

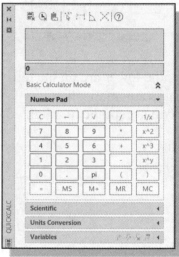

❖ Note that the QuickCalc option brings up the AutoCAD calculator, which can be used to perform a full range of mathematical, scientific, and geometric calculations. We can also use QuickCalc to create and use variables, as well as to convert units of measurement.

3. Click the **Measure Distance** icon, which is located on the top section of the *QuickCalc* calculator pad.

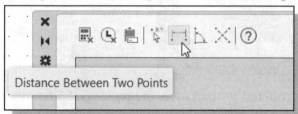

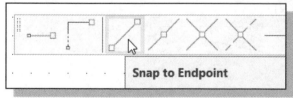

4. Pick **Snap to Endpoint** in the *Object Snap* toolbar.

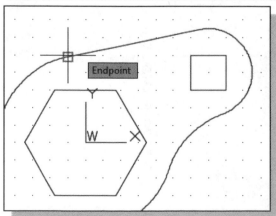

5. Select the **top tangent line**, near the lower endpoint, as shown.

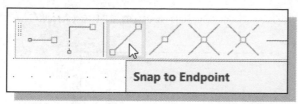

6. Pick **Snap to Endpoint** in the *Object Snap* toolbar.

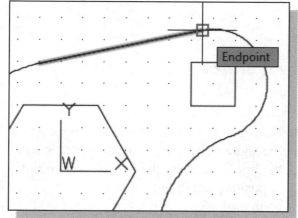

7. Select the tangent line, near the upper endpoint, as shown.

❖ The length of the line is displayed in the *QuickCalc* calculator as shown.

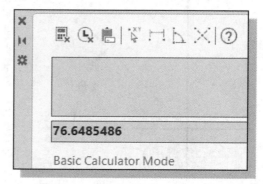

8. Click the **Clear** icon to remove the number displayed.

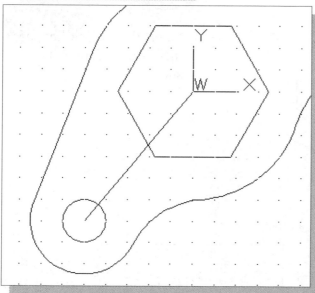

9. On your own, repeat the above steps and measure the center to center distance of the lower region of the design as shown. (Hint: use the Snap to Center option.)

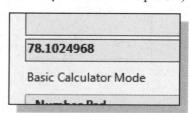

10. Click the **Measure Angle** icon, which is located on the top section of the *QuickCalc* calculator pad.

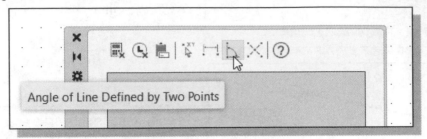

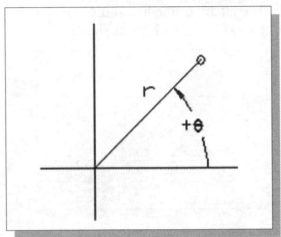

❖ Note that this option allows us to measure the angle between the horizontal axis and the line formed by the two selected points. A positive angle indicates a counterclockwise direction.

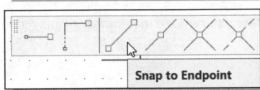

11. Pick **Snap to Endpoint** in the *Object Snap* toolbar.

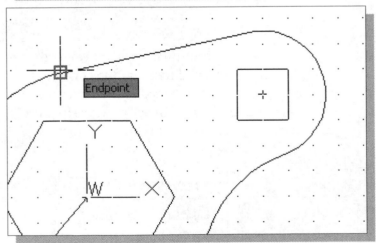

12. Select the tangent line, near the lower endpoint, as shown.

13. Pick **Snap to Endpoint** in the *Object Snap* toolbar.

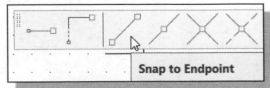

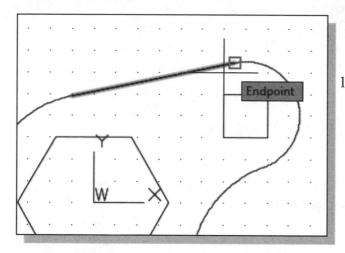

14. Select the tangent line, near the upper endpoint, as shown.

➤ The measured angle is displayed in the calculator pad as shown.

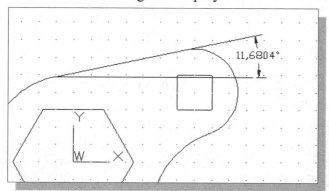

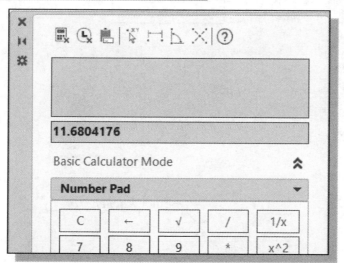

15. On your own, experiment with the available **Get Coordinates** options.

➤ Note also that the QuickCalc calculator can remain active while you are using other AutoCAD commands.

Saving the CAD File

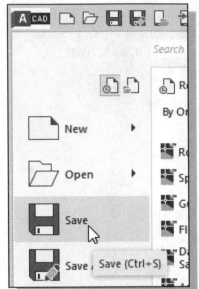

1. In the *Application Menu*, select:

   [Application] → [Save]

   ❖ Note the command can also be activated with quick-key combination of [**Ctrl**]+[**S**].

2. In the *Save Drawing As* dialog box, select the folder in which you want to store the CAD file and enter **RockerArm** in the *File name* box.

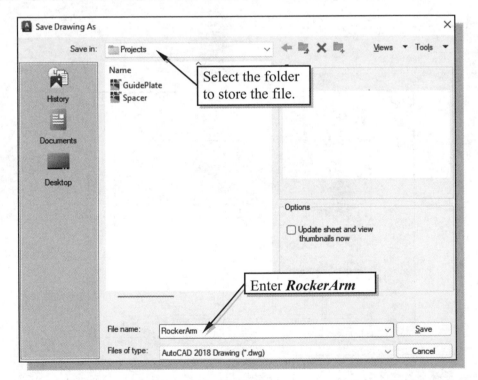

3. Pick **Save** in the *Save Drawing As* dialog box to accept the selections and save the file.

Exit AutoCAD

To exit **AutoCAD 2023**, select **Exit AutoCAD** from the *Application Menu* or type **QUIT** at the command prompt.

AutoCAD Quick Keys

Besides using the mouse and/or AutoCAD menu systems, AutoCAD also provides quick keys to access the more commonly used commands, with many of them being a single key stroke.

Q   QSAVE / Saves the current drawing.

A   ARC / Creates an arc.

Z   ZOOM / Increases or decreases the magnification of the view in the current viewport.

W   WBLOCK / Writes objects or a block to a new drawing file.

S   STRETCH / Stretches objects crossed by a selection window or polygon.

X   EXPLODE / Breaks a compound object into its component objects.

E   ERASE / Removes objects from a drawing.

D   DIMSTYLE / Creates and modifies dimension styles.

C   CIRCLE / Creates a circle.

R   REDRAW / Refreshes the display in the current viewport.

F   FILLET / Rounds and fillets the edges of objects.

V   VIEW / Saves and restores named views, camera views, layout views, and preset views.

T   MTEXT / Creates a multiline text object.

G   GROUP / Creates and manages saved sets of objects called groups.

B   BLOCK / Creates a block definition from selected objects.

H   HATCH / Fills an enclosed area or selected objects with a hatch pattern, solid fill, or gradient fill.

J   JOIN / Joins similar objects to form a single, unbroken object.

M   MOVE / Moves objects a specified distance in a specified direction.

I   INSERT / Inserts a block or drawing into the current drawing.

O   OFFSET / Creates concentric circles, parallel lines, and parallel curves.

L   LINE / Creates straight line segments.

P   PAN / Adds a parameter with grips to a dynamic block definition.

For a more detailed description on the quick keys, refer to the following Autodesk website:
http://www.autodesk.com/store/autocad-shortcuts

Review Questions: (Time: 20 minutes)

1. Describe the procedure to activate the AutoCAD Startup option.

2. List and describe three options in the AutoCAD *Object Snap* toolbar.

3. Which AutoCAD command can we use to remove a portion of an existing entity?

4. Describe the difference between the *circumscribed* and *inscribed* options when using the AutoCAD Polygon command.

5. Create the following triangle and fill in the blanks: Length = ____, Angle = ____. (Dimensions are in inches.)

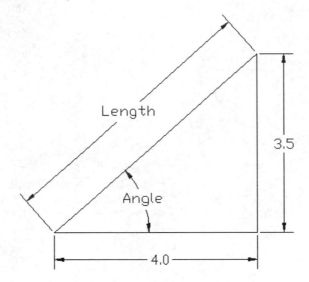

6. Create the following drawing; line **AB** is tangent to both circles. Fill in the blanks: Length = _____, Angle = _____. (Dimensions are in inches.)

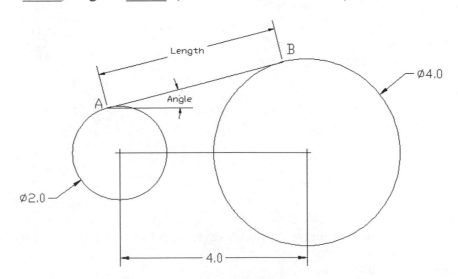

Exercises:

(Unless otherwise specified, dimensions are in inches.) (Time: 120 minutes.)

1. Adjustable Support

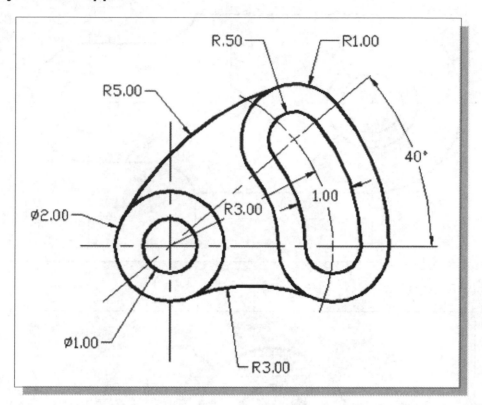

2. V-Slide Plate (The design has two sets of parallel lines with implied tangency.)

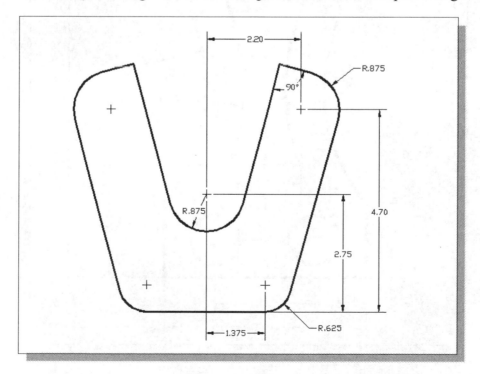

3. Swivel Base (Dimensions are in Millimeters.)

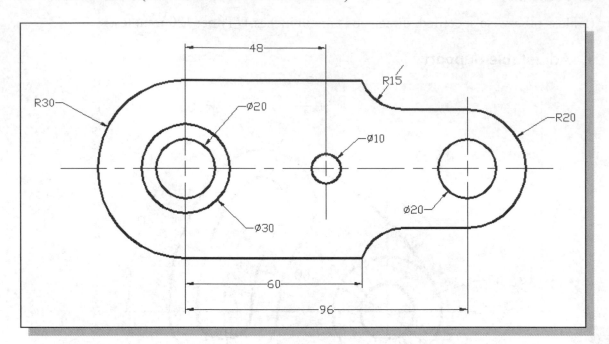

4. Sensor Mount

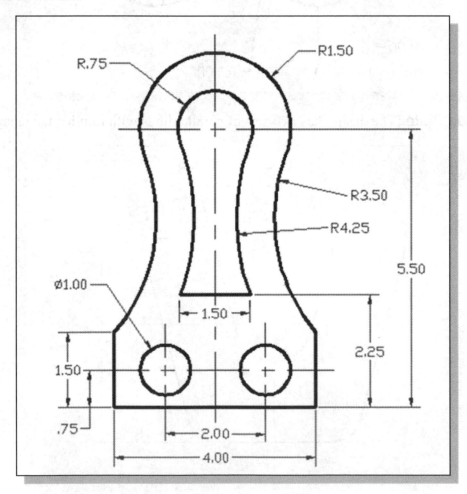

5. Flat Hook (Dimensions are in Millimeters. Thickness: 25 mm.)

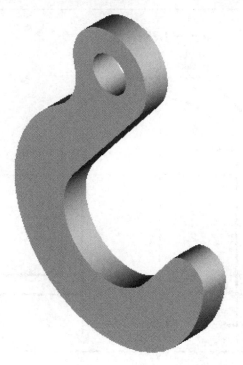

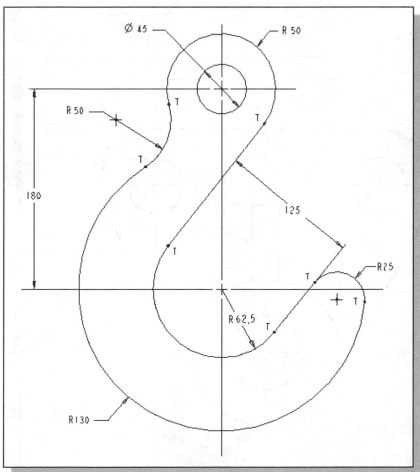

6. Journal Bracket

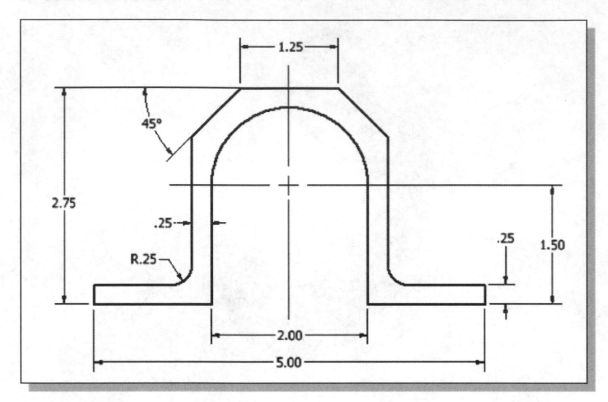

7. Support Base

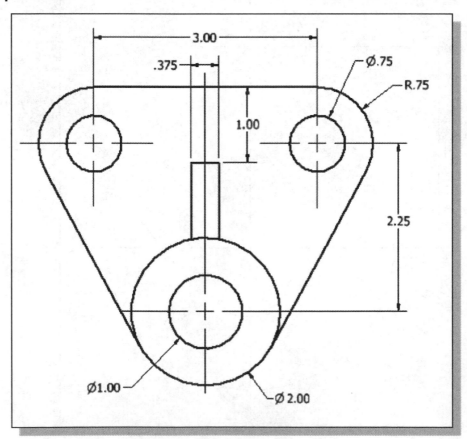

8. Pivot Arm

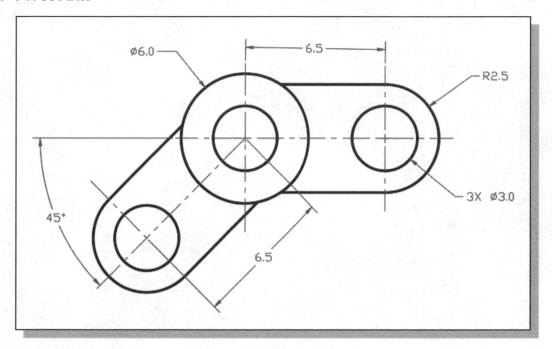

9. Extruder Cover (Dimensions are in Millimeters)

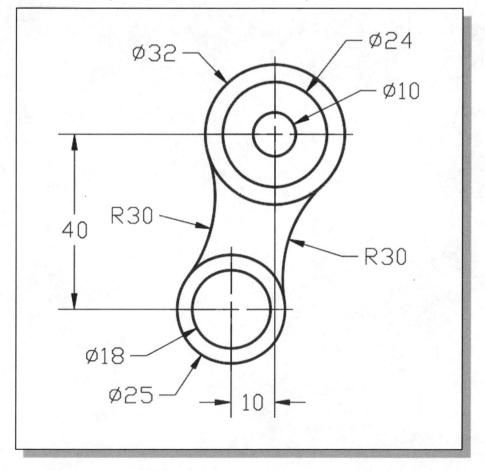

Notes:

Chapter 3
Geometric Construction and Editing Tools

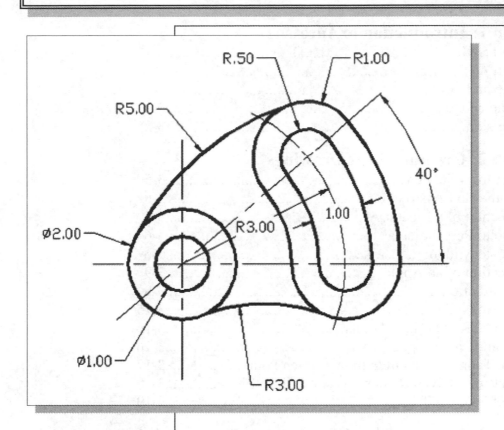

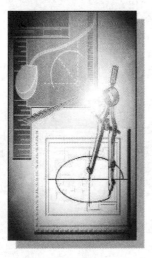

Learning Objectives

- ♦ **Set up the display of Drawing Units**
- ♦ **Display AutoCAD's toolbars**
- ♦ **Set up and use OBJECT SNAPS**
- ♦ **Edit using EXTEND and TRIM**
- ♦ **Use the FILLET command**
- ♦ **Create parallel geometric entities**
- ♦ **Using the PEDIT command**
- ♦ **Use the EXPLODE command**

AutoCAD Certified User Examination Objectives Coverage

This table shows the pages on which the objectives of the Certified User Examination are covered in Chapter 3.

Certified User Reference Guide

Tips on Taking the AutoCAD Certified User Examination

1. **Study**: The first step to maximize your potential on an exam is to sufficiently prepare for it. You need to be familiar with the AutoCAD package, and this can only be achieved by doing drawings and exploring the different commands available. The AutoCAD Certified User exam is designed to measure your familiarity with the AutoCAD software. You must be able to perform the given task and answer the exam questions correctly and quickly.

2. **Make Notes**: Take notes of what you learn either while attending classroom sessions or going through study material. Use these notes as a review guide before taking the actual test.

3. **Time Management**: Manage the time you spend on each question. Always remember you do not need to score 100% to pass the exam. Also keep in mind that some questions are weighed more heavily and may take more time to answer.

4. **Be Cautious**: Devote some time to ponder and think of the correct answer. Ensure that you interpret all the options correctly before selecting from available choices.

5. **Use Common Sense**: If you are unable to get the correct answer and unable to eliminate all distracters, then you need to select the best answer from the remaining selections. This may be a task of selecting the best answer from amongst several correct answers, or it may be selecting the least incorrect answer from amongst several poor answers.

6. **Take Your Time**: The examination has a time limit. If you encounter a question you cannot answer in a reasonable amount of time, use the Save As feature to save a copy of the data file, and mark the question for review. When you review the question, open your copy of the data file and complete the performance task. After you verify that you have entered the answer correctly, unmark the question so it no longer appears as marked for review.

7. **Don't Act in Haste**: Don't go into panic mode while taking a test. Always read the question carefully before you look out for choices in hand. Use the *Review* screen to ensure you have reviewed all the questions you may have marked for review. When you are confident that you have answered all questions, end the examination to submit your answers for scoring. You will receive a score report once you have submitted your answers.

8. **Relax before exam**: In order to avoid last minute stress, make sure that you arrive 10 to 15 minutes early and relax before taking the exam.

Certified User Reference Guide

Geometric Constructions

The creation of designs usually involves the manipulation of geometric shapes. Traditionally, manual graphical construction used simple hand tools like a T-square, straightedge, scales, triangles, compass, dividers, pencils, and paper. The manual drafting tools are designed specifically to assist in the construction of geometric shapes. For example, the T-square and drafting machine can be used to construct parallel and perpendicular lines very easily and quickly. Today, modern CAD systems provide designers much better control and accuracy in the construction of geometric shapes.

In technical drawings, many of the geometric shapes are constructed with specific geometric properties, such as perpendicularity, parallelism and tangency. For example, in the drawing below, quite a few implied geometric properties are present.

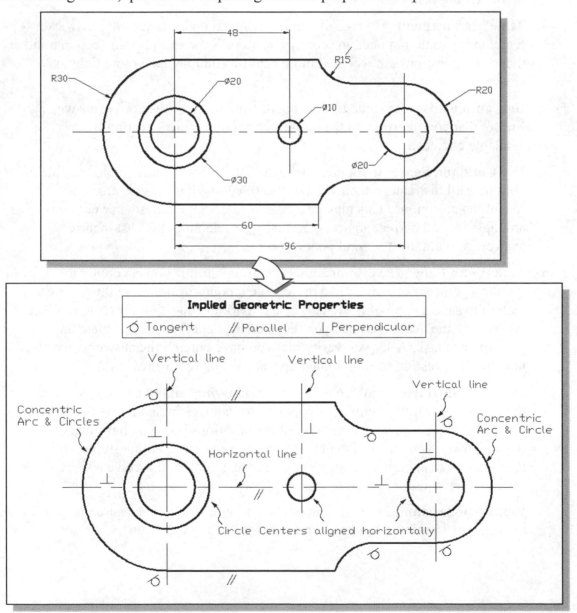

Starting Up AutoCAD 2023

1. Select the AutoCAD 2023 option on the *Program* menu or select the AutoCAD 2023 icon on the *Desktop*. Once the program is loaded into the memory, the **AutoCAD 2023** drawing screen will appear on the screen.

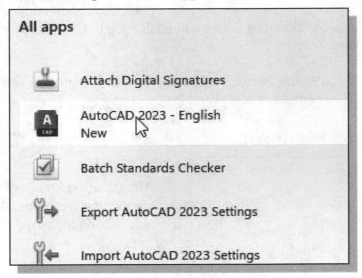

2. In the *Startup* window, select **Start from Scratch**, as shown in the figure below.

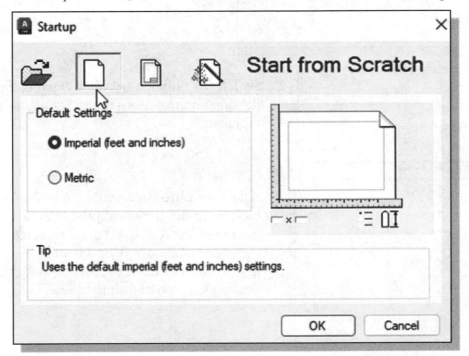

3. In the *Default Settings* section, pick **Imperial (feet and inches)** as the drawing units.

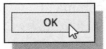

4. Pick **OK** in the *Startup* dialog box to accept the selected settings.

Geometric Construction – CAD Method

The main characteristic of any CAD system is its ability to create and modify 2D/3D geometric entities quickly and accurately. Most CAD systems provide a variety of object construction and editing tools to relieve the designer of the tedious drudgery of this task, so that the designer can concentrate more on design content. A good understanding of the computer geometric construction techniques will enable the CAD users to fully utilize the capability of the CAD systems.

➢ Note that with CAD systems, besides following the classic geometric construction methods, quite a few options are also feasible.

- ## Bisection of a Line or Arc

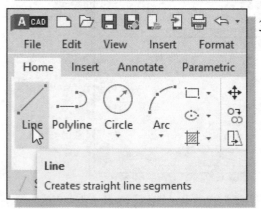

1. Create an arbitrary arc **AB** at any angle, and create line **AB** by connecting the two endpoints of the arc.

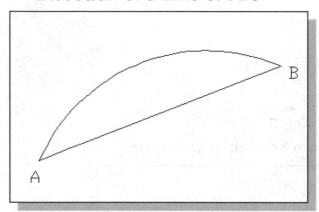

2. Switch **ON** only the *Dynamic Input* option by clicking on the buttons in the *Status Bar* area as shown.

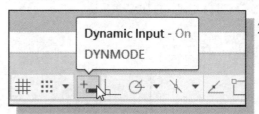

3. Select the **Line** command icon in the *Draw* toolbar. In the command prompt area, near the bottom of the AutoCAD drawing screen, the message "*Line Specify first point:*" is displayed. AutoCAD expects us to identify the starting location of a straight line.

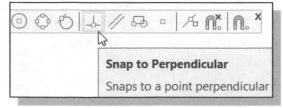

4. Pick **Snap to Perpendicular** in the *Object Snap* toolbar. In the command prompt area, the message "*_per to*" is displayed.

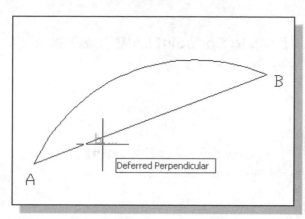

5. Select line **AB** at any position.

❖ Note the displayed tooltip ***Deferred Perpendicular*** indicates the construction is deferred until all inputs are completed.

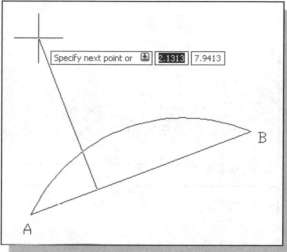

6. Select an arbitrary point above the line as shown.

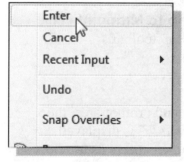

7. Inside the *Drawing Area*, **right-click** to activate the option menu and select **Enter** with the left-mouse-button to end the Line command.

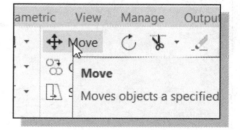

8. Select **Move** in the *Modify* toolbar as shown.

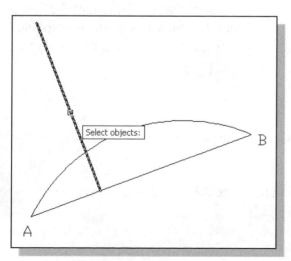

9. Select the **perpendicular line** we just created.

❖ In the *command prompt area*, the message "*Specify the base point or [Displacement]*" is displayed. AutoCAD expects us to select a reference point as the base point for moving the selected object.

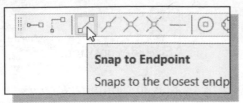

10. Pick **Snap to Endpoint** in the *Object Snap* toolbar.

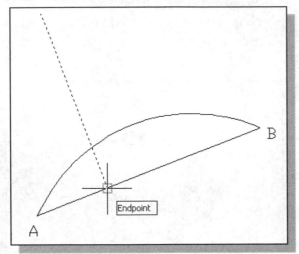

11. Select the lower **Endpoint** of the selected line as shown.

12. Move the cursor inside the Drawing Area, and notice the line is moved to the new cursor location on the screen.

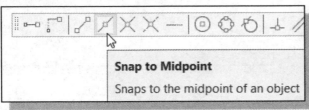

13. Pick **Snap to Midpoint** in the *Object Snap* toolbar.

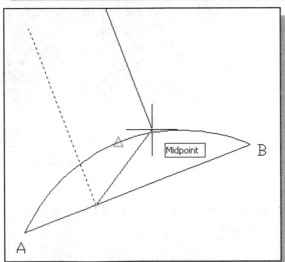

❖ In the command prompt area, the message "_mid to" is displayed. AutoCAD now expects us to select an existing arc or line on the screen.

14. Select the arc to move the line to the midpoint of the arc. Note that the midpoint of an arc or a line is displayed when the cursor is on top of the object.

15. On your own, repeat the above process and move the perpendicular line to the midpoint of line **AB**.

➢ The constructed bisecting line is perpendicular to line **AB** and passes through the midpoint of the line or arc **AB**.

- **Bisection of an Angle**

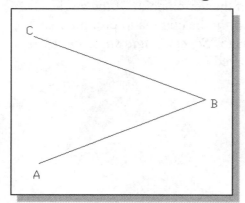

1. Create an arbitrary angle **ABC** as shown in the figure.

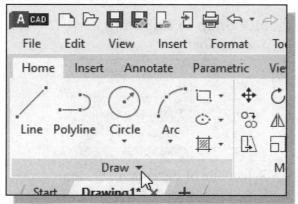

2. Click once with the left-mouse-button on the small triangle in the titlebar of the *Draw* toolbar as shown.

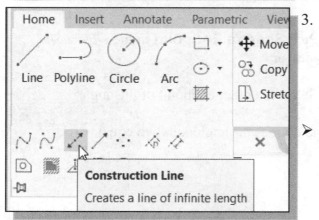

3. Select the **Construction Line** icon in the *Draw* toolbar. In the *command prompt area*, the message "*_xline Specify a point or [Hor/Ver/Ang/Bisect/Offset]:*" is displayed.

➢ *Construction lines* are lines that extend to infinity. Construction lines are usually used as references for creating other objects.

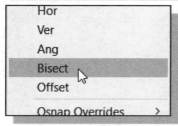

4. Inside the *Drawing Area*, **right-click** once to bring up the option menu.

5. Select **Bisect** from the option list as shown. In the command prompt area, the message "*Specify angle vertex point:*" is displayed.

6. Pick **Snap to Endpoint** in the *Object Snap* toolbar.

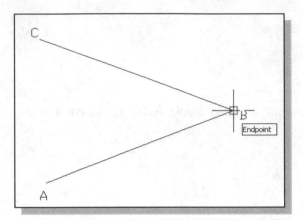

7. Select the vertex point of the angle as shown. In the command prompt area, the message "*Specify angle start point:*" is displayed.

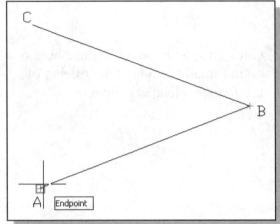

8. Pick **Snap to Endpoint** in the *Object Snap* toolbar.

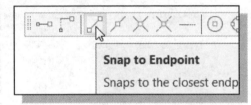

9. Select one of the endpoints, point A or C, of the angle.

10. Pick **Snap to Endpoint** in the *Object Snap* toolbar.

11. Select the other endpoint of the angle.

➢ Note that the constructed bisection line divides the angle into two equal parts.

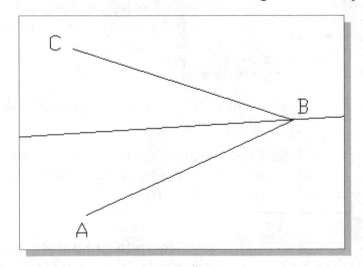

- ## **Transfer of an Angle**

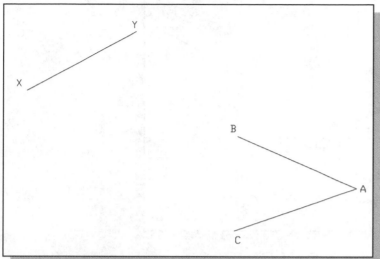

1. Create an arbitrary angle **ABC** and a separate line **XY** as shown in the figure.

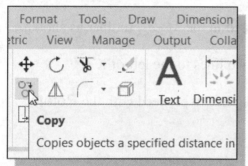

2. Select **Copy** in the *Modify* toolbar as shown.

3. Select line **AB** and **AC** as the objects to be copied.

4. **Right-click** once to proceed with the Copy command.

5. Pick **Snap to Endpoint** in the *Object Snap* toolbar.

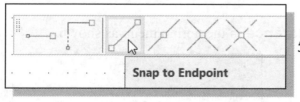

6. Select point **A** as the base point.

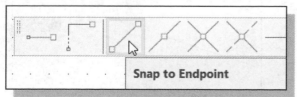

7. Pick **Snap to Endpoint** in the *Object Snap* toolbar.

8. Select point **X** to place the copied lines.

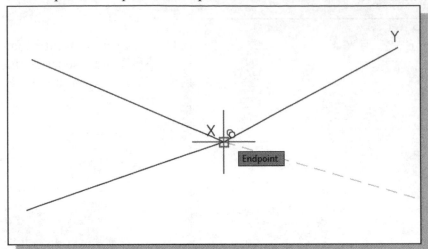

9. Hit the **ENTER** key once to end the Copy command.

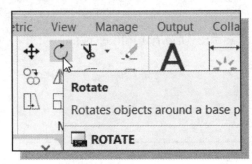

10. Select **Rotate** in the *Modify* toolbar as shown.

11. Select the two lines to the left of **X**.

12. **Right-click** once to proceed with the Rotate command.

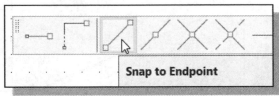

13. Pick **Snap to Endpoint** in the *Object Snap* toolbar.

14. Select point **X** as the base point for the rotation.

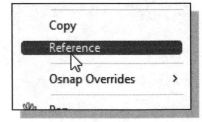

15. Inside the *Drawing Area*, **right-click** once to bring up the option menu.

16. Select **Reference** in the option menu as shown.

17. Select point **X** as the first point.

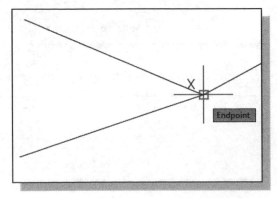

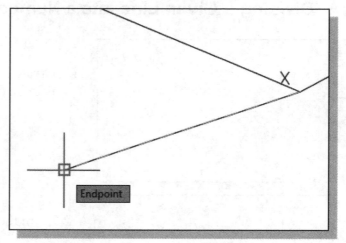

18. Select the left end point of the bottom line as shown.

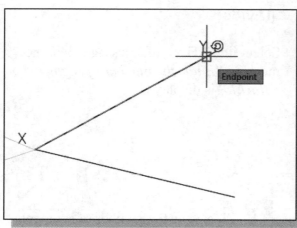

19. Select **Point Y** to align the lines as shown.

➢ Note the **angle** has been transferred to align to the XY line.

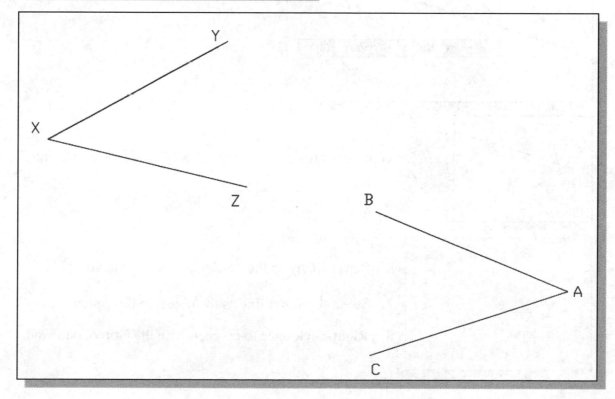

- **Dividing a Given Line into a Number of Equal Parts**

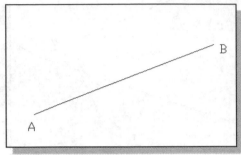

1. Create a line **AB** at an arbitrary angle; the line is to be divided into five equal parts.

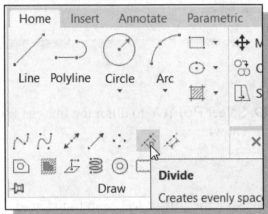

2. In the *Ribbon* toolbar, under *Draw*, select:

   [Divide]

3. Select line **AB**. In the message area, the message "*Enter the number of segments or [Block]:*" is displayed.

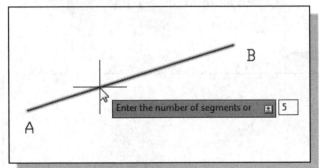

4. Enter **5** as the number of segments needed.

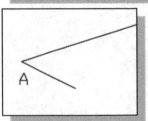

5. On your own, create an arbitrary short line segment at point **A**.

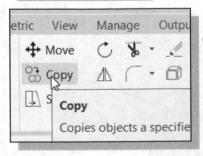

6. Select **Copy** in the *Modify* toolbar as shown.

7. Select the **short line** as the object to be copied.

8. **Right-click** once to proceed with the Copy command.

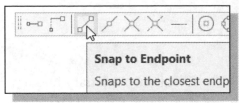

9. Pick **Snap to Endpoint** in the *Object Snap* toolbar.

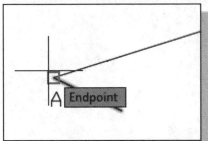

10. **Select** point **A** as the base reference point.

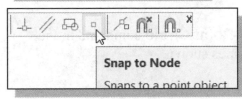

11. Pick **Snap to Node** in the *Object Snap* toolbar.

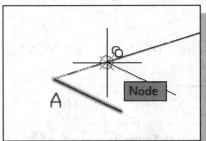

12. Move the cursor along line **AB**, and select the next node point as shown.

13. Repeat the above steps and create 3 additional lines at the nodes indicating the division of the line into five equal parts.

➢ We can also change the display of the created Points.

14. In the *Menu Bar*, select:

   [Format] → [Point Style]

15. In the *Point Style* window, choose the 4th icon in the second row, as shown.

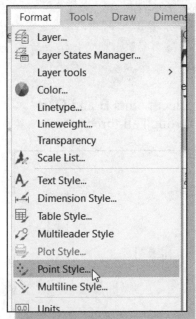

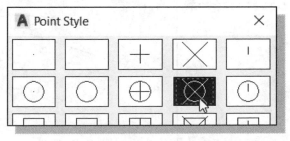

16. Click **OK** to accept the selection and adjust the point style.

- ## Circle through Three Points

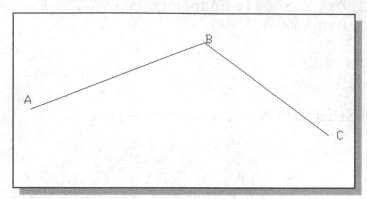

1. Create two arbitrary line segments, **AB** and **BC**, as shown.

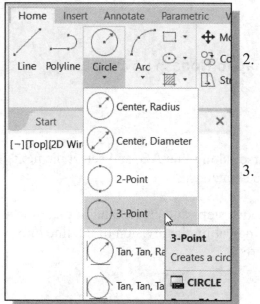

2. Select the **3-Point Circle** command in the *Draw* toolbar as shown.

3. Pick **Snap to Endpoint** in the *Object Snap* toolbar.

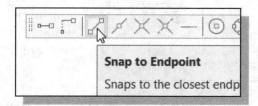

Snap to Endpoint

Snaps to the closest endp

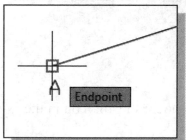

4. Select the first point, point **A**.

5. Repeat the above steps and select points **B** and **C** to create the circle that passes through all three points.

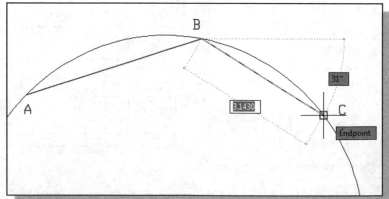

• Line Tangent to a Circle from a Given Point

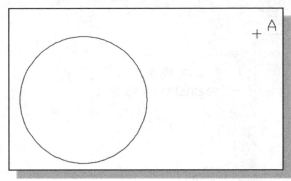

1. Create a circle and a point **A**. (Use the Point command to create point **A**.)

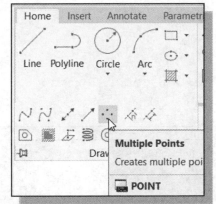

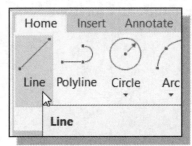

2. Select the **Line** command icon in the *Draw* toolbar. In the command prompt area, near the bottom of the AutoCAD drawing screen, the message "*Line Specify first point:*" is displayed. AutoCAD expects us to identify the starting location of a straight line.

3. Pick **Snap to Node** in the *Object Snap* toolbar.

4. Select point **A** as the starting point of the new line.

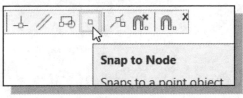

5. Pick **Snap to Tangent** in the *Object Snap* toolbar.

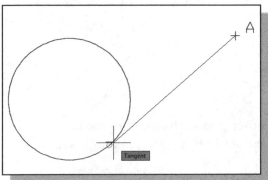

6. Move the cursor on top of the circle and notice the **Tangent** symbol is displayed.

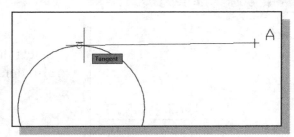

➢ Note that we can create two tangent lines, one to the top and one to the bottom of the circle, from point **A**.

- ## **Circle of a Given Radius Tangent to Two Given Lines**

Option I: TTR circle

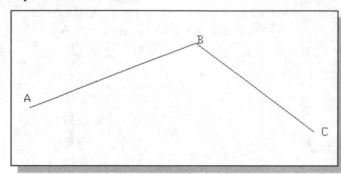

1. Create two arbitrary line segments as shown.

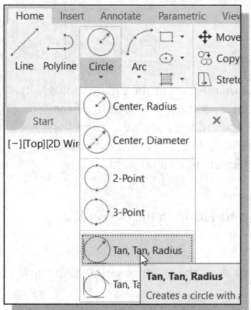

2. Select **TTR Circle** in the *Draw* toolbar as shown.

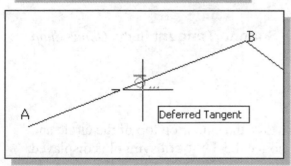

3. Select one of the line segments; note the tangency is deferred until all inputs are completed.

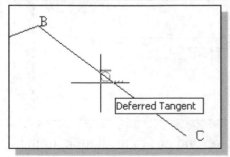

4. Select the other line segment; note the tangency is also deferred until all inputs are completed.

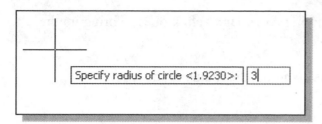

5. Enter **3** as the radius of the circle.

➤ The circle is constructed exactly tangent to both lines.

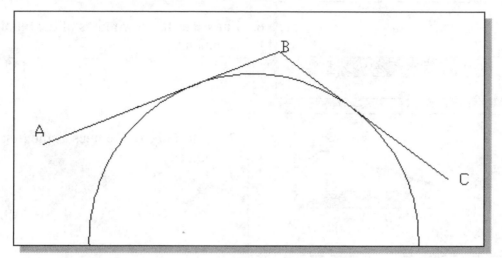

Option II: Fillet Command

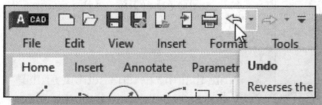

1. Select the **Undo** icon in the *Standard* toolbar as shown. This will undo the last step, the circle.

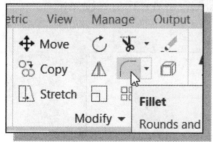

2. Select **Fillet** in the *Modify* toolbar as shown.

3. In the command prompt area, the message "*Select first object or [Undo/Polyline/Radius/Trim/Multiple]*" is displayed. By default, *Mode* is set to **Trim** and the current arc *Radius* is set to **0**.

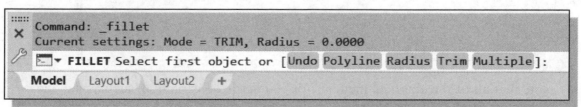

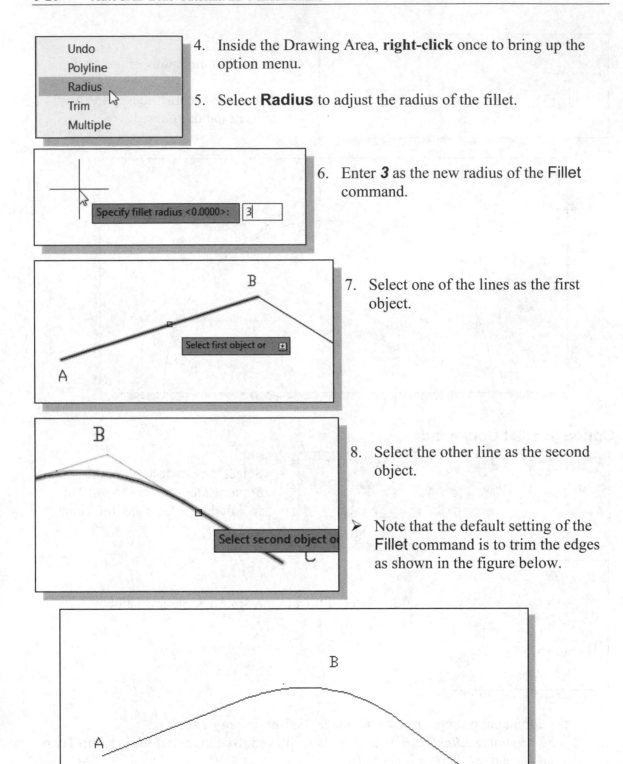

| Undo |
| Polyline |
| **Radius** |
| Trim |
| Multiple |

4. Inside the Drawing Area, **right-click** once to bring up the option menu.

5. Select **Radius** to adjust the radius of the fillet.

Specify fillet radius <0.0000>:   3

6. Enter **3** as the new radius of the **Fillet** command.

B

Select first object or

A

7. Select one of the lines as the first object.

B

Select second object o

C

8. Select the other line as the second object.

➢ Note that the default setting of the **Fillet** command is to trim the edges as shown in the figure below.

B

A

C

• Note that all of the classical methods for geometric construction, such as the one shown on page 3-14, are also applicable in CAD systems.

The Gasket Design

Exploring the possibilities of a CAD system can be very exciting. For persons who have board drafting experience, the transition from the drafting board to the computer does require some adjustment. However, the essential skills required to work in front of a computer are not that much different from those needed for board drafting. In fact, many of the basic skills acquired in board drafting can also be applied to a computer system. For example, the geometric construction techniques that are typically used in board drafting can be used in AutoCAD. The main difference between using a CAD system over the traditional board drafting is the ability to create and modify geometric entities very quickly and accurately. As was illustrated in the previous sections, a variety of object construction and editing tools, which are available in AutoCAD, are fairly easy to use. It is important to emphasize that a good understanding of the geometric construction fundamentals remains the most important part of using a CAD system. The application of the basic geometric construction techniques is one of the main tasks in using a CAD system.

In the following sections, we will continue to examine more of the geometric construction and editing tools provided by **AutoCAD 2023**. We will be looking at the geometric construction tools, such as Trim, Extend, Edit Polyline and Offset, which are available in **AutoCAD 2023**.

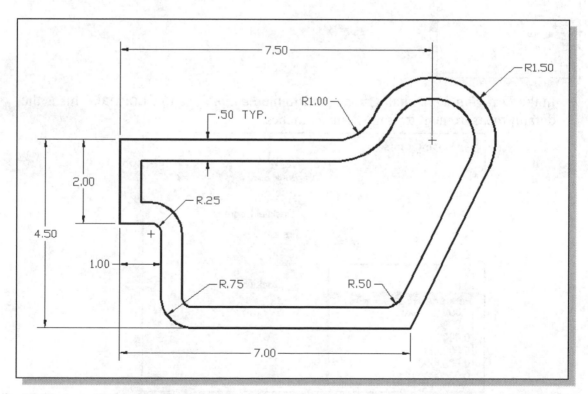

❖ Before continuing to the next page, on your own make a rough sketch showing the steps that can be used to create the design. Be aware that there are many different approaches to accomplishing the same task.

Drawing Units Display Setup

Before creating the first geometric entity, the value of the units within the CAD system should be determined. For example, in one drawing, a unit might equal one millimeter of the real-world object. In another drawing, a unit might equal an inch. The unit type and number of decimal places for object lengths and angles can be set through the UNITS command. These *drawing unit settings* control how AutoCAD interprets the coordinate and angle entries and how it displays coordinates and units in the *Status Bar* and in the dialog boxes.

1. In the *Menu Bar* select:

    [Format] → [Units]

2. In the *Drawing Units* dialog box, confirm the *Length Type* to **Decimal**. This is the default measurement to English units, inches.

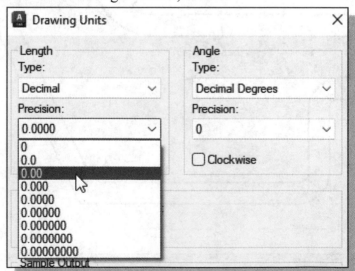

3. Set the *Precision* to **two digits** after the decimal point.

4. Pick **OK** to exit the *Drawing Units* dialog box.

GRID and *SNAP* Intervals Setup

1. In the *Menu Bar*, select:

    [Tools] → [Drafting Settings]

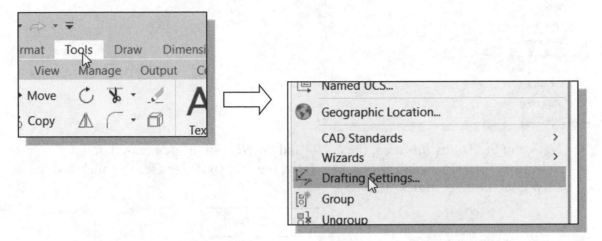

2. In the *Drafting Settings* dialog box, select the **Snap and Grid** tab if it is not the page on top.

3. Change *Grid Spacing* to **1.00** for both X and Y directions.

4. Switch **ON** the *Grid On* and *Snap On* options as shown.

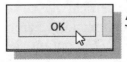

5. Pick **OK** to exit the *Drawing Units* dialog box.

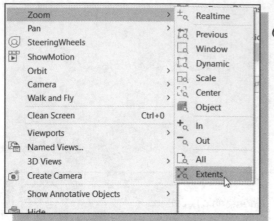

6. On your own, use the **Zoom Extents** command, under the **View** pull-down menu, to reset the display.

❖ Notice in the *Status Bar* area, the *GRID* and *SNAP* options are pressed down indicating they are switched *ON*. Currently, the grid spacing is set to 1 inch and the snap interval is set to 0.5 inch.

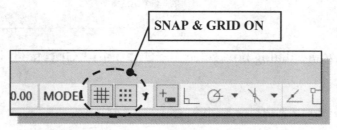

Using the Line Command with ORTHO option

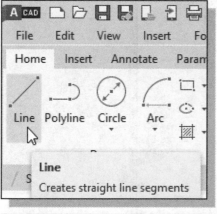

1. Select the **Line** command icon in the *Draw* toolbar. In the command prompt area, near the bottom of the AutoCAD drawing screen, the message "*Line Specify first point:*" is displayed. AutoCAD expects us to identify the starting location of a straight line.

2. In the Drawing Area, move the cursor to **world coordinates (4,6)**. **Left-click** to position the starting point of the line at that location.

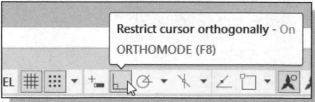

3. We will next turn *off* the *dynamic input* option and switch *ON* the *ORTHO* option by toggling on the **ORTHO** button in the *Status Bar* area.

❖ The *ORTHO* option constrains cursor movement to the horizontal or vertical directions, relative to the current coordinate system. With the Line command, we are now restricted to creating only horizontal or vertical lines with the *ORTHO* option.

4. Move the graphics cursor below the last point we selected on the screen and create a vertical line that is **two units** long (*Y coordinate: 4.00*).

5. Move the graphics cursor to the right of the last point and create a horizontal line that is **one unit** long (*X coordinate: 5.00*).

6. Move the graphics cursor below the last point and create a vertical line that is **2.5 units** long (*Y coordinate: 1.50*).

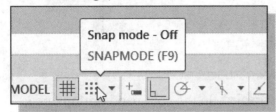

7. Turn **OFF** the *SNAP* option in the *Status Bar* area.

8. Move the graphics cursor to the right of the last point and create a horizontal line that is about seven units long (near *X coordinate: 12.00*). As is quite common during the initial design stage, we might not always know all of the dimensions at the beginning.

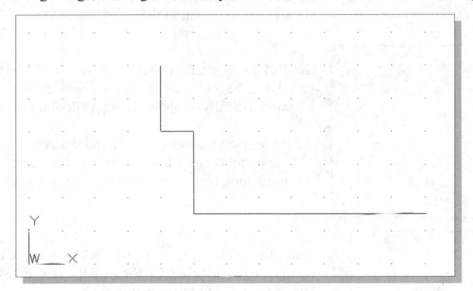

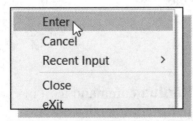

9. Inside the *Drawing Area*, **right-click** to activate the option menu and select **Enter** with the left-mouse-button to end the Line command.

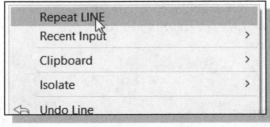

10. Activate the **Line** command by picking the icon in the *Draw* toolbar or **right-click** to activate the option menu and select **Repeat Line**.

Object Snap Toolbar

Object Snap is an extremely powerful construction tool available on most CAD systems. During an entity's creation operations, we can snap the cursor to points on objects such as endpoints, midpoints, centers, and intersections. For example, we can quickly draw a line to the center of a circle, the midpoint of a line segment, or the intersection of two lines.

1. Bring up the *Object Snap* toolbar through the **[Tools]** ➔ **[Toolbars]** menu.

2. In the *Object Snap* toolbar, pick **Snap to Endpoint**. In the command prompt area, the message "*_endp of*" is displayed. AutoCAD now expects us to select a geometric entity on the screen.

❖ The **Snap to Endpoint** option allows us to snap to the closest endpoint of objects such as lines or arcs. AutoCAD uses the midpoint of the entity to determine which end to snap to.

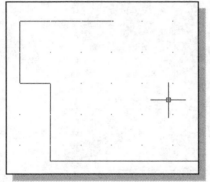

3. Pick the **top left vertical line** by selecting a location above the midpoint of the line. Notice AutoCAD automatically snaps to the top endpoint of the line.

4. Move the graphics cursor to the right of the last point and create a **horizontal line** that is about **three units** long (near *X coordinate: 7.00*).

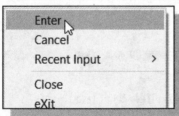

5. Inside the Drawing Area, **right-click** to activate the option menu and select **Enter** with the left-mouse-button to end the Line command.

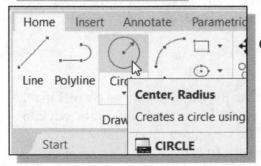

6. Select the **Circle-Radius** command icon in the *Draw* toolbar. In the command prompt area, the message "*Specify center point for circle or [3P/2P/Ttr (tan tan radius)]:*" is displayed.

7. In the *Status Bar* area, switch *ON* the *SNAP* option.

8. In the *Drawing Area*, move the cursor to world coordinates (**11.5,6**). **Left-click** to position the center point of the circle at this location.

9. Move the graphics cursor to world coordinates (**13,6**). **Left-click** at this location to create a circle (radius **1.5** inches).

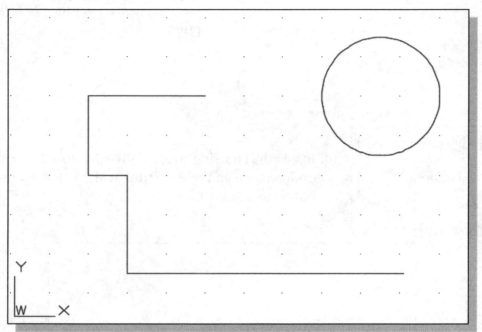

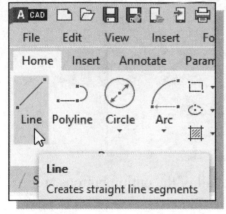

10. Select the **Line** command icon in the *Draw* toolbar. In the command prompt area, the message "*Line Specify first point:*" is displayed.

11. In the *Drawing Area*, move the cursor to world coordinates (**11,1.5**). **Left-click** to position the first point of a line at this location.

12. Pick **Snap to Tangent** in the *Object Snap* toolbar. In the command prompt area, the message "*_tan to*" is displayed. AutoCAD now expects us to select a circle or an arc on the screen.

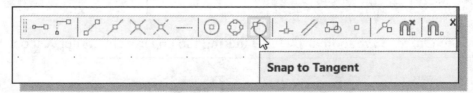

❖ The **Snap to Tangent** option allows us to snap to the point on a circle or arc that, when connected to the last point, forms a line tangent to that object.

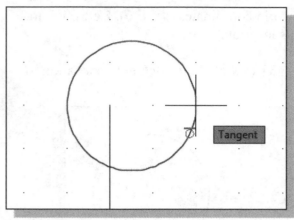

13. Pick a location on the right side of the **circle** and create the line tangent to the circle. Note that the *Object Snap* options take precedence over the *ORTHO* option.

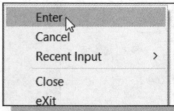

14. Inside the Drawing Area, **right-click** to activate the option menu and select **Enter** with the left-mouse-button to end the Line command.

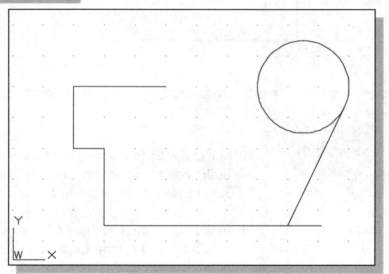

15. In the *Status Bar* area, reset the option buttons so that all of the option buttons are switched *OFF*.

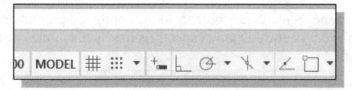

16. Close the *Object Snap* toolbar by **left-clicking** the upper right corner **X** icon.

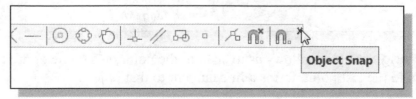

Using the Extend Command

The Extend command lengthens an object so that it ends precisely at a selected boundary.

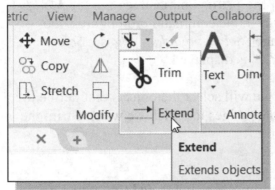

1. Select the **Extend** command icon in the *Modify* toolbar. In the command prompt area, the message "*Select boundary edges... Select objects:*" is displayed.

❖ First, we will use the **Boundary edges option** to select the objects we want to extend an object to.

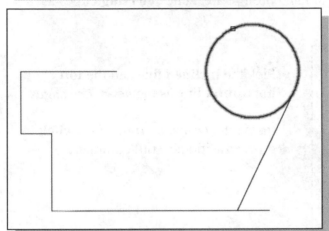

2. In the command prompt area, click **Boundary edges** as shown.

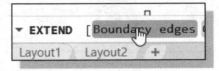

3. Pick the **circle** as the *boundary edge*.

4. Inside the *Drawing Area*, **right-click** to proceed with the Extend command.

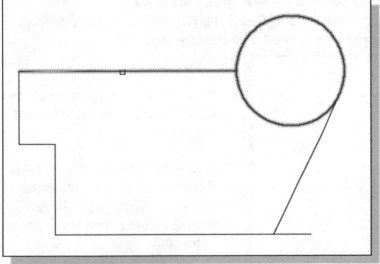

5. The message "*Select object to extend or shift-select object to trim or [Project/Edge/Undo]:*" is displayed in the command prompt area. Extend the **horizontal line** that is to the left side of the circle by clicking near the right endpoint of the line.

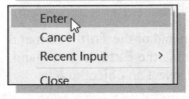

6. Inside the Drawing Area, **right-click** to activate the option menu and select **Enter** with the left-mouse-button to end the Extend command.

Using the Trim Command

The Trim command shortens an object so that it ends precisely at a selected boundary.

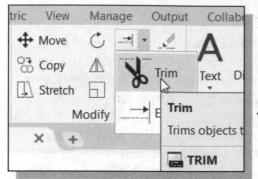

1. Select the **Trim** command icon in the *Modify* toolbar. In the command prompt area, the message "*Select boundary edges... Select objects:*" is displayed.

❖ First, we will select the objects that define the boundary edges to which we want to trim the object.

2. In the command prompt area, click **cuTting edges** as shown.

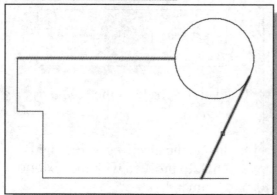

3. Pick the **inclined line** and the **top horizontal line** as the *boundary edges*.

4. Inside the *Drawing Area*, **right-click** to proceed with the Trim command.

5. The message "*Select object to trim or shift-select object to extend or [Fence/Crossing/Project/Edge/eRase/Undo]:*" is displayed in the command prompt area. Pick near the **right endpoint** of the bottom horizontal line.

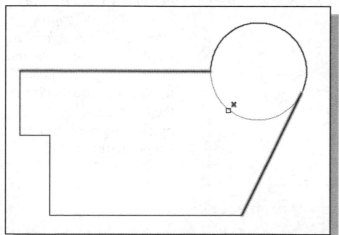

6. Pick the **bottom of the circle** by clicking on the lower portion of the circle.

7. Inside the *Drawing Area*, **right-click** to activate the option menu and select **Enter** with the left-mouse-button to end the Trim command.

• Note that in **AutoCAD 2023**, we can use the Extend command or the Trim command for trimming or extending an object. For example, when using the **Extend** command, we can select an object to *extend* or hold down **SHIFT** and select an object to *trim*.

Creating a TTR Circle

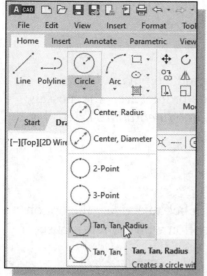

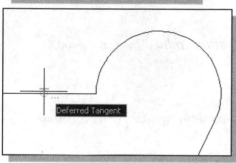

1. In the *Draw* toolbar, click the triangle next to the **Circle** icon to show the additional options.

2. In the displayed list, select the **Tan, Tan, Radius** option. This option allows us to create a circle that is tangent to two objects.

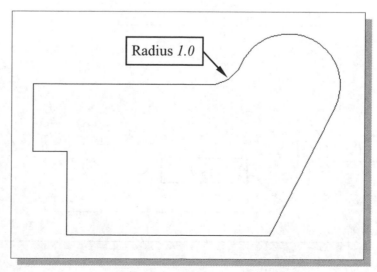

3. Pick the **top horizontal line** that is to the left side of the arc. We will create a circle that is tangent to this line and the circle.

4. Pick the **arc** by selecting a location that is above the right endpoint of the horizontal line. AutoCAD interprets the location we selected as being near the tangency.

5. In the command prompt area, the message "*Specify radius of circle <1.50>*" is displayed.
   Specify radius of circle <1.50>: **1.0** [ENTER]

➤ On your own, use the **Trim** command and trim the circle, the horizontal line, and the arc as shown.

Radius *1.0*

Using the Fillet Command

The Fillet command can be used to round or fillet the edges of two arcs, circles, elliptical arcs, or lines with an arc of a specified radius.

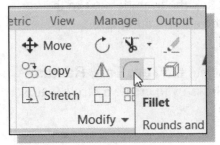

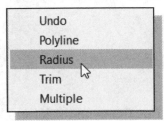

1. Select the **Fillet** command icon in the *Modify* toolbar. In the command prompt area, the message *"Select first object or [Polyline/Radius/Trim]:"* is displayed.

2. Inside the Drawing Area, right-click to activate the option menu and select the **Radius** option with the left-mouse-button to specify the radius of the fillet.

3. In the command prompt area, the message *"Specify fillet radius:"* is displayed.

   Specify fillet radius: **0.75 [ENTER]**

4. Pick the **bottom horizontal line** and the **adjacent vertical line** to create a rounded corner as shown.

➢ On your own, use the Fillet command and create a radius **0.25** fillet at the corner as shown.

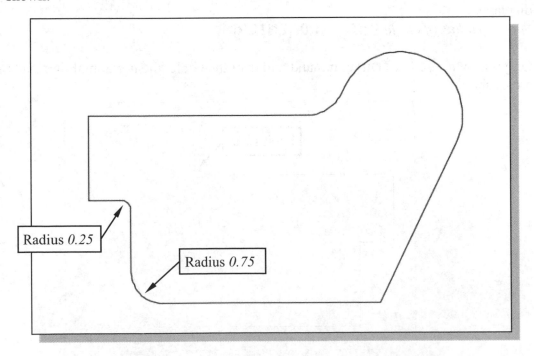

Radius *0.25*

Radius *0.75*

Converting Objects into a Polyline

The next task in our project is to use the Offset command and create a scaled copy of the constructed geometry. Prior to using the Offset command, we will simplify the procedure by converting all objects into a **compound object – a *polyline*.**

A *polyline* in AutoCAD is a 2D line of adjustable width composed of line and arc segments. A polyline is treated as a single object with definable options.

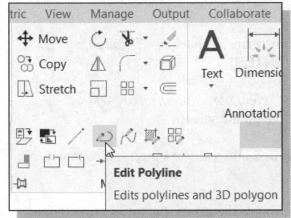

1. In the *Ribbon Toolbars*, select:

   [Modify] → [Edit Polyline]

2. The message "*Select polyline:*" is displayed in the command prompt area. Select **any** of the objects on the screen.

3. The message "*Object selected is not a polyline, Do you want to turn it into one? <Y>*" is displayed in the command prompt area. **Right-click** to accept the *Yes* default.

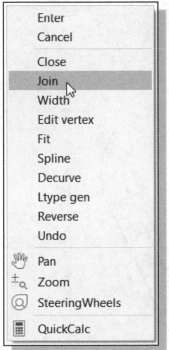

4. Inside the *Drawing Area*, **right-click** once to activate the option menu and select the **Join** option with the left-mouse-button to add objects to the polyline.

5. **Pick all objects** by enclosing them inside a *selection window* (select the top left corner first with a single click).

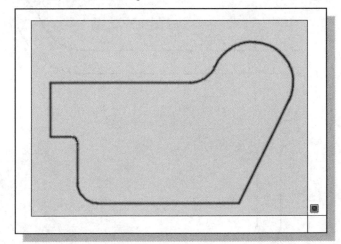

6. Inside the *Drawing Area*, **right-click** once to accept the selected objects.

7. Inside the *Drawing Area*, **right-click** once to activate the option menu and select **Enter** to end the Edit Polyline command.

Using the Offset Command

The Offset command creates a new object at a specified distance from an existing object or through a specified point.

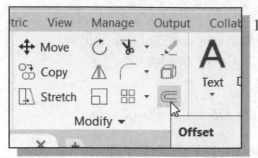

1. Select the **Offset** command icon in the *Modify* toolbar. In the *command prompt area*, the message "*Specify offset distance or [Through]:*" is displayed.

    Specify offset distance or [Through]:
    0.5 [ENTER]

2. In the *command prompt area*, the message "*Select object to offset or <exit>:*" is displayed. Select **any segment** of the polyline on the screen.

➤ Since all the lines and arcs have been converted into a single object, all segments are now selected.

3. AutoCAD next asks us to identify the direction of the offset. Pick a location that is *inside* the polyline.

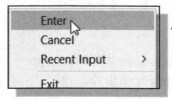

4. Inside the *Drawing Area*, **right-click** and select **Enter** to end the **Offset** command.

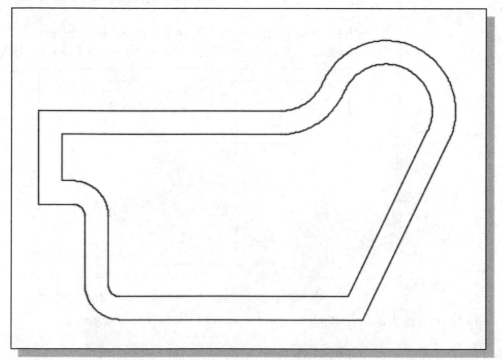

Using the Area Inquiry Tool to Measure Area and Perimeter

AutoCAD also provides several tools that will allow us to measure distance, area, perimeter, and even mass properties. With the use of polylines, measurements of areas and perimeters can be done very quickly.

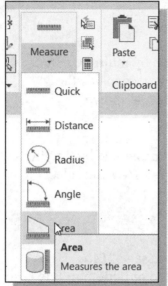

1. In the *Ribbon* tabs area, left-click once on the **Measure** title in the *Utilities* toolbar as shown.

2. In the *Inquiry* toolbar, click on the **Area** icon to activate the Calculates the area and perimeter of selected objects command.

• Note the different **Measure** options that are available in the list.

3. In the command prompt area, the message *"Specify first corner point or [Object/Add/Subtract]:"* is displayed. By default, AutoCAD expects us to select points that will form a polygon. The area and perimeter of the polygon will then be calculated.

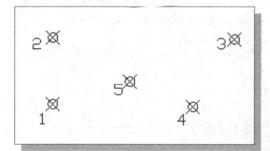

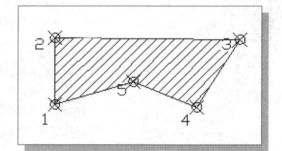

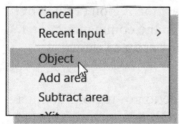

4. AutoCAD can also calculate the area and perimeter of objects that define closed regions. For example, a circle or a rectangle can be selected as both of these objects define closed regions. We can also select a region defined by a *polyline*. To activate this option, **right-click** once inside the Drawing Area and select **Object** as shown.

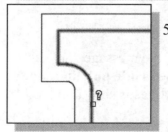

5. In the command prompt area, the message *"Select objects:"* is displayed. Pick the **inside polyline** and the associated area and perimeter information are shown in the prompt area.

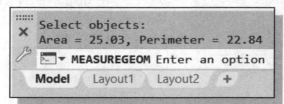

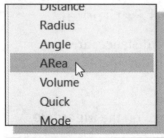

6. Inside the *Drawing Area*, right-click once to bring up the option menu and select **Area** as shown.

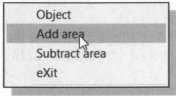

7. We can also select a region defined by multiple *polylines*. To activate this option, **right-click** once inside the Drawing Area and select **Add area** as shown.

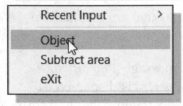

8. To also activate selection of regions *defined* by polylines, **right-click** once inside the *Drawing Area* and select **Object** as shown.

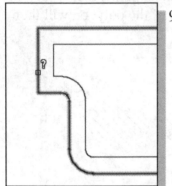

9. In the command prompt area, the message "*[Add Mode] Select objects:*" is displayed. Pick the **outside polyline**. The associated area and perimeter information are shown in the prompt area. **Right-click** once to end the selection.

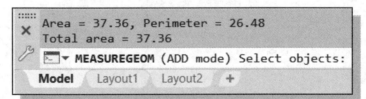

```
Area = 37.36, Perimeter = 26.48
Total area = 37.36
MEASUREGEOM (ADD mode) Select objects:
Model  Layout1  Layout2  +
```

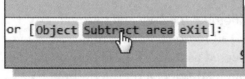

10. Next, we will subtract the region defined by the inside polyline. In the command prompt area, select **Subtract area** as shown.

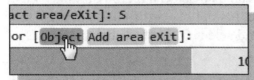

11. To also activate selection of regions defined by polylines, select **Object** in the command prompt area as shown.

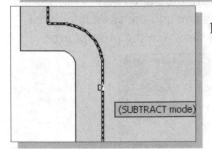

12. In the *command prompt area*, the message "*Select objects:*" is displayed. Pick the **inside polyline** and the area between the two polylines is shown in the *command prompt area*.

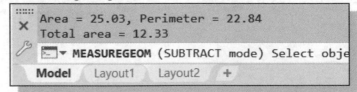

```
Area = 25.03, Perimeter = 22.84
Total area = 12.33
MEASUREGEOM (SUBTRACT mode) Select obje
Model  Layout1  Layout2  +
```

Using the Explode Command

The Explode command breaks a compound object into its component objects.

1. Select the **Explode** command icon in the *Modify* toolbar. In the command prompt area, the message "*Select objects:*" is displayed.

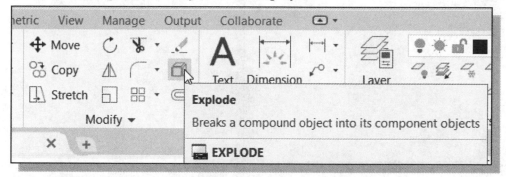

2. Pick the *inside polyline* that we created using the Offset command.

3. Inside the Drawing Area, **right-click** to end the Explode command.

Create another Fillet

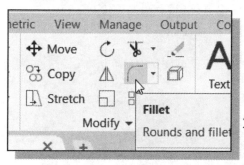

1. Select the **Fillet** command icon in the *Modify* toolbar. In the command prompt area, the message "*Select first object or [Polyline/Radius/Trim]:*" is displayed.

2. On your own, set the *Fillet Radius* to **0.5.** *Specify fillet radius:* **0.5 [ENTER]**

3. Pick the **horizontal line** and the **adjacent inclined line** to create a rounded corner as shown.

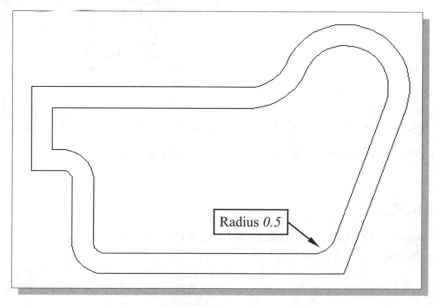

Radius *0.5*

Saving the CAD File

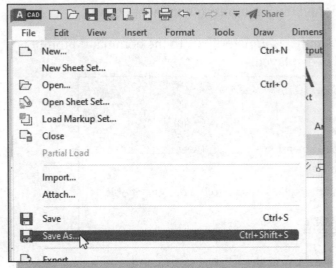

1. In the pull-down menus, select:

    [File] → [Save As]

2. In the *Save Drawing As* dialog box, select the folder in which you want to store the CAD file and enter **Gasket** in the *File name* box.

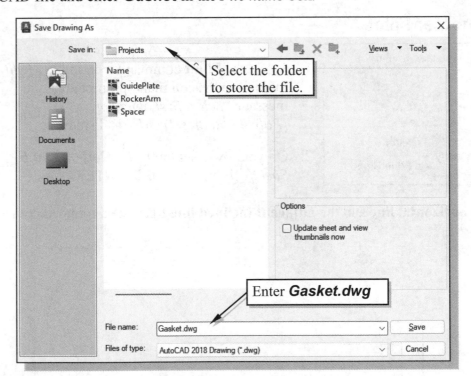

3. Pick **Save** in the *Save Drawing As* dialog box to accept the selections and save the file.

Exit AutoCAD

To exit **AutoCAD 2023**, select **Exit AutoCAD** from the *Application Menu* or type **QUIT** at the command prompt.

Review Questions: (Time: 20 minutes)

1. Describe when and why you would use the AutoCAD **ORTHO** option.

2. What is the difference between a *line* and a *polyline* in AutoCAD?

3. Which AutoCAD command can we use to break a compound object, such as a polyline, into its component objects?

4. Which AutoCAD command can we use to quickly calculate the area and perimeter of a closed region defined by a polyline?

5. Describe the procedure to calculate the area and perimeter of a closed region defined by a polyline?

6. What does the **Offset** command allow us to do?

7. Create the following triangle and measure the area and perimeter of the triangle.

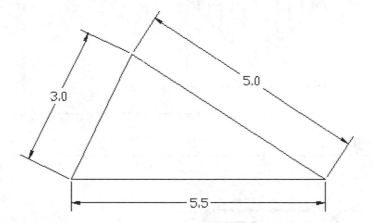

8. Create the following triangle and measure the area and perimeter of the triangle. Also find the angle Θ.

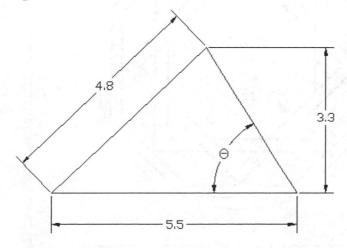

Exercises:

(Unless otherwise specified, dimensions are in inches.) (Time: 150 minutes)

1. Lines & Squares Pattern

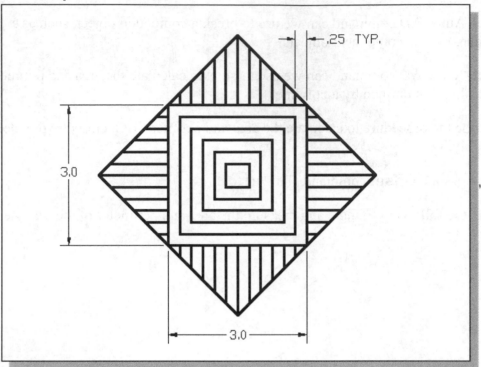

2. Interlacement Design

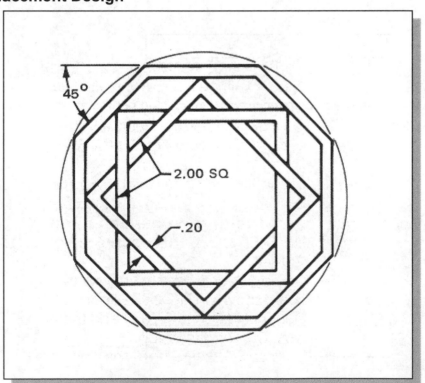

3. Polygons & Circles Pattern (Dimensions are in inches.)

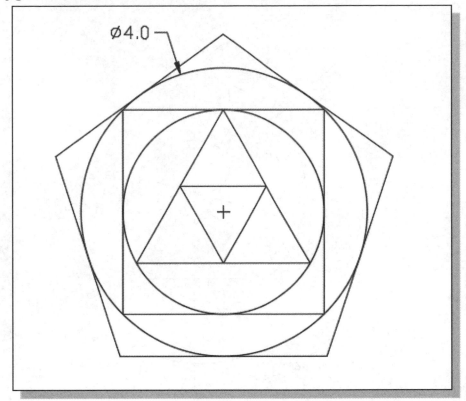

4. Positioning Spacer (Dimensions are in inches.)

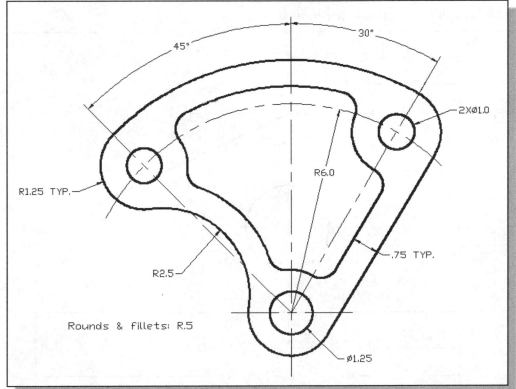

5. Indexing Base (Dimensions are in inches.)

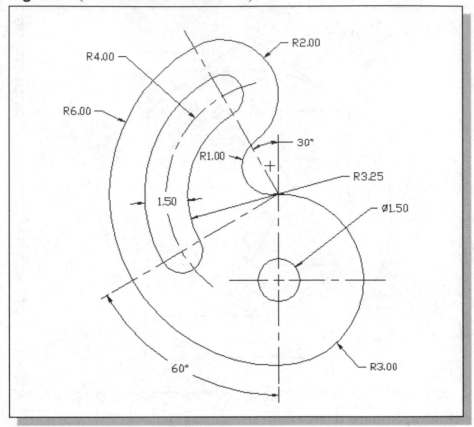

6. Guide Block (Dimensions are in inches.)

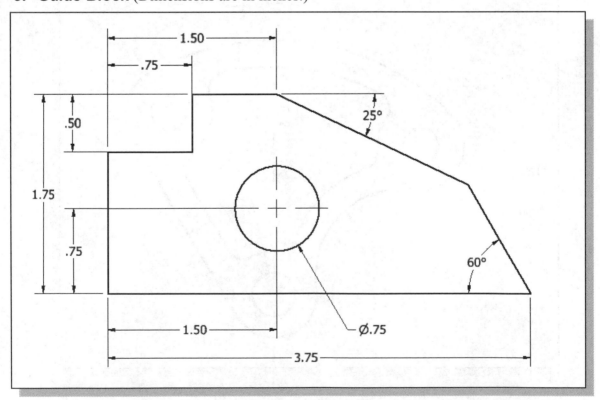

7. Slider Guide (Dimensions are in inches.)

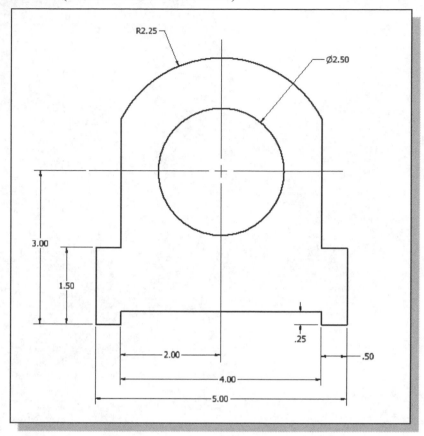

8. Vent Cover

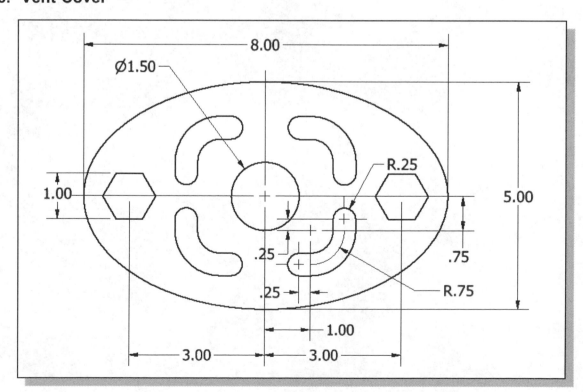

Notes:

Chapter 4
Object Properties and Organization

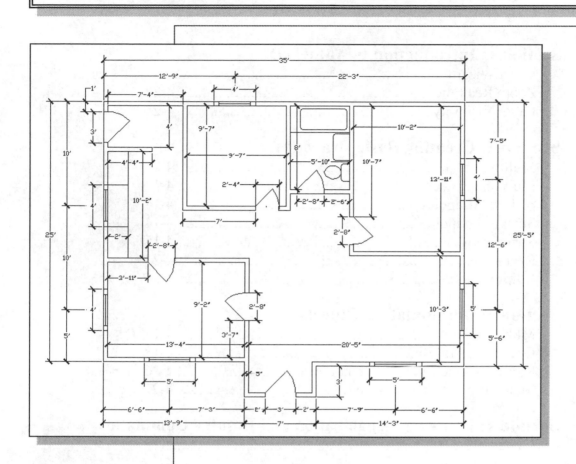

Learning Objectives

- ♦ **Using the AutoCAD Quick Setup Wizard**
- ♦ **Create new Multiline Styles**
- ♦ **Draw using the MULTILINE command**
- ♦ **Use the Multiline Editing commands**
- ♦ **Create new layers**
- ♦ **Pre-selection of objects**
- ♦ **Controlling Layer Visibility**
- ♦ **Moving objects to a different layer**

AutoCAD Certified User Examination Objectives Coverage

This table shows the pages on which the objectives of the Certified User Examination are covered in Chapter 4.

Certified User Reference Guide

Introduction

The CAD database of a design may contain information regarding the hundreds of CAD entities that are used to create the CAD model. One of the advantages of using a CAD system is its ability to organize and manage the database so that the designer can access the information quickly and easily. Typically, CAD entities that are created to describe one feature, function, or process of a design are perceived as related information and therefore are organized into the same group. In AutoCAD, the Layer command is used extensively for this purpose. For example, an architectural drawing typically will show walls, doors, windows, and dimensions. Using layers, we can choose to display or hide sub-systems for clarity; we can also change object properties, such as colors and linetypes, quickly and easily.

In this chapter, we will continue to explore the different construction and editing tools that are available in **AutoCAD 2023**. We will demonstrate the use of the Limits, Mline, Medit, and Layer commands. As you become proficient with the CAD tools and understand the underlying CAD modeling concepts, you are encouraged to experiment with new ideas in using the CAD tools and develop your own style of using the system.

The Floor Plan Design

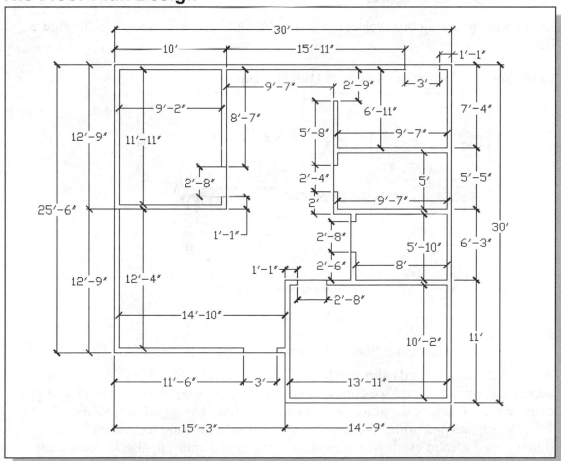

Starting Up AutoCAD 2023

1. Select the **AutoCAD 2023** option on the *Program* menu or select the **AutoCAD 2023** icon on the *Desktop*. Once the program is loaded into the memory, the **AutoCAD 2023** drawing screen will appear on the screen.

Using the Setup Wizard

1. In the *Startup* dialog box, select the **Use a Wizard** option as shown in the figure below.

2. In the *Select a Wizard* section, pick **Quick Setup**.

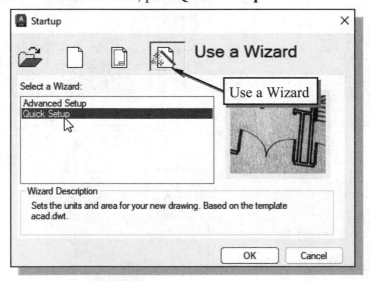

- AutoCAD setup wizards allow us to customize several of the AutoCAD settings depending on the wizard we choose. The *Quick Setup* wizard sets the units and grid display area. Choices for units include *Decimal, Engineering, Architectural, Fractional,* and *Scientific*. We can also specify the width and length of a two-dimensional area to establish the extents of the *grid* displayed, also known as the *limits* of the working area.

Drawing Units Setup

1. In the *Quick Setup Units* option, select **Architectural**.

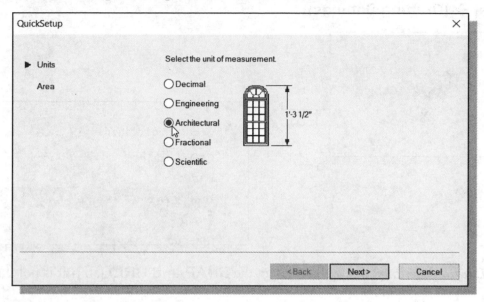

2. Pick **Next** to continue with the *Quick Setup* settings.

Reference Area Setup

1. In the *Quick Setup Area* option, enter **60′** and **40′** for the width and length.

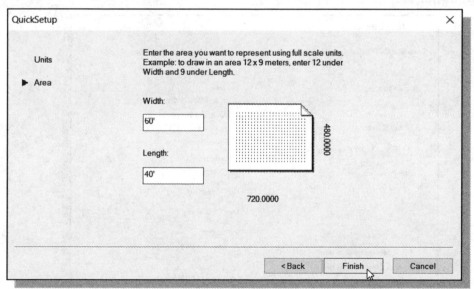

- The two-dimensional area we set up in the *Quick Setup* is the *drawing limits* in AutoCAD.

2. Pick **Finish** to accept the settings and end the *Quick Setup* wizard.

GRID and *SNAP* Intervals Setup

1. In the *Menu Bar*, select:
    [Tools] → [Drafting Settings]

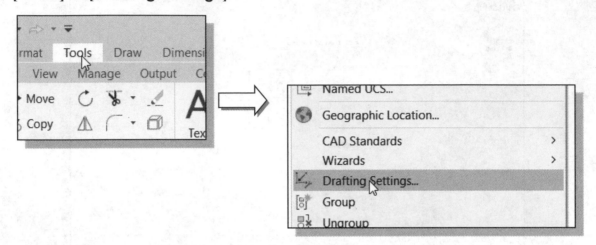

2. In the *Drafting Settings* dialog box, select the **SNAP and GRID** tab if it is not the page on top.

3. Change *Grid Spacing* to **6″** for both X and Y directions.

4. Also adjust the *Snap Spacing* to **6″** for both X and Y directions.

5. Turn **OFF** the **Adaptive Grid** option. (This switch is used to limit the grid display when zooming.)

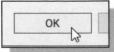   6. Pick **OK** to exit the *Drafting Settings* dialog box.

Using the Zoom Extents Command in the Navigation Bar

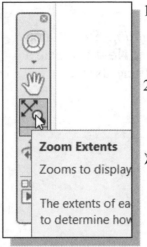

1. Move the cursor inside the *Drawing Area* and notice that, although we have set the *limits* to 40' by 60', the default display is still not adjusted to the new settings.

2. In the *Navigation* toolbar, select **Zoom Extents** by clicking the left-mouse-button on the icon as shown. We can also click on the triangle icon to select other *Zoom* options.

➢ The *navigation bar* is a user interface element that provides quick access to display related tools, such as **Zoom**, **Pan** and **3D rotation.**

The AutoCAD Multiline Command

The **Multiline** command in **AutoCAD 2023** is used to create multiple parallel lines. This command is very useful for creating designs that contain multiple parallel lines, such as walls for architectural designs and for highway designs in civil engineering. The Multiline command creates a sct of parallel lines (up to 16 lines) and all line segments are grouped together to form a single *multiline object*, which can be modified using Multiline Edit and Explode commands. We will first create a new ***multiline style*** for our floor plan design.

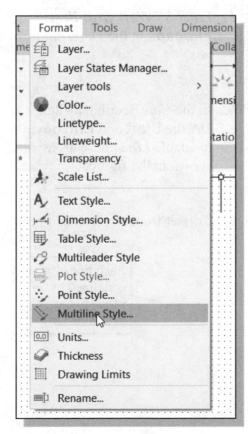

1. In the *Menu Bar*, select:

   [Format] → [Multiline Style]

❖ The default AutoCAD multiline style is called *STANDARD*, and it consists of two elements (two parallel lines) with an offset distance of 1.0 inch.

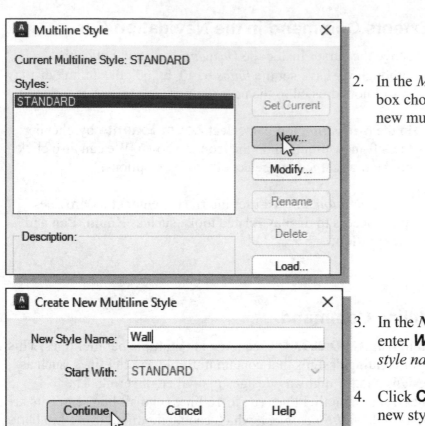

2. In the *Multiline Style* dialog box choose **New** to create a new multiline style.

3. In the *New Style Name* box, enter ***Wall*** as the new *multiline style name*.

4. Click **Continue** to create the new style.

5. Enter ***5" Wall with line endcaps*** in the *Description* box.

6. In the *Caps* section, switch ***ON*** the **Start** and **End** boxes to enable *Line* end-caps as shown in the figure.

❖ All line elements in the multiline style are defined by an offset from a reference line, the *multiline origin*.

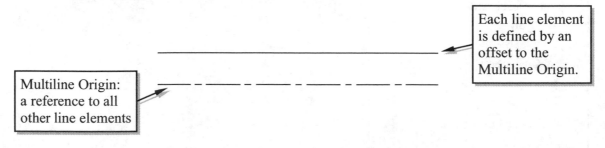

Each line element is defined by an offset to the Multiline Origin.

Multiline Origin: a reference to all other line elements

❖ Note that in the *Elements* section, all the line elements are listed in descending order with respect to their offsets. We will create two line elements representing a five-inch wall (offsetting on both sides of the reference location).

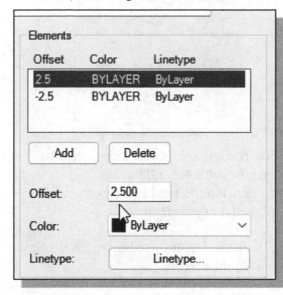

7. In the *Elements* section, highlight the first element in the list, and change the *Offset* to **2.5**.

8. Pick the *second element* in the *Elements* section and change the *Offset* to **-2.5**.

9. Choose **OK** to exit the *Element Properties* dialog box.

❖ Notice the **Add** and **Delete** options are also available, which allow us to create or remove additional elements.

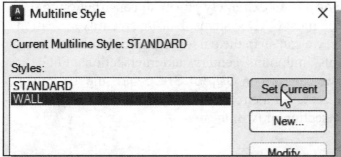

10. With the **Wall** style highlighted, click **Set Current** as shown.

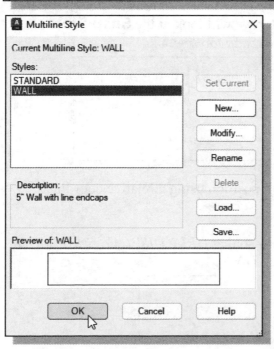

11. Click the **OK** button to end the **Multiline Style** command. (Note: The *Preview* section shows the current **WALL** style.)

❖ Note that the **Save** button will save a multiline style to the library of multiline styles. By default, AutoCAD saves the multiline styles information to a file called **acad.mln**. The **Load** button allows users to retrieve multiline styles from a library.

Object Snap Toolbar

1. Move the cursor to the *Menu Bar* area and choose **[Tools]** → **[Toolbars]** → **[AutoCAD]**.

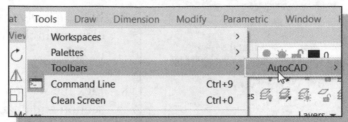

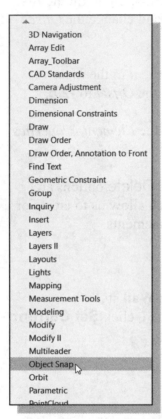

❖ AutoCAD provides 50+ predefined toolbars for access to frequently used commands, settings, and modes. A *checkmark* (next to the item) in the list identifies the toolbars that are currently displayed on the screen.

2. Select **Object Snap**, with the left-mouse-button, to display the *Object Snap* toolbar on the screen.

❖ **Object Snap** is an extremely powerful construction tool available on most CAD systems. During an entity's creation operations, we can snap the cursor to points on objects such as endpoints, midpoints, centers, and intersections. For example, we can turn on **Object Snap** and quickly draw a line to the center of a circle, the midpoint of a line segment, or the intersection of two lines.

3. In the previous chapter, we used several of the object snap options to quickly locate positions on existing geometry. In this chapter we will look at the **Snap From** option, which is the second icon in the *Object Snap toolbar*.

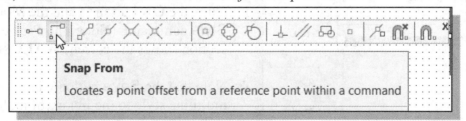

Snap From

Locates a point offset from a reference point within a command

➢ The **Snap From** option allows us to locate a position using a relative coordinate system with respect to a selected position.

4. In the *Status Bar* area, reset the option buttons so that *GRID DISPLAY, SNAP MODE, DYNAMIC INPUT*, and *ORTHO* are switched **ON**.

Drawing Multilines

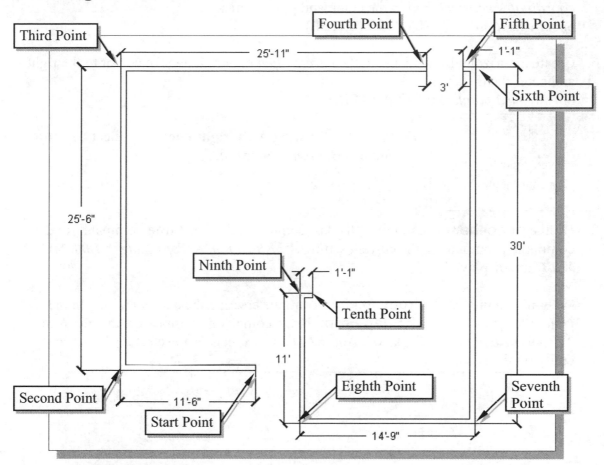

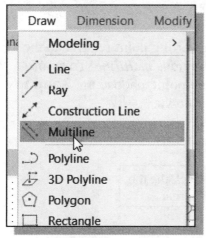

1. Select the **Multiline** command icon in the *Draw* pull-down menu, through the *Menu Bar,* as shown. In the command prompt area, the current settings, such as "*Justification = **Top**, Scale = **1.00**, Style = **Wall**"* are displayed.

2. On your own, confirm the *Scale* is set to **1.00**, the *Styles* to **Wall** and the *Justification* to **Top**. You can use the **right-mouse-button** to bring up the option list and adjust the settings if needed.

3. In the command prompt area, the message "*Specify start point or [Justification/Scale/Style]:*" is displayed. Select a location **near** the bottom center of the *Drawing Area* as the **start point** of the multiline.

4. Create a horizontal line by using the *Dynamic Input* option or the *relative rectangular coordinates entry method* in the command prompt area:
   Specify next point: **@-11'6",0 [ENTER]**

5. Create a vertical line by using the *Dynamic Input* option or the *relative rectangular coordinates entry method* in the command prompt area:
    Specify next point: **@25'6"<90 [ENTER]**

6. Create a horizontal line by using the *Direct Input* option; move the cursor to the right and enter the distance:
    Specify next point: **25'11" [ENTER]**

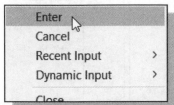

7. Inside the Drawing Area, **right-click** and select **Enter** to end the *Multiline* command.

8. Hit the **spacebar** once to repeat the last command, the **Multiline** command. In the command prompt area, the current settings "*Justification = Top, Scale = 1.00, Style = Wall*" are displayed.

9. We will use the **Snap From** option to continue creating the exterior walls. In the *Object Snap* toolbar, pick **Snap From**. In the command prompt area, the message "*_from Base point*" is displayed. AutoCAD now expects us to select a geometric entity on the screen.

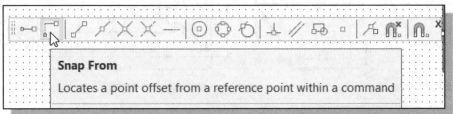

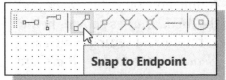

10. We will position the starting point relative to the last position of the previous *multiline*. To assure the selection of the endpoint, choose the **Snap to Endpoint** option as shown.

11. Pick the **upper corner** of the top horizontal multiline as shown.

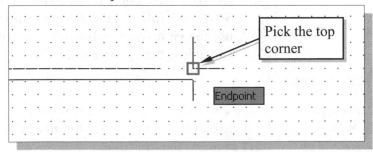

➢ Note that it is feasible to stack snap options for precise positioning of geometry.

12. The position of the starting point of the new *multiline* segments is 3′ to the right of the reference point we just picked. At the command prompt, enter @**3',0"** [**ENTER**].

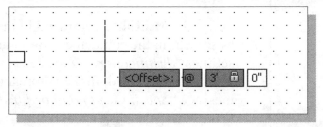

13. Now enter @**1'1",0**, to define the top corner of the exterior wall.

14. On your own, complete the multiline segments by specifying the rest of the corners using the dimensions as shown in the figure below.

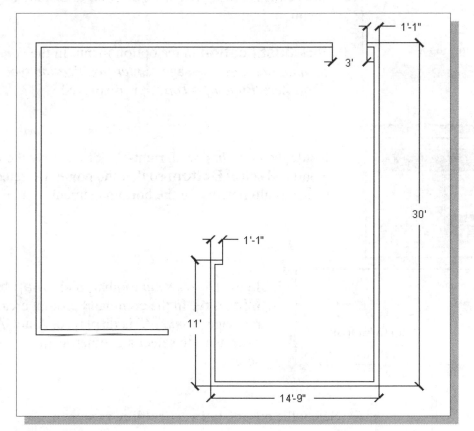

• Note that the points we specified are defining the outside corners of the floor plan design.

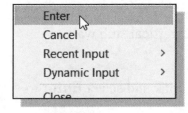

15. Inside the Drawing Area, right-click and select **Enter** to end the Multiline command.

Creating Interior Walls

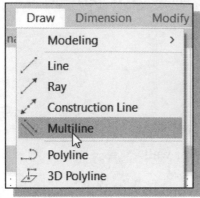

1. Select the **Multiline** command in the ***Draw*** *Menu Bar* as shown. In the command prompt area, the current settings "*Justification* = ***Top***, *Scale* = ***1.00***, *Style* = ***Wall***" are displayed. In the command prompt area, the message "*Specify start point or [Justification/ Scale/Style]:*" is displayed.

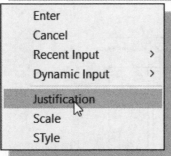

2. Inside the *Drawing Area*, **right-click** to display the option menu.

3. Pick **Justification** in the option menu. In the *command prompt area*, the message "*Enter justification type [Top/Zero/Bottom] <Top>:*" is displayed.

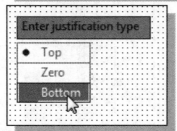

4. Inside the *Drawing Area*, right-click to display the option menu and select **Bottom** so that the points we select will be set as alignments for the bottom element.

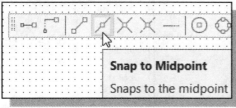

5. In the *Object Snap* toolbar, pick **Snap to Midpoint**. In the command prompt area, the message "*_mid of*" is displayed. AutoCAD now expects us to select a geometric entity on the screen.

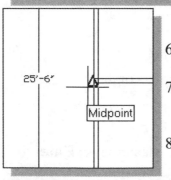

6. Select the outside **left vertical line** as shown.

7. Using the *Dynamic Input* option, create a ***10'*** inside wall toward the right.

8. Now enter ***@0,1'1"*** to define the vertical stub wall.

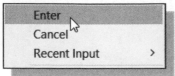

9. Inside the *Drawing Area*, right-click and select **Enter** to end the **Multiline** command.

- Next, we will create a vertical wall right above the last corner.

10. Hit the **spacebar** once to repeat the last command, the **Multiline** command. Since the last position used is directly below the new location, we will just enter the relative coordinates. At the command prompt, enter **@0,2'8"** [ENTER].

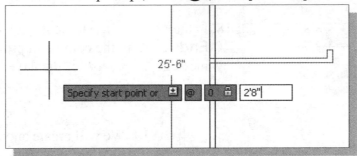

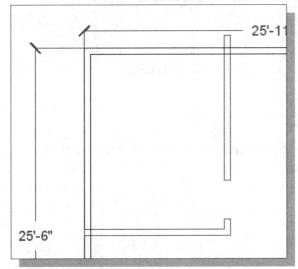

11. Place the other end above the top horizontal line as shown in the figure below. In the next section, we will use the **Multiline Edit** tools to adjust these constructions.

12. Inside the Drawing Area, right-click and select **Enter** to end the Multiline command.

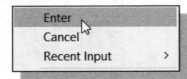

13. Hit the **spacebar** once to repeat the last command, the **Multiline** command. In the text window, the current settings "*Justification = Bottom, Scale = 1.00, Style = Wall*" are displayed. In the command prompt area, the message "*Specify start point or [Justification/Scale/Style]:*" is displayed.

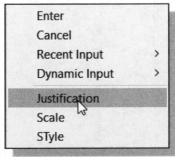

14. Inside the *Drawing Area*, **right-click** to display the option menu.

15. Pick **Justification** in the option menu. In the command prompt area, the message "*Enter justification type [Top/Zero/Bottom] <Bottom>:*" is displayed.

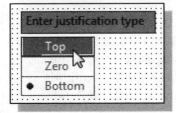

16. Inside the *Drawing Area*, right-click to display the *option menu* and select **Top** so that the points we select will align to the top element.

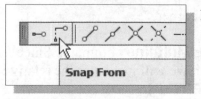

17. In the *Object Snap* toolbar, pick **Snap From**. In the command prompt area, the message "*_from Base point*" is displayed. AutoCAD now expects us to select a geometric entity on the screen.

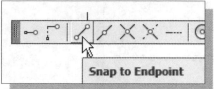

18. In the *Object Snap* toolbar, pick **Snap to Endpoint**. In the command prompt area, the message "*_endp of*" is displayed.

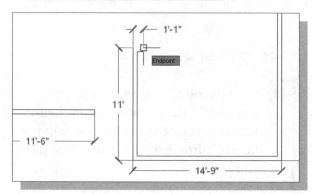

19. We will create another inside wall on the right side. Pick the corner as shown.

20. At the command prompt, enter **@2′8″,0** [ENTER].

21. Pick a location that is to the right of the right-vertical exterior wall. The drawing should appear as shown in the figure below.

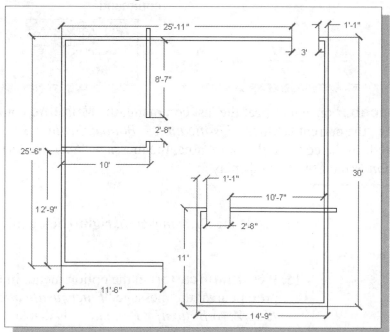

❖ One of the main advantages of using a CAD system to create drawings is the ability to create and/or modify geometric entities quickly, using many of the available tools. Unlike traditional board drafting, where typically only the necessary entities are created, CAD provides a much more flexible environment that requires a slightly different way of thinking, as well as taking a different view of the tasks at hand.

Joining the Walls Using Multiline Edit

1. In the *Menu Bar*, select **[Modify]** → **[Object]**→ **[Multiline]**.

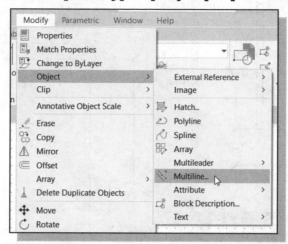

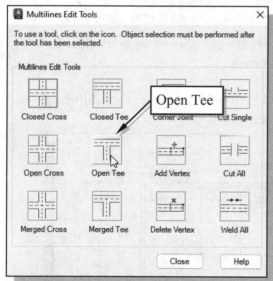

➤ The *Multiline Edit Tools* dialog box appears. Select the **Help** button to see the description of the available *Multiline Edit Tools*.

2. Pick the **Open Tee** option in the dialog box.

➤ We will need to select two multilines for this option: first, select the multiline to trim or extend; and second, select the intersecting multiline.

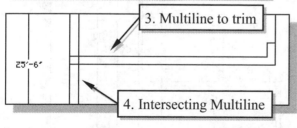

3. Pick the **horizontal multiline** as the first object, *multiline to trim,* as shown.

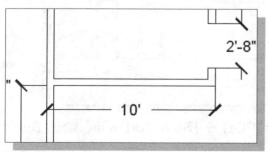

4. Pick the **vertical multiline** as the 2nd object, *intersecting multiline.*

➤ The **Open Tee** option automatically trims the lines to form the proper shapes.

5. Repeat the above steps and modify the connection of the other two inside walls.

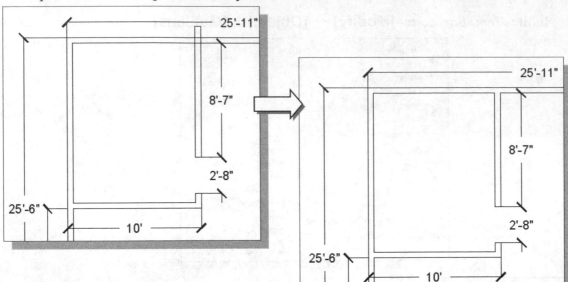

6. Inside the Drawing Area, right-click once and select **Enter** to end the **Mledit** command.

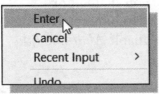

7. Using the *Multiline/MEdit* options, create the additional walls and doorways as shown.

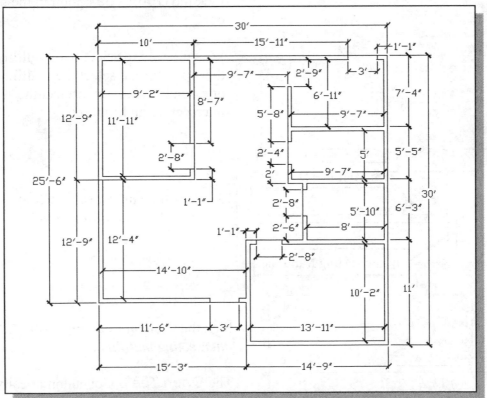

❖ Now is a good time to save the design. Select **[File] → [Save As]** in the *Menu Bar* and use ***FloorPlan*** as the *File name.*

Using Layers and Object Properties

In **AutoCAD 2023**, *layers* can be thought of as transparent overlays on which we organize different kinds of design information. Typically, CAD entities that are created to describe one feature or function of a design are considered as related information and therefore can be organized into the same group. The objects we organized into the same group will usually have common properties such as colors, linetypes, and lineweights. Color helps us visually distinguish similar elements in our designs. Linetype helps us identify easily the different drafting elements, such as centerlines or hidden lines. Lineweight increases the legibility of an object through width. Consider the floor plan we are currently working on. The floor plan can be placed on one layer, electrical layout on another, and plumbing on a third layer. Organizing layers and the objects on layers makes it easier to manage the information in our designs. Layers can be used as a method to control the visibility of objects. We can temporarily switch *ON* or *OFF* any layer to help construction and editing of our designs.

AutoCAD allows us to create an infinite number of layers. In general, twenty to thirty layers are sufficient for most designs. Most companies also require designers and CAD operators to follow the company standards in organizing objects in layers.

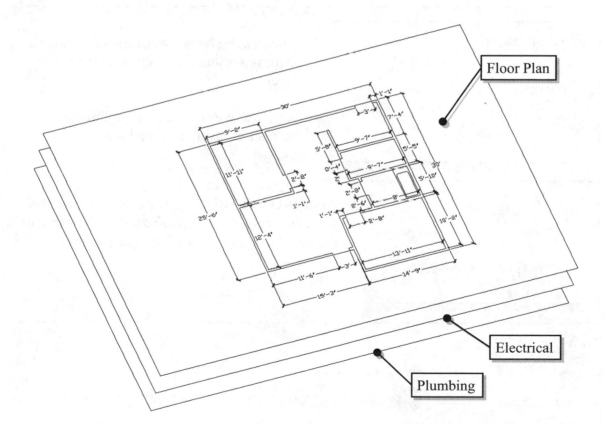

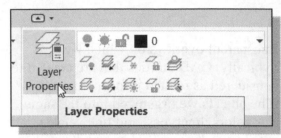

1. Pick **Layers Properties** in the *Layers* toolbar as shown.

❖ The *Layer Properties Manager* dialog box appears. AutoCAD creates a default layer, *layer 0*, which we cannot rename or delete. Note that *Layer 0* has special properties that are used by the system.

❖ In AutoCAD, we always construct entities on a layer. It may be the default layer or a layer that we create. Each layer has associated properties such as the visibility setting, color, linetype, lineweight, and plot style.

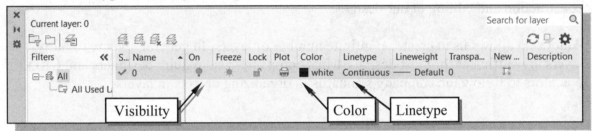

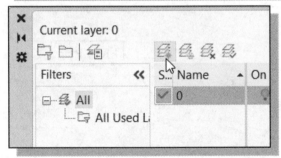

2. Click on the **New Layer** button. Notice a layer is automatically added to the list of layers.

➢ Note that we can create an unlimited number of layers in a drawing.

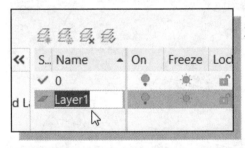

3. AutoCAD will assign a generic name to the new layer (*Layer1*). Enter **BathRoom** as the name of the new layer as shown in the figure below.

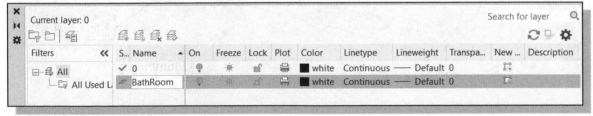

❖ Layer properties can be adjusted by clicking on the icon or name of a property. For example, clicking on the *light-bulb* icon toggles the visibility of the layer *ON* or *OFF*.

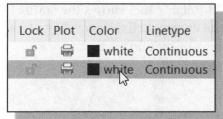

4. Pick the color swatch or the color name (**White**) of the *BathRoom* layer. The *Select Color* dialog box appears.

5. Pick **Cyan** *(Index color: 4)* in the *Standard Colors* section. Notice the current color setting is displayed at the bottom of the dialog box.

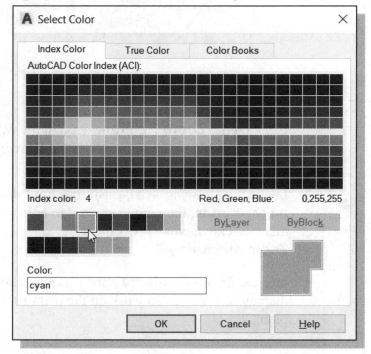

6. Click on the **OK** button to accept the color assignment.

7. Click on the **Set Current** button to make *BathRoom* the *Current Layer*. There can only be one *Current Layer*, and new entities are automatically placed on the layer that is set to be the *Current Layer*.

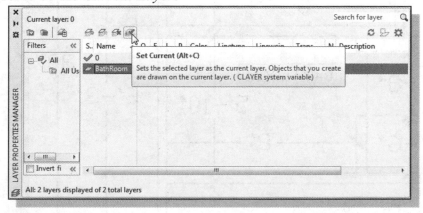

8. Click on the **Close** button, at the upper left corner of the dialog box, to accept the settings and exit the *Layer Properties Manager* dialog box.

❖ The *Layer Control* toolbar in the top of the AutoCAD toolbar panel shows the status of the active layer. The *BathRoom* layer is shown as the current active layer. Note that this *Layer* toolbar can also be used to control the settings of individual layers.

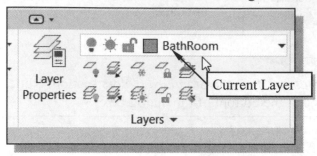

Use the Zoom Realtime option

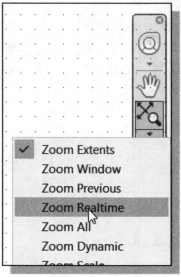

1. Click on the **Zoom Realtime** icon in the *Navigate* toolbar located to the right side of the Drawing Area.

2. Inside the *Drawing Area*, **push and hold down the left-mouse-button**, then move upward to enlarge the current display scale factor. (Press the [**Esc**] key to exit the Zoom command.)

3. Use the **Zoom Realtime** option to reposition the display so that we can work on the bathroom of the floor plan.

➢ Note that the mouse-wheel can also be used to Zoom Realtime; turning the wheel forward will enlarge the current display scale factor.

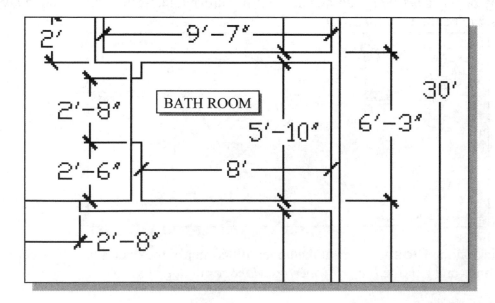

Modeling the Bathroom

1. In the *Status Bar* area, reset the option buttons so that all of the buttons are switched *OFF*.

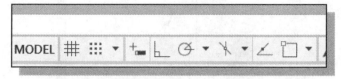

2. Click on the **Rectangle** command icon in the *Draw* toolbar. In the *command prompt area*, the message "*Specify first corner point:*" is displayed.

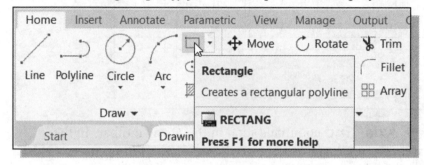

3. In the *Object Snap* toolbar, pick **Snap to Endpoint**. In the command prompt area, the message "*_endp of*" is displayed. AutoCAD now expects us to select a geometric entity on the screen.

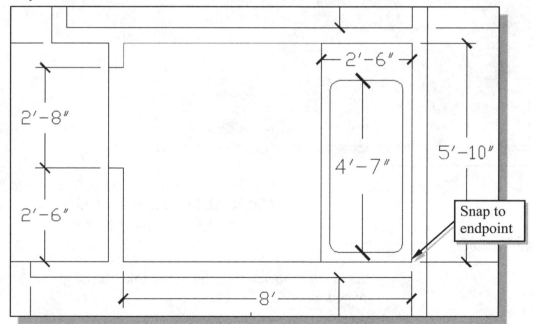

4. Use the *Dynamic Input* options and create the outer rectangle of the tub (**2'-6" × 5'-10"**).

5. Complete the inner shape by creating a rectangle with a distance of 3" from the outer rectangle and rounded corners of 3" radius.

6. Create two rectangles (*10″ × 20″* and *20″ × 30″*) with rounded corners (radius **3″**) and position them as shown.

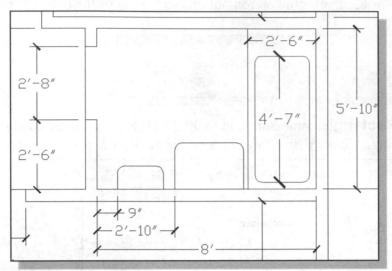

7. Select the **Ellipse → Axis, End** command icon in the *Draw* toolbar. In the command prompt area, the message "*Specify axis endpoint of ellipse or [Arc/Center]:*" is displayed.

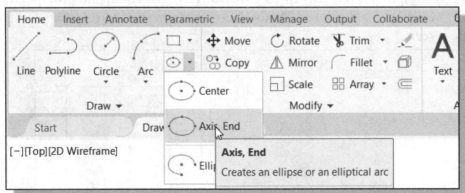

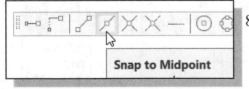

8. In the *Object Snap* toolbar, pick **Snap to Midpoint**. In the command prompt area, the message "*_mid of*" is displayed.

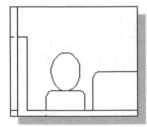

9. Pick the **top horizontal line** of the small rectangle we just created.

10. For the second point location, enter **@0,20″**[ENTER].

11. For the third point, enter **@7.5″,0** [ENTER].

❖ An ellipse has a major axis, the longest distance between two points on the ellipse, and a minor axis, the shorter distance across the ellipse. The three points we specified identify the two axes.

Controlling Layer Visibility

AutoCAD does not display or plot the objects that are on invisible layers. To make layers invisible, we can *freeze* or *turn off* those layers. Turning off layers only temporarily removes the objects from the screen; the objects remain active in the CAD database. Freezing layers will make the objects invisible and also disable the objects in the CAD database. Freezing layers will improve object selection performance and reduce regeneration time for complex designs. When we *thaw* a frozen layer, AutoCAD updates the CAD database with the screen coordinates for all objects in the design.

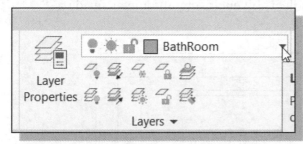

1. On the *Layers* toolbar panel, choose the triangle next to the **Layer Control** box with a click of the left-mouse-button.

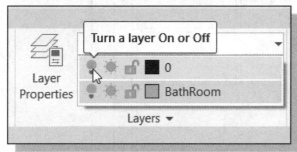

2. Move the cursor over the **lightbulb** icon for *layer 0*. The tool tip *"Turn a layer On or Off"* appears.

3. **Left-click once** and notice the icon color is changed to a dark color, representing the layer (*Layer 0*) is turned *OFF*.

4. Move the cursor into the Drawing Area and **left-click once** to accept the layer control settings.

➢ On your own, practice turning on *Layer 0* and freezing/thawing *Layer 0*. What would happen if we turn off all layers?

Adding a New Layer

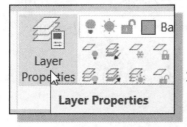

1. Pick **Layer Properties Manager** in the *Layers* toolbar panel. The *Layer Properties Manager* dialog box appears.

2. Create a new layer (layer name: **Walls**) and change the layer color to **Green**.

3. Turn **ON** the *0* layer, turn **OFF** the *BathRoom* layer, and set the **Walls** layer as the *Current Layer*. Click on the **Close** button to exit *Layer Properties*.

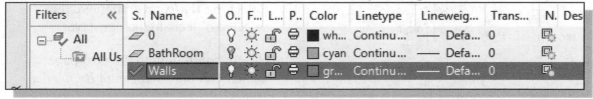

Filters	«	S..	Name	▲	O..	F...	L...	P..	Color	Linetype	Lineweig...	Trans...	N.	Des
⊟ 🗐 All		⬦	0		💡	☼	🔓	🖶	■ wh...	Continu...	—— Defa...	0	🗗	
🗀 All Us		⬦	BathRoom		💡	☼	🔓	🖶	□ cyan	Continu...	—— Defa...	0	🗗	
		✓	Walls		💡	☼	🔓	🖶	□ gr...	Continu...	—— Defa...	0	🗗	

Moving Objects to a Different Layer

AutoCAD 2023 provides a flexible graphical user interface that allows users to select graphical entities BEFORE the command is selected (*pre-selection*), or AFTER the command is selected (*post-selection*). The procedure we have used so far is the *post-selection* option. We can pre-select one or more objects by clicking on the objects at the command prompt (**Command:**). To deselect the selected items, press the **[Esc]** key twice.

1. Inside the Drawing Area, pre-select all objects by enclosing all objects inside a **selection window** as shown.

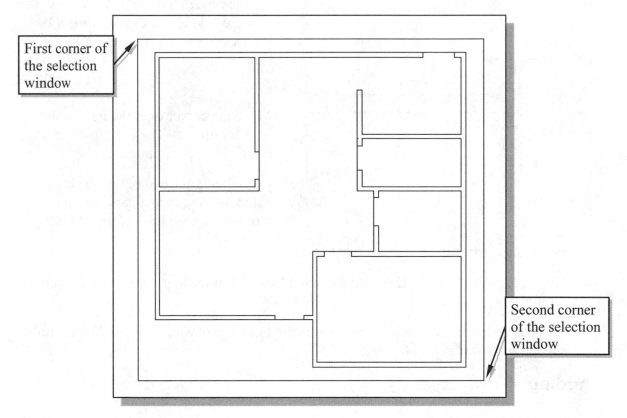

First corner of the selection window

Second corner of the selection window

2. On the *Object Properties* toolbar, choose the ***Layer Control*** box with the left-mouse-button.

❖   Notice the layer name displayed in the *Layer Control* box is the selected object's assigned layer and layer properties.

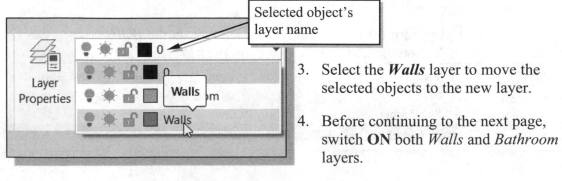

Selected object's layer name

3. Select the ***Walls*** layer to move the selected objects to the new layer.

4. Before continuing to the next page, switch **ON** both *Walls* and *Bathroom* layers.

Matching Layer Properties

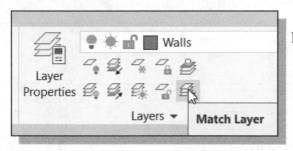

1. Pick **Match** in the *Layer control* toolbar panel. In the command prompt area, the message "*Select objects to be changed:*" is displayed.

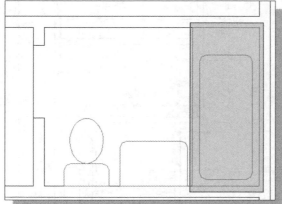

2. Select the **bathtub** using a selection window as shown.

3. Inside the Drawing Area, **right-click** once to accept the selection.

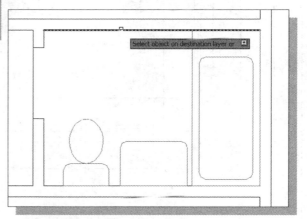

4. Sclect one of the **Walls** as the object properties to match.

➤ In the *command prompt area,* notice the selected objects have been moved to the Walls layer.

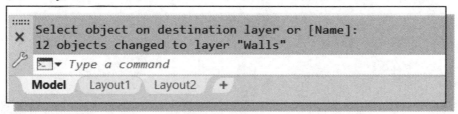

```
::::::
 X   Select object on destination layer or [Name]:
     12 objects changed to layer "Walls"
     ▼ Type a command
   Model   Layout1   Layout2   +
```

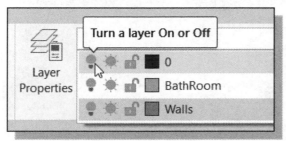

5. On your own, switch **on** and **off** of the ***Walls and bathroom*** layers to examine the results of the *Match Layer Properties* command.

➢ On your own, complete the floor plan by creating the 4′ and 5′ windows in a new
   layer *Windows*. The dimensions are as shown in the figure below.

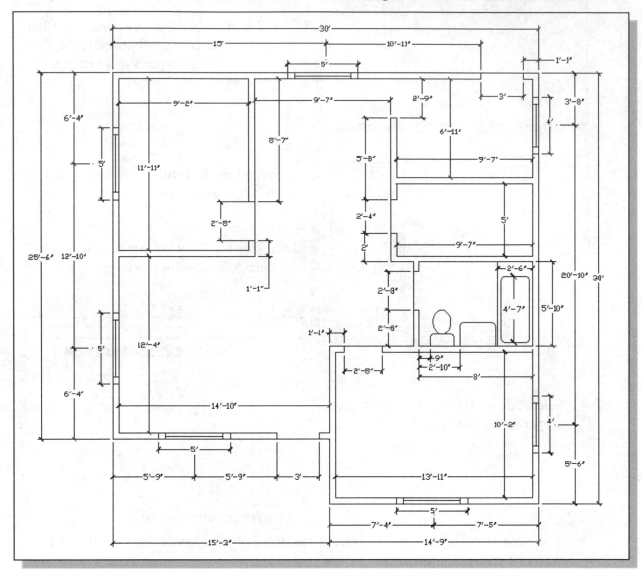

Review Questions: (Time: 25 minutes)

1. List some of the advantages of using *layers*.

2. List two methods to control the *layer visibility* in **AutoCAD 2023**.

3. Describe the procedure to move objects from one layer to another.

4. When and why should you use the Multiline command?

5. Is there a limitation to how many layers we can set up in AutoCAD?

6. List and describe the two options available in AutoCAD to create ellipses.

7. Is there a limitation to how many parallel lines we can set up when using AutoCAD Multiline objects?

8. What is the name of the layer that AutoCAD creates as the default layer (the layer that we cannot rename or delete)?

9. When and why would you use the Match Properties command?

10. A **chamfer** connects two objects with an angled line. A chamfer is usually used to represent a beveled edge on a corner. Construct the following corners by using the **Chamfer** command.

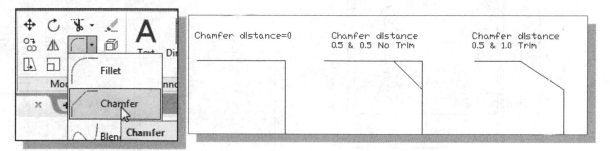

11. List the commands you would use to create the following multilines in a drawing.

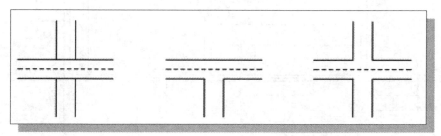

Exercises: (Time: 120 minutes)

1. Floor Plan A (Wall thickness: 5 inch)

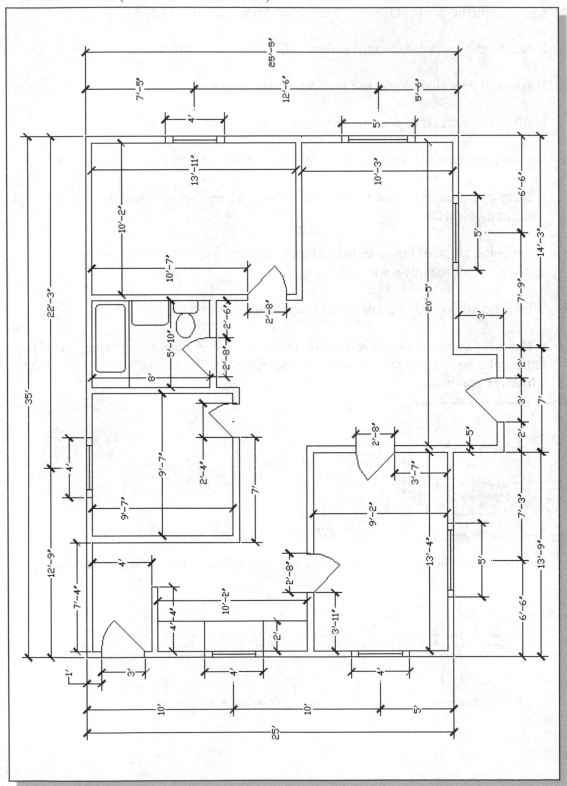

2. **Floor Plan B** (Wall thickness: 5 inch)

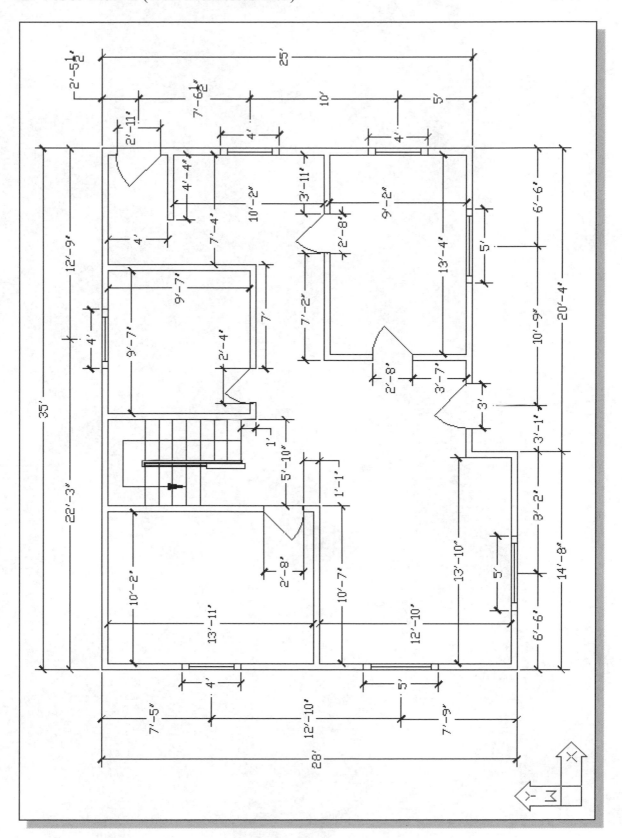

Notes:

Chapter 5
Orthographic Views in Multiview Drawings

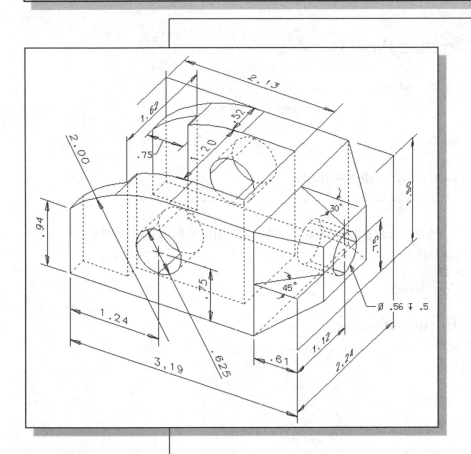

Learning Objectives

♦ **Create 2D orthographic views using AutoCAD**

♦ **Using the Construction Line command to draw**

♦ **Using Running Object Snaps**

♦ **Use AutoCAD's AutoSnap and AutoTrack features**

♦ **Create a Miter line to transfer dimensions**

♦ **Using Projection lines between orthographic views**

♦ **Use the Polar Tracking option**

AutoCAD Certified User Examination Objectives Coverage

This table shows the pages on which the objectives of the Certified User Examination are covered in Chapter 5.

Certified User Reference Guide

Introduction

Most drawings produced and used in industry are ***multiview drawings***. Multiview drawings are used to provide accurate three-dimensional object information on two-dimensional media, a means of communicating all of the information necessary to transform an idea or concept into reality. The standards and conventions of multiview drawings have been developed over many years, which equip us with a universally understood method of communication. The age of computers has greatly altered the design process, and several CAD methods are now available to help generate multiview drawings using CAD systems.

Multiview drawings usually require several orthographic views to define the shape of a three-dimensional object. Each orthographic view is a two-dimensional drawing showing only two of the three dimensions of the three-dimensional object. Consequently, no individual view contains sufficient information to completely define the shape of the three-dimensional object. All orthographic views must be looked at together to comprehend the shape of the three-dimensional object. The arrangement and relationship between the views are therefore very important in multiview drawings. In this chapter, the common methods of creating two-dimensional orthographic views with AutoCAD are examined.

The Locator Design

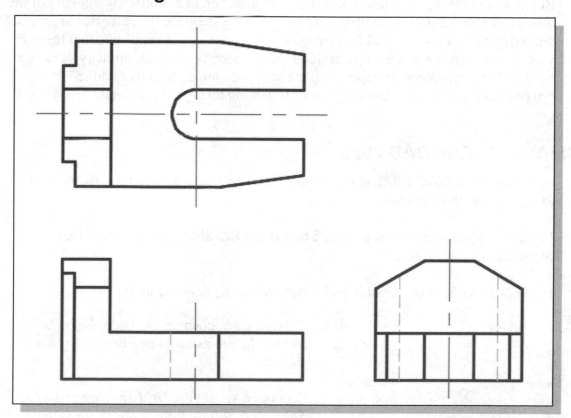

The Locator Part

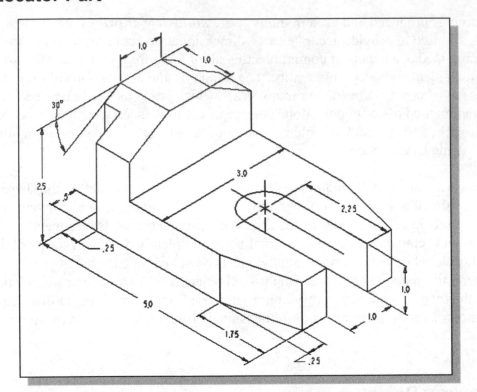

➤ Before going through the tutorial, make a rough sketch of a multiview drawing of the part. How many 2D views will be necessary to fully describe the part? Based on your knowledge of **AutoCAD 2023** so far, how would you arrange and construct these 2D views? Take a few minutes to consider these questions and do preliminary planning by sketching on a piece of paper. You are also encouraged to construct the orthographic views on your own prior to following through the tutorial.

Starting Up AutoCAD 2023

1. Select the **AutoCAD 2023** option on the *Program* menu or select the **AutoCAD 2023** icon on the *Desktop*.

2. In the *Startup* dialog box, select the **Start from Scratch** option with a single click of the left-mouse-button.

3. In the *Default Settings* section, pick **Imperial** as the drawing units.

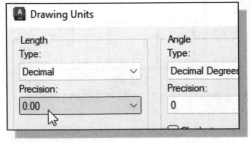

4. On your own, open up the *Units* dialog box, and set the precision to **two decimal places** as shown.

5. On your own, set the *Grid Spacing* and *Snap Spacing* to **0.5** for both X and Y directions.

Layers Setup

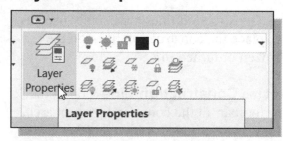

1. Pick **Layer Properties Manager** in the *Layers* toolbar.

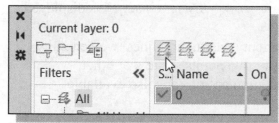

2. Click on the **New Layer** icon to create new layers.

3. Create two **new** layers with the following settings:

Layer	Color	Linetype
Construction	**White**	**Continuous**
Object	**Blue**	**Continuous**

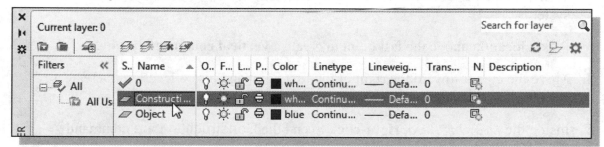

4. Highlight the layer *Construction* in the list of layers.

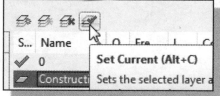

5. Click on the **Current** button to set layer *Construction* as the *Current Layer*.

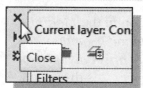

6. Click on the **Close** button to accept the settings and exit the *Layer Properties Manager* dialog box.

7. In the *Status Bar* area, reset the option buttons so that only *SNAP Mode* and *GRID Display* are switched **ON**.

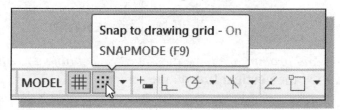

Drawing Construction Lines

Construction lines are lines that extend to infinity. Construction lines are usually used as references for creating other objects. We will also place the construction lines on the *Construction* layer so that the layer can later be frozen or turned off.

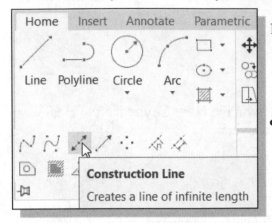

1. Select the **Construction Line** icon in the *Draw* toolbar. In the command prompt area, the message "*_xline Specify a point or [Hor/Ver/Ang/Bisect/ Offset]:*" is displayed.

- To orient construction lines, we generally specify two points. Note that other orientation options are also available.

2. Select a location near the **lower left corner** of the *Drawing Area*. It is not necessary to align objects to the origin of the world coordinate system. CAD systems provide us with many powerful tools to manipulate geometry. Our main goal is to use the CAD system as a flexible and powerful tool and to be very efficient and effective with the systems.

3. Pick a location above the last point to create a **vertical construction line**.

4. Move the cursor toward the right of the first point and pick a location to create a **horizontal construction line**.

5. Inside the *Drawing Area*, **right-click** to end the Construction Line command.

6. In the *Status Bar* area, turn **OFF** the *SNAP* option.

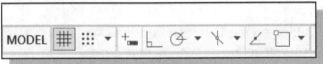

Using the Offset Command

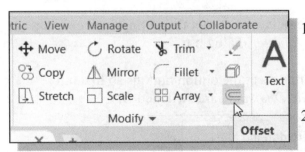

1. Select the **Offset** icon in the *Modify* toolbar. In the *command prompt area*, the message "*Specify offset distance or [Through/Erase/Layer]:*" is displayed.

2. In the *command prompt area*, enter **5.0** [**ENTER**].

3. In the *command prompt area*, the message "*Select object to offset or <exit>:*" is displayed. Pick the **vertical line** on the screen.

4. AutoCAD next asks us to identify the direction of the offset. Pick a location that is to the **right** of the vertical line.

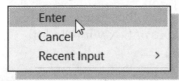

5. Inside the *Drawing Area*, **right-click** and choose **Enter** to end the **Offset** command.

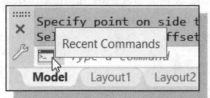

6. In the *command prompt area*, click on the small **triangle icon** to access the list of recent commands.

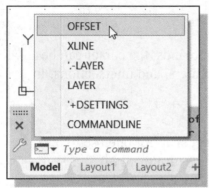

7. Select **Offset** in the pop-up list, to repeat the **Offset** command.

8. In the *command prompt area*, enter **2.5** [ENTER].

9. In the command prompt area, the message "*Select object to offset or <exit>:*" is displayed. Pick the **horizontal line** on the screen.

10. AutoCAD next asks us to identify the direction of the offset. Pick a location that is **above** the horizontal line.

11. Inside the Drawing Area, **right-click** to end the **Offset** command.

12. Repeat the **Offset** command and create the offset lines as shown.

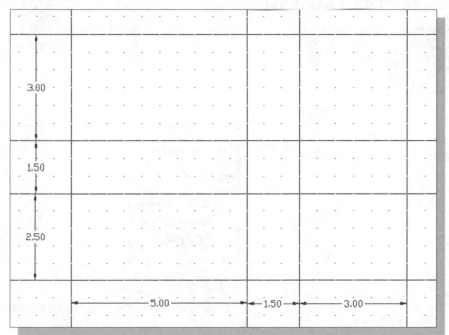

Set Layer Object as the Current Layer

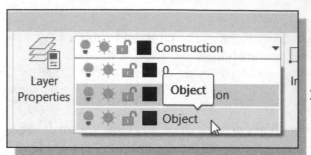

1. On the *Layers* toolbar panel, choose the **Layer Control** box with the left-mouse-button.

2. Move the cursor over the name of the layer **Object**. The tool tip "*Object*" appears.

3. **Left-click once** on the layer *Object* to set it as the *Current Layer*.

Using the Running Object Snaps

In **AutoCAD 2023**, while using geometry construction commands, the cursor can be placed to points on objects such as endpoints, midpoints, centers, and intersections. In AutoCAD, this tool is called the **Object Snap**.

Object snaps can be turned on in one of two ways:
• **Single Point (or override) Object Snaps**: Sets an object snap for one use.
• **Running Object Snaps**: Sets object snaps *active* until we turn them off.

The procedure we have used so far is the *Single Point Object Snaps* option, where we select the specific object snap from the *Object Snap* toolbar for one use only. The use of the *Running Object Snaps* option to assist the construction is illustrated next.

1. In the *Menu Bar*, select:

   [Tools] → [Drafting Settings]

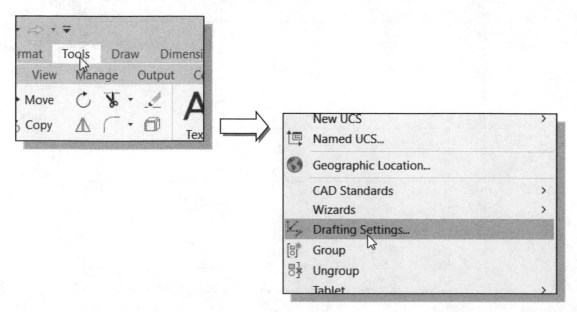

2. In the *Drafting Settings* dialog box select the **Object Snap** tab.

- The *Running Object Snap* options can be turned on or off by clicking the different options listed. Notice the different symbols associated with the different *Object Snap* options.

3. Turn ***ON*** the *Running Object Snap* by clicking the **Object Snap On** box, or hit the **[F3]** key once.

4. Confirm the ***Intersection***, ***Endpoint*** and ***Extension*** options are switched ***ON*** and click on the **OK** button to accept the settings and exit from the *Drafting Settings* dialog box.

❖ Notice in the *Status Bar* area the ***OSNAP*** button is switched ***ON***. We can toggle the *Running Object Snap* option on or off by clicking the *OSNAP* button.

5. Press the **[F3]** key once and notice the *OSNAP* button is switched ***OFF*** in the *Status Bar* area.

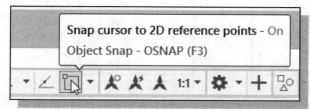

6. Press the **[F3]** key again and notice the *OSNAP* button is now switched ***ON*** in the *Status Bar* area.

➤ **AutoCAD 2023** provides many input methods and shortcuts; you are encouraged to examine the different options and choose the option that best fits your own style.

Creating Object Lines

We will define the areas for the front view, top view and side view by adding object lines using the *Running Object Snap* option.

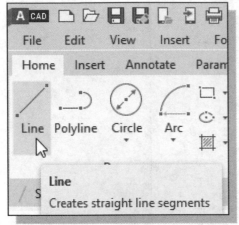

1. Select the **Line** command icon in the *Draw* toolbar. In the command prompt area, the message "*Line Specify first point:*" is displayed.

2. Move the cursor to the **intersection** of any two lines and notice the visual aid automatically displayed at the intersection.

3. Pick the four intersection points closest to the lower left corner to create the four sides of the area of the front view.

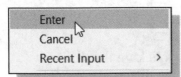

4. Inside the *Drawing Area*, **right-click** once to activate the option menu and select **Enter** with the left-mouse-button to end the Line command.

5. Repeat the **Line** command to define the top view and side view as shown.

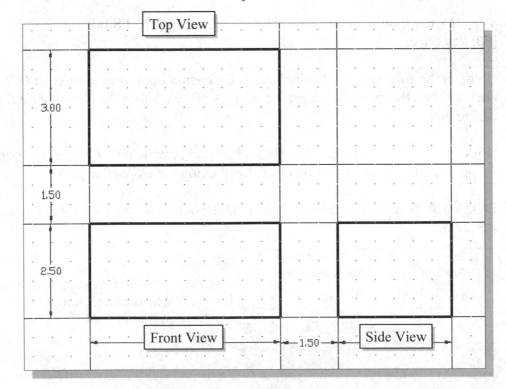

Turn Off the Construction Lines Layer

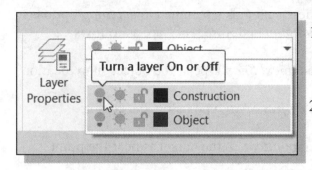

1. On the *Object Properties* toolbar, choose the **Layer Control** box with the left-mouse-button.

2. Move the cursor over the lightbulb icon for layer **Construction**. The tool tip *"Turn a layer On or Off"* appears.

3. **Left-click once** on the *lightbulb* icon and notice the icon color is changed to gray color, representing the layer (layer *Construction*) is turned **OFF**.

Adding More Objects in the Front View

1. Use the **Offset** command and create the two parallel lines in the front view as shown.

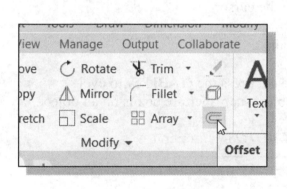
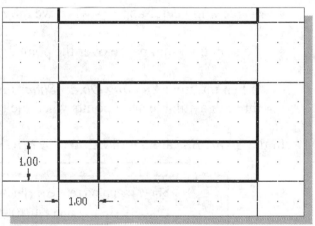

2. Use the **Trim** command and modify the front view as shown.

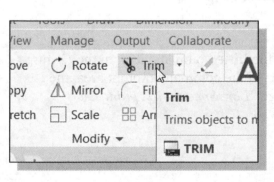
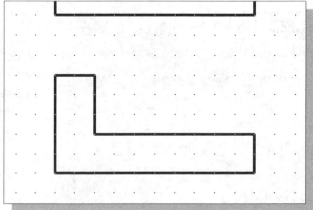

AutoCAD's AutoSnap™ and AutoTrack™ Features

AutoCAD's *AutoSnap* and *AutoTrack* provide visual aids when the *Object Snap* options are switched *on*. The main advantages of *AutoSnap* and *AutoTrack* are as follows:

- **Symbols**: Automatically displays the *Object Snap* type at the object snap location.

- **Tooltips**: Automatically displays the *Object Snap* type below the cursor.

- **Magnet**: Locks the cursor onto a snap point when the cursor is near the point.

With **Object Snap Tracking,** the cursor can track along alignment paths based on other object snap points when specifying points in a command. To use *Object Snap Tracking*, one or more object snaps must be switched on. The basic rules of using the **Object Snap Tracking** option are as follows:

- To track from a *Running Object Snap* point, pause over the point while in a command.

- A tracking vector appears when we move the cursor.

- To stop tracking, pause over the point again.

- When multiple *Running Object Snaps* are on, press the **[TAB]** key to cycle through available snap points when the object snap aperture box is on an object.

1. In the *Status Bar* area, turn **ON** the *OTRACK/AUTOSNAP* option.

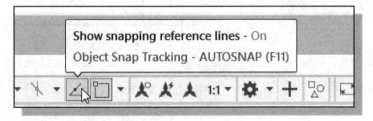

2. Select the **Line** command icon in the *Draw* toolbar. In the command prompt area, the message "*Line Specify first point:*" is displayed.

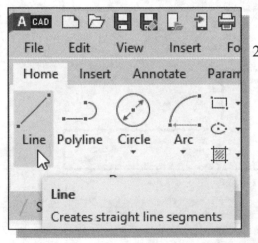

3. Move the cursor near the top right corner of the vertical protrusion in the front view. Notice that *AUTOSNAP* automatically locks the cursor to the corner and displays the ***Endpoint*** symbol.

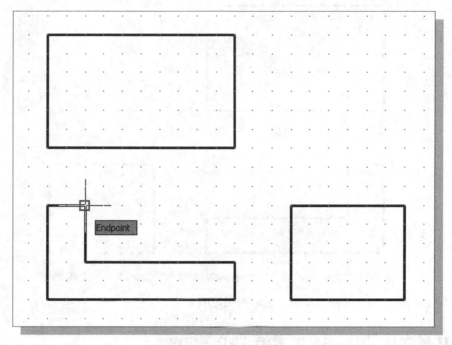

4. Move the cursor upward and notice that *Object Tracking* displays a dashed line, showing the alignment to the top right corner of the vertical protrusion in the front view. Move the cursor near the top horizontal line of the top view and notice that *AUTOSNAP* displays the intersection point.

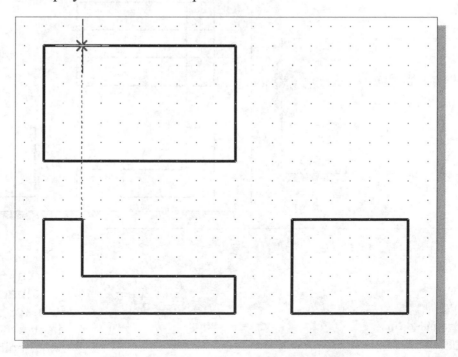

5. Left-click to place the starting point of a line at the intersection.

6. Move the cursor to the top left corner of the front view to activate the tracking feature.

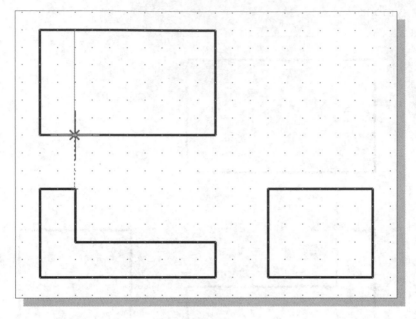

7. Create the line as shown in the above figure.

Adding More Objects in the Top View

1. Use the **Offset** command and create the two parallel lines in the top view as shown.

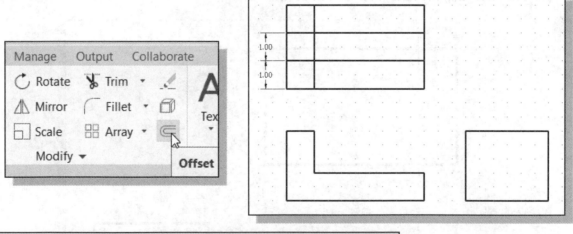

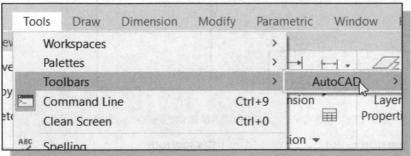

2. On your own, select **[Tools]** → **[Toolbars]** → **[Object Snap]** to display the *Object Snap* toolbar on the screen.

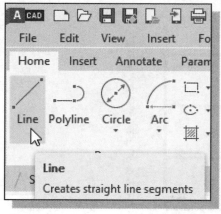

3. Select the **Line** command icon in the *Draw* toolbar. In the command prompt area, the message "*Line Specify first point:*" is displayed.

4. In the *Object Snap* toolbar, pick **Snap From**. In the command prompt area, the message "*_from Base point*" is displayed. AutoCAD now expects us to select a geometric entity on the screen.

➤ The ***Single Point Object Snap*** overrides the ***Running Object Snap*** option.

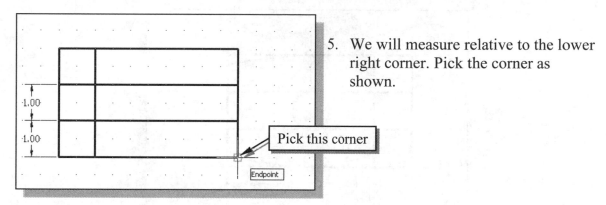

5. We will measure relative to the lower right corner. Pick the corner as shown.

6. In the command prompt area, enter **@0,0.25** [ENTER].

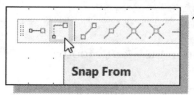

7. In the *Object Snap* toolbar, pick **Snap From**. Pick the lower right corner of the top view again.

8. In the command prompt area, enter **@-1.75,0** [ENTER].

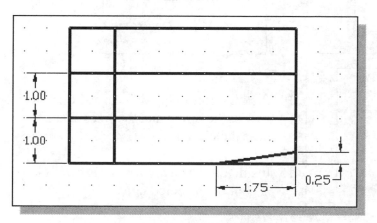

9. Inside the Drawing Area, right-click to activate the option menu and select **Enter** with the left-mouse-button to end the Line command.

10. Repeat the procedure and create the line on the top right corner as shown.

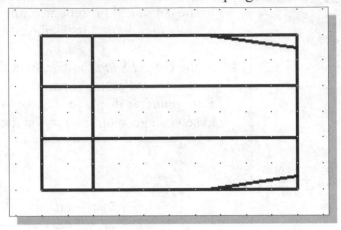

11. Using the *Snap From* option, create the circle (diameter *1.0*) as shown.

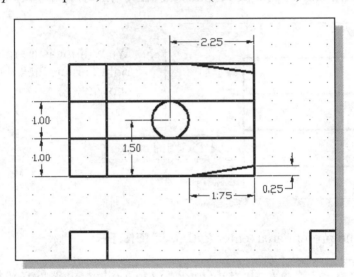

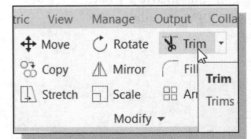

12. Select the **Trim** icon in the *Modify* toolbar.

13. In the command prompt area, click **cuTting edges** as shown.

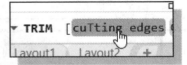

14. Pick the following objects as boundary edges: the circle and the lines that are near the circle.

15. Inside the *Drawing Area,* **right-click** to accept the selected objects.

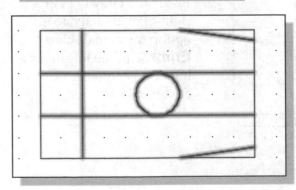

16. Select the unwanted portions and modify the objects as shown.

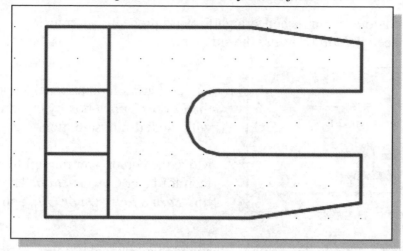

17. On your own, use the **Offset** and **Trim** commands and modify the top view as shown.

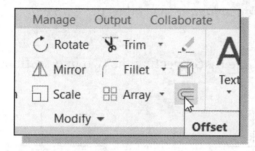
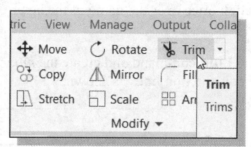

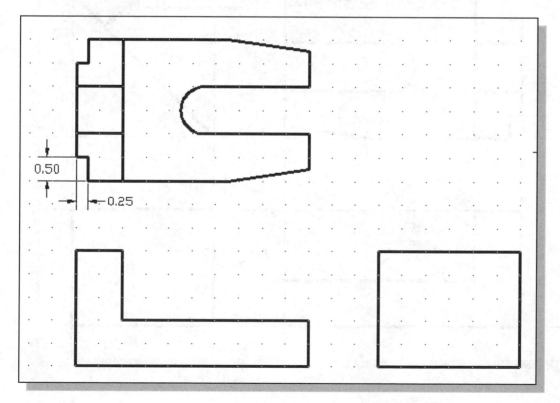

Drawing Using the Miter Line Method

The *45° miter line* method is a simple and straightforward procedure to transfer measurements in between the top view and the side view.

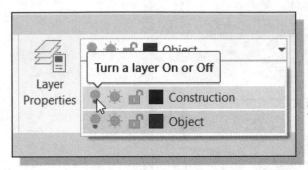

1. On the *Layers* toolbar panel, choose the **Layer Control** box by clicking once with the left-mouse-button.

2. Move the cursor over the **light-bulb** icon for layer **Construction**. The tool tip *"Turn a layer On or Off"* appears.

3. **Left-click once** and notice the icon color is changed to a light color, representing the layer (layer *Construction*) is turned *ON*.

4. **Left-click once** over the name of the layer **Construction** to set it as the *Current Layer*.

5. Use the **Line** command and create the *miter line* by connecting the two intersections of the construction lines as shown.

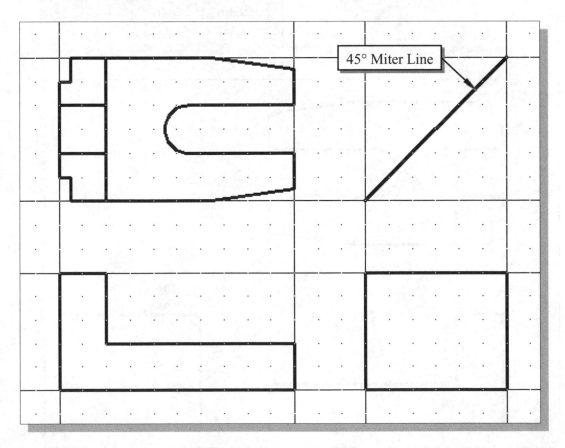

45° Miter Line

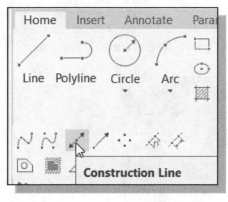

6. Select the **Construction Line** command in the *Draw* toolbar as shown.

7. In the *command prompt area*, select the **Horizontal** option as shown.

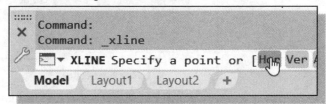

8. On your own, create horizontal projection lines through all the corners in the top view as shown.

9. Use the **Trim** command and trim the projection lines as shown in the figure below.

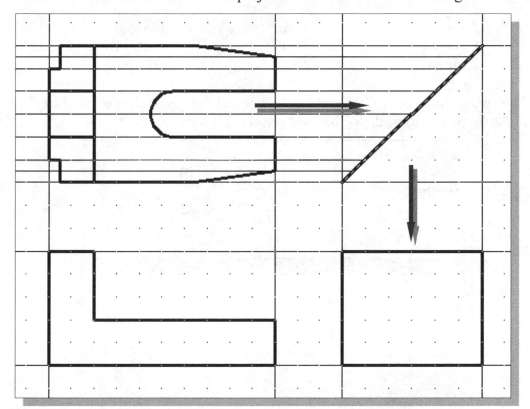

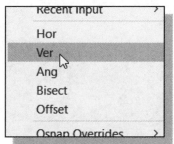

10. On your own, create additional **Construction Lines** (use the *vertical* option) through all the intersection points that are on the *miter line*.

More Layers Setup

1. Pick *Layer Properties Manager* in the *Layers* toolbar panel as shown in the figure below.

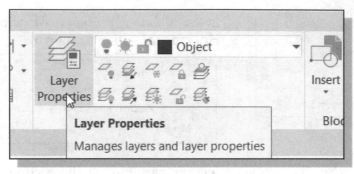

2. Click on the **New** icon to create new layers.

3. Create two **new layers** with the following settings:

Layer	Color	Linetype
Center	Red	CENTER
Hidden	Cyan	HIDDEN

• The default linetype is *Continuous*. To use other linetypes, click on the **Load** button in the *Select Linetype* dialog box and select the desired linetypes.

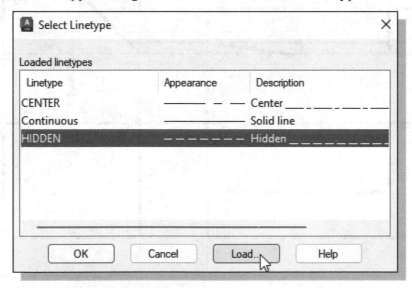

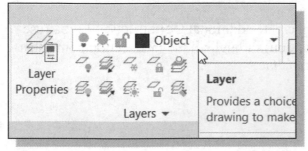

4. On your own, set the layer *Object* as the *Current Layer*.

Top View to Side View Projection

1. Using the *Running Object Snaps*, create the necessary **object-lines** in the side view.

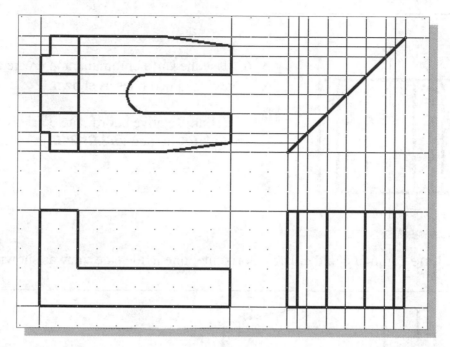

2. Set layer **Hidden** as the *Current Layer* and create the two necessary hidden lines in the side view.

3. Set layer **Center** as the *Current Layer* and create the necessary centerlines in the side view.

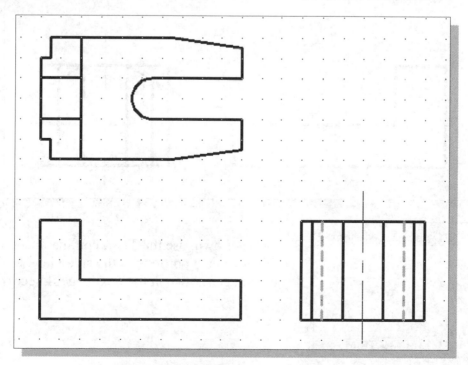

4. In the *Layer Control* box, turn **OFF** the **construction lines**.

5. Set layer **Object** as the *Current Layer*.

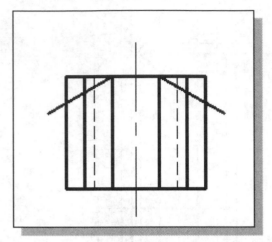

6. Use the **Line** command and create the two 30° inclined lines as shown.

    (Hint: Relative coordinate entries of **@2.0<-30** and **@2.0<210**.)

7. Use the **Line** command and create a horizontal line in the side view as shown.

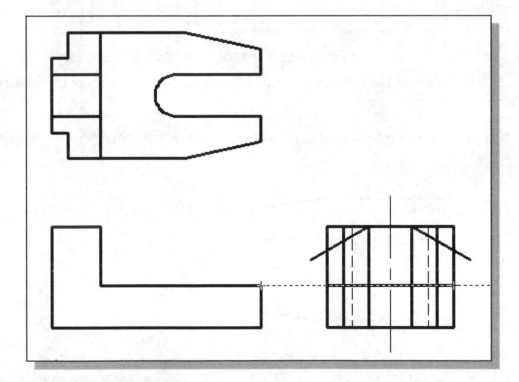

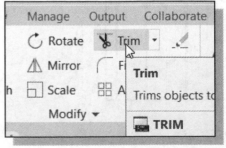

8. On your own, use the **Trim** command and remove the unwanted portions in the side view. Refer to the image shown on the next page if necessary.

Completing the Front View

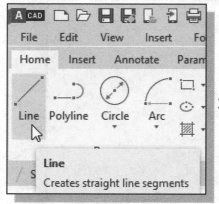

1. Select the **Line** command icon in the *Draw* toolbar. In the command prompt area, the message "*Line Specify first point:*" is displayed.

2. Move the cursor to the top left corner in the side view and the bottom left corner in the top view to activate the *Object Tracking* option to both corners.

3. Left-click once when the cursor is aligned to both corners as shown.

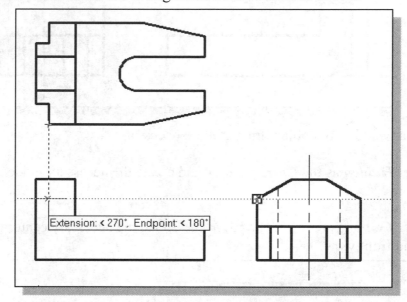

4. Create the **horizontal line** as shown.

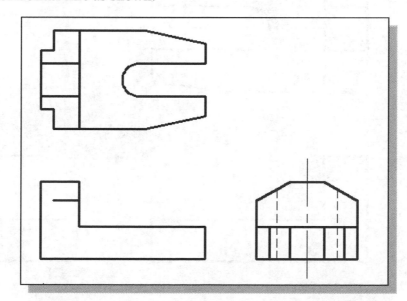

5. Repeat the procedure and create the lines in the front view as shown.

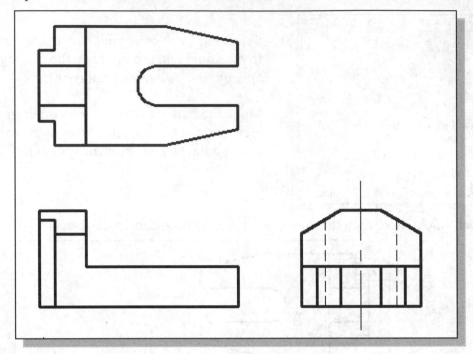

6. Add in any additional object lines that are necessary.

7. Set layer **Hidden** as the *Current Layer* and create the necessary hidden lines in the front view.

8. Set layer **Center** as the *Current Layer* and create the necessary centerlines in the top view and front view.

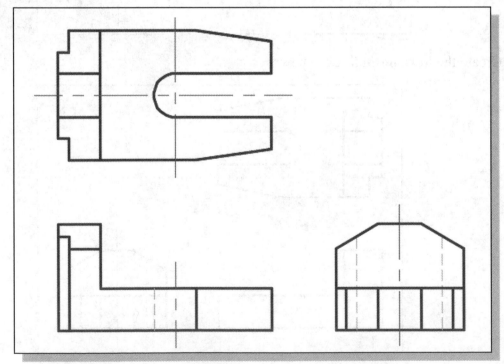

Object Information Using the List Command

AutoCAD provides several tools that will allow us to get information about constructed geometric objects. The **List** command can be used to show detailed information about geometric objects.

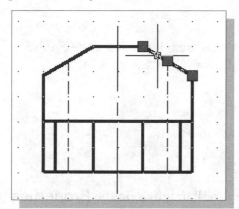

1. Move the cursor to the side view and select the inclined line on the right, as shown in the figure.

2. In the *Properties* toolbar, click on the **List** icon to activate the command.

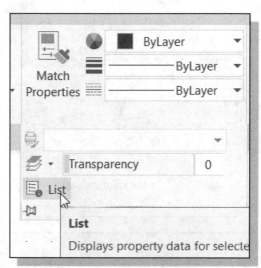

- Note the information regarding the selected object is displayed in the *AutoCAD Text Window* as shown.

```
AutoCAD Text Window - Drawing1.dwg                    —    □    ✕

 Edit

Specify next point or [Close/eXit/Undo]:
Specify next point or [Close/eXit/Undo]:
Specify next point or [Close/eXit/Undo]:

Command:
Command:
Command:
Command: _list 1 found

              LINE      Layer: "Object"
                        Space: Model space
                Handle = bf
        from point, X= 11.5000  Y=  6.0000  Z=   0.0000
          to point, X= 12.5000  Y=  5.4226  Z=   0.0000
    Length =   1.1547,  Angle in XY Plane =    330
          Delta X =   1.0000, Delta Y =  -0.5774, Delta Z =   0.0000

Command:
```

- Note the **List** command can be used to show detailed information about the selected line. The angle and length of the line, as well as the X, Y and Z components between the two endpoints are all listed in a separate window.

3. Press the [**F2**] key once to close the *AutoCAD Text Window*.

4. Press the [**Esc**] key once to deselect the selected line.

Object Information Using the Properties Command

AutoCAD also provides tools that allow us to display and change properties of constructed geometric objects. The **Properties** command not only provides the detailed information about geometric objects—modifications can also be done very quickly.

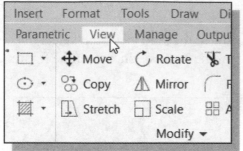

1. In the *Ribbon* tabs area, left-click once on the **View** tab as shown.

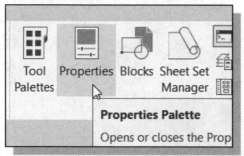

2. In the *Palettes* toolbar, click on the **Properties** icon to activate the command.

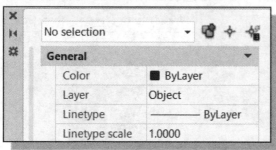

3. Note the *Properties* panel appears on the screen. The "*No selection*" on top of the panel indicates no object has been selected.

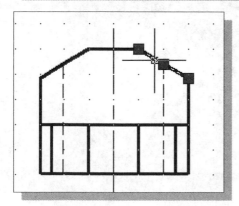

4. Move the cursor to the side view and select the inclined line on the right, as shown in the figure.

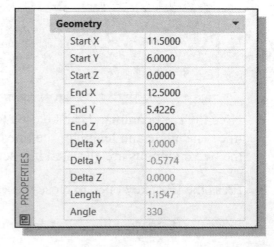

5. The geometry information is listed at the bottom section. Note the line length is *1.1547* and at the angle of *330* degrees.

Geometry	
Start X	11.5000
Start Y	6.0000
Start Z	0.0000
End X	12.5000
End Y	5.4226
End Z	0.0000
Delta X	1.0000
Delta Y	-0.5774
Delta Z	0.0000
Length	1.1547
Angle	330

Review Questions: (Time: 20 minutes)

1. Explain what an orthographic view is and why it is important to engineering graphics.

2. What does the *Running Object Snaps* option allow us to do?

3. Explain how a *miter line* can assist us in creating orthographic views.

4. Describe the AutoCAD *AUTOSNAP* and *AutoTrack* options.

5. List and describe two AutoCAD commands that can be used to get geometric information about constructed objects.

6. List and describe two options you could use to quickly create a 2-inch line attached to a 2-inch circle, as shown in the below figure.

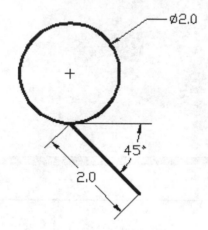

7. What are the length and angle of the inclined line, highlighted in the figure below, in the top view of the *Locator* design?

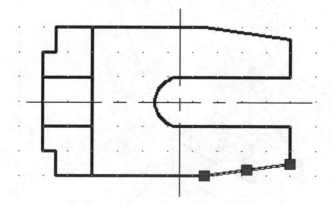

Exercises:

(Unless otherwise specified, dimensions are in inches.) (Time: 175 minutes)

1. Saddle Bracket

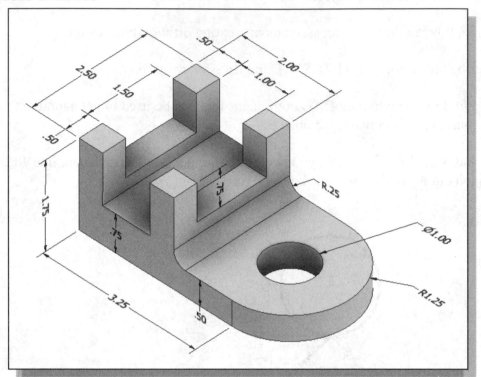

2. Anchor Base

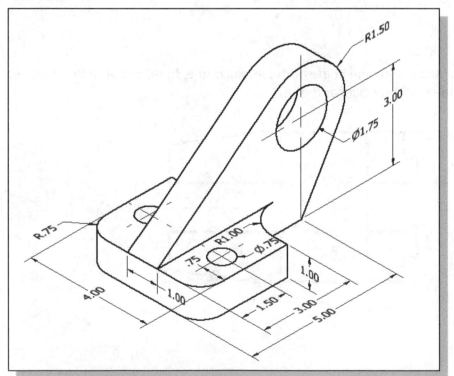

3. Bearing Base

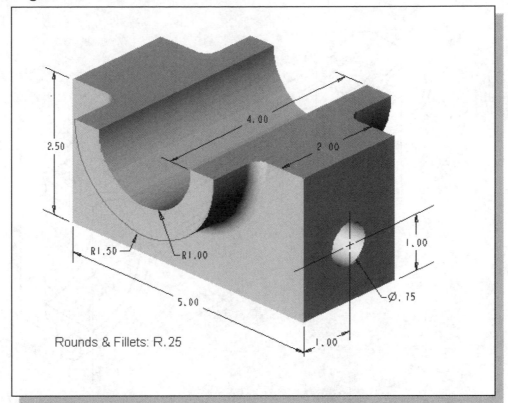

4. Shaft Support (Dimensions are in Millimeters. Note the two R40 arcs at the base share the same center.)

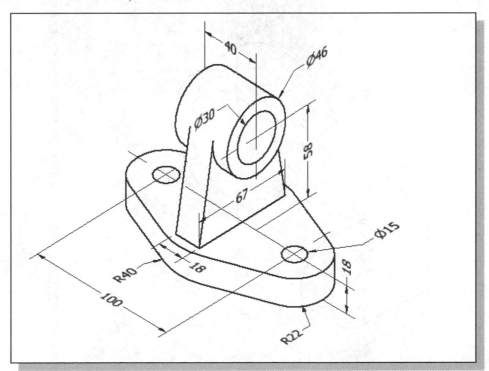

5. Connecting Rod

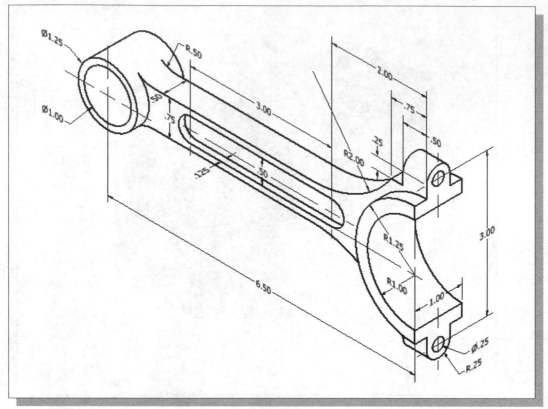

6. Tube Hanger

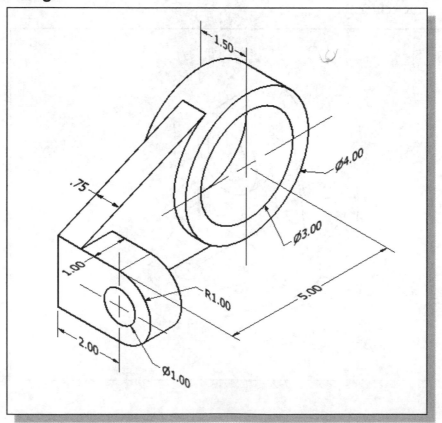

7. Tube Spacer

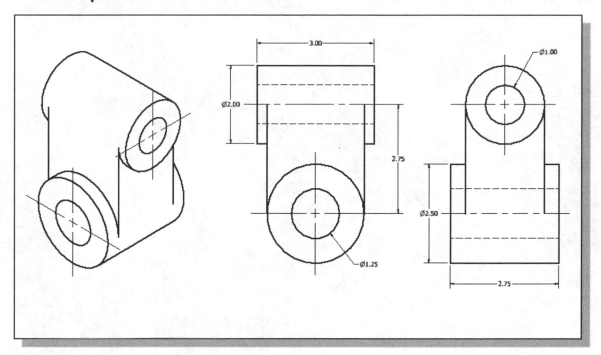

8. Slider (Dimensions are in Millimeters.)

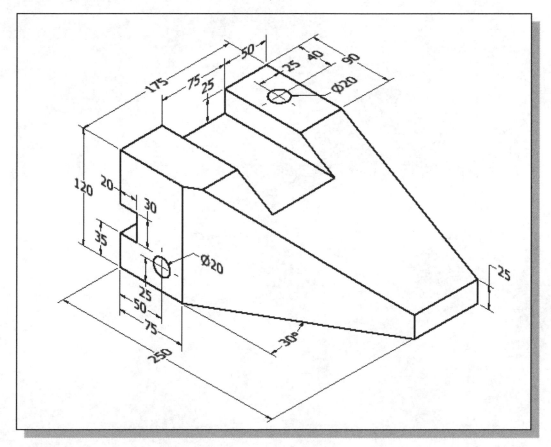

Notes:

Chapter 6
AutoCAD 2D Isometric Drawings

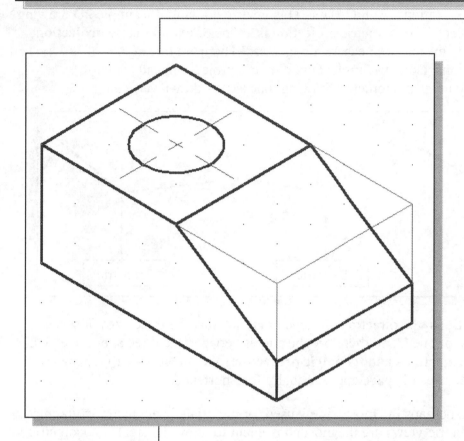

Learning Objectives

- ◆ **Understand the Axonometric projections**
- ◆ **Use the AutoCAD 2D Isometric Drawing tools**
- ◆ **Understand and Use the AutoCAD Isoplane and Isodraft commands**
- ◆ **Create the AutoCAD Isocircles**
- ◆ **Use the Object Snap options**
- ◆ **Control Objects with the Parametric Geometry Constraint tools**

Introduction

Although AutoCAD has full 3D solid modeling capabilities, it makes sense to use isometric drawing when all that is needed is some simple isometric views for design concepts and /or presentations. AutoCAD's 2D Isometric drawing is a simple 2D drawing to represent a 3D object. The Isometric projection is a type of **axonometric projection**; the word *axonometric* means "to measure along axes." There are three types of axonometric projections: isometric projection, dimetric projection, and trimetric projection. Typically in an axonometric drawing, one axis is drawn vertically.

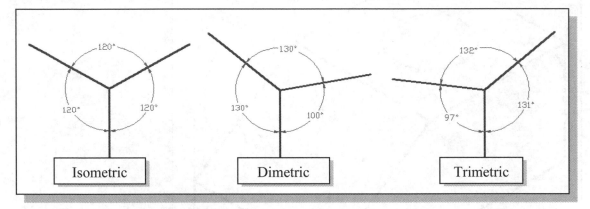

In **isometric projections**, the direction of viewing is such that the three axes of space appear equally foreshortened, and therefore the angles between the three axes are equal, while in **dimetric projections** and **trimetric projections** the directions of viewing are such that not all three axes of space appear equally foreshortened.

Isometric drawing is perhaps the most widely used for pictorials in engineering graphics, mainly because isometric views are the most convenient to draw. To create an isometric drawing, the visible portions of the individual 2D-views are constructed on the corresponding sides of the box. Adjustments of the locations of surfaces are then made, by moving the edges, to complete the isometric view.

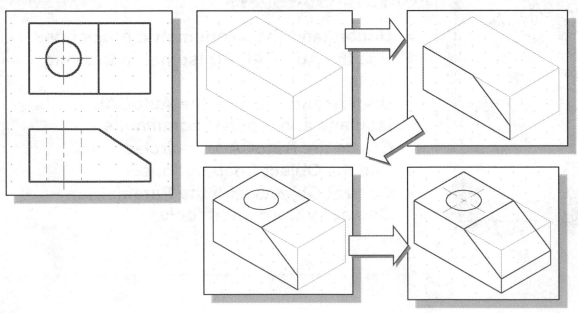

In an isometric drawing, cylindrical or circular shapes appear as ellipses. It can be confusing drawing the ellipses in an isometric view; one simple rule to remember is the **major axis** of the ellipse is always **perpendicular** to the **center axis** of the cylinder as shown in the figures below.

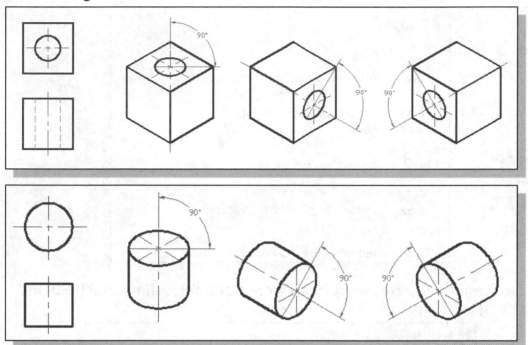

The general procedure to constructing an AutoCAD 2D isometric drawing is illustrated in the following sections. Note that this is a fairly simple and effective way of representing a three-dimensional object in two dimensional space.

The Angle Support Design

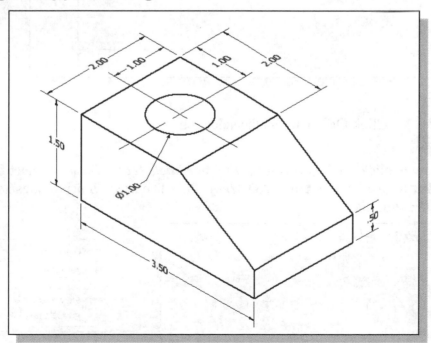

Starting Up AutoCAD 2023

1. Select the **AutoCAD 2023** option on the *Program* menu or select the **AutoCAD 2023** icon on the *Desktop*.

2. In the *Startup dialog box*, select **[Start from Scratch]** → **[Imperial (feet and inches)]** as shown.

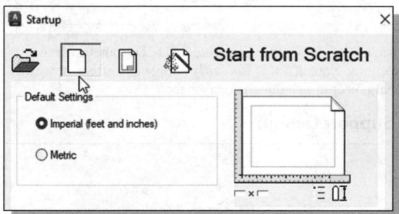

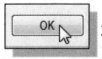 3. Click **OK** to accept the selection.

4. On your own, click on the down-arrow in the *Quick Access Bar* and select **Show Menu Bar** to display the **AutoCAD** *Menu Bar*. The *Menu Bar* provides access to all AutoCAD commands.

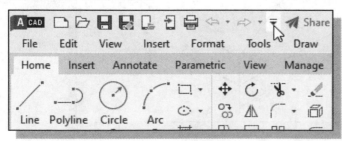 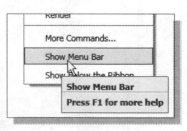

AutoCAD Isoplane and Isodraft Commands

AutoCAD uses the Isoplane and Isodraft commands to control the alignment of the cursor to the three isometric axes. The following settings can be used when drawing 2D isometric representations of 3D models:

- Ortho directions
- Snap orientation
- Grid orientation and style
- Polar tracking angles
- Orientation of isometric circles

The isometric plane affects the cursor movement only when the snap style is set to *Isometric* in the *Drafting Settings* dialog box. When the *snap option* is set to *Isometric*, the *Ortho* mode will use the appropriate axis paired to 30, 90, and/or 150 degrees.

Left Plane

 The cursor is aligned to the left-facing plane of the isometric axes.

Top Plane

 The cursor is aligned to the top-facing plane of the isometric axes.

Right Plane

 The cursor is aligned to the right-facing plane of the isometric axes.

➢ The three isometric planes can be selected by toggling the [**F5**] or [**Ctrl+E**] keys or using the **Isodraft** command.

The **Isodraft** command can be accessed through the icon in the *Status Bar* area. The primary advantage of *Isodraft* is that when it is toggled on, all the related settings are automatically set as well, so the cursor will be aligned to the isometric axes of the active *Isoplane*.

Layers Setup

1. Click **Layer Properties Manager** in the *Layers* toolbar.

❖ In AutoCAD, we always construct entities on a layer. It may be the default layer or a layer that we create. Each layer has associated properties such as the visibility setting, color, linetype, lineweight, and plot style.

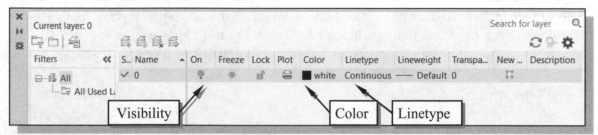

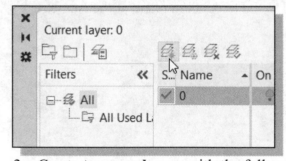

2. Click on the **New Layer** button. Notice a layer is automatically added to the list of layers.

3. Create **two new layers** with the following settings:

Layer	Color	Linetype	Lineweight
Construction	Grey	Continuous	Default
Object	Cyan	Continuous	0.30mm

4. Highlight the layer *Construction* in the list of layers.

5. Click on the **Current** icon to set layer *Construction* as the *Current Layer*.

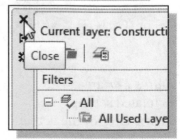

6. Click on the **Close** button to accept the settings and exit the *Layer Properties Manager* dialog box.

Create the Base Box of the Design

We will first use construction geometry to define the outer edges of the design.

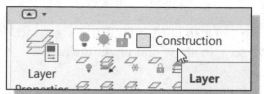

1. In the *Layer Control* box, confirm that layer *Construction* is set as the *Current Layer*.

2. Click on the *Customization* icon, and switch **on** the **Isometric Drafting** option as shown.

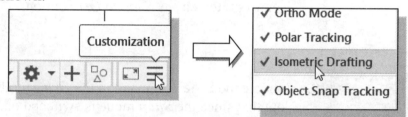

3. In the *Status Bar* area, reset the option buttons so that only the *GRID DISPLAY, Dynamic Input, Object Snap Tracking* and *Object Snap* options are switched *ON.*

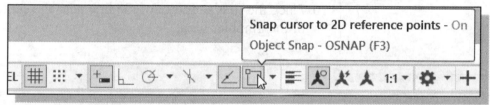

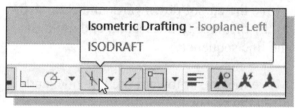

4. In the *Status Bar* area, click on the **Isodraft** icon to switch **on** the *Isoplane* option. Note the default isoplane is set to the *Left plane*.

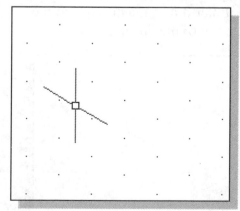

5. Inside the graphics area, notice the cursor is set to align to the **Left Isoplane**. The *Grid Display* is also aligned to the isoplane.

6. Hit the [**F5**] key or [**Ctrl+E**] once and notice the isoplane is now set to the *Top plane* as shown.

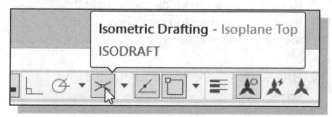

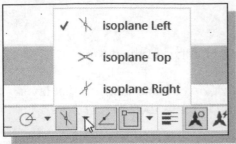

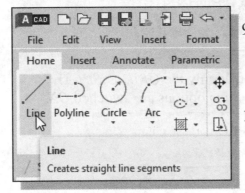

7. Hit the [**F5**] key or [**Ctrl+E**] twice and notice the isoplane is now set back to the *Left plane* as shown.

➢ Note that we can also click on the arrow next to the *Isodraft* icon to show and/or set the active *isoplane*.

8. In the *Status Bar* area, click on the **Snap Mode** icon to switch **on** the *Snap to Grid* option.

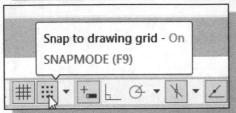

9. Activate the **Line** command in the *Draw* toolbar. Note that since the *Snap* mode is switched on, the cursor is aligned to the grid points in the directions of the left isoplane.

10. Place the first endpoint of the line near the center of the graphics area.

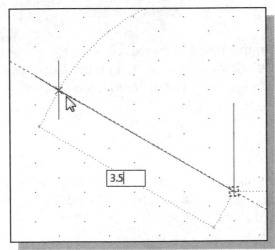

11. Move the cursor downward, and enter **1.5** as the distance to the second end point of the line sequence.

12. Move the cursor toward the left side, and align the cursor in the 150 degrees direction. Enter **3.5** as the distance to the third end point of the line sequence.

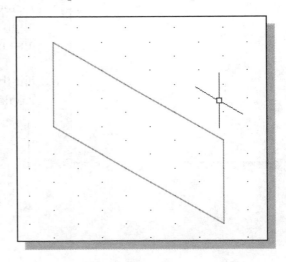

13. Repeat the above steps and create the front face of the isometric box for the *Angle Support design*.

14. Hit the [**Esc**] key once to end the line sequence.

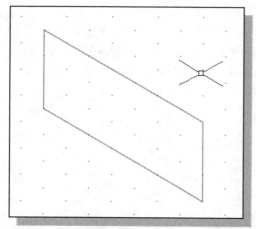

15. Hit the [**F5**] key or [**Ctrl+E**] once and notice the isoplane is now set to the *Top plane* as shown.

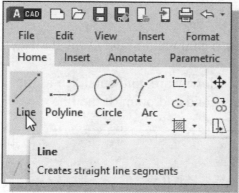

16. Activate the **Line** command in the *Draw* toolbar.

17. Snap to the **top left corner** of the polygon we just created.

18. Move the cursor toward the right side, and align the cursor in the 30 degrees direction. Enter **2.0** as the distance to the second end point of the line sequence.

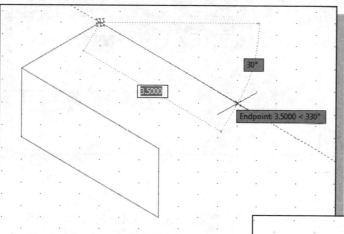

19. Move the cursor toward the right side, and align the cursor in the -30 degrees direction. Enter **3.5** as the distance to the third end point of the line sequence.

20. Snap to the **top right corner** of the front face of the isometric box to form the top face of the isometric box.

21. Hit the [**Esc**] key once to end the line sequence.

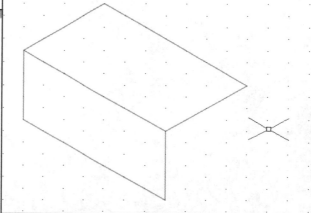

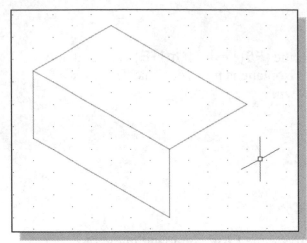

22. Hit the [**F5**] key or [**Ctrl+E**] once and notice the isoplane is now set back to the *Right plane* as shown.

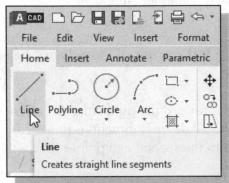

23. Activate the **Line** command in the *Draw* toolbar.

24. Move the cursor downward, and enter **1.5** as the distance to the second end point of the line sequence.

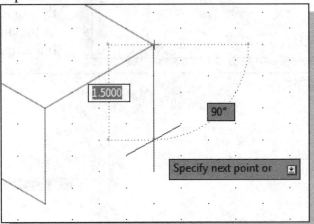

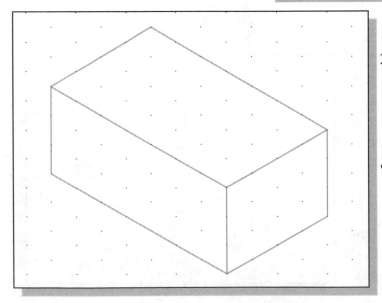

25. Snap to the **lower right corner** of the front face of the isometric box to form the right face of the isometric box.

● We will use the isometric box to help create the *Angle Support design*. Note that the entire process is done in two-dimensional space to represent a 3D design.

Create the Design inside the Base Box

We will now create the design using the base isometric box.

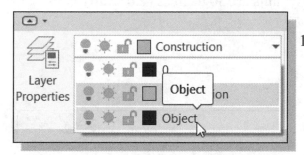

1. In the *Layer Control* box, set the layer *Object* as the *Current Layer*.

2. In the *Menu Bar* select **[Tools]** → **[Toolbars]** → **[AutoCAD]** → **[Object Snap]**.

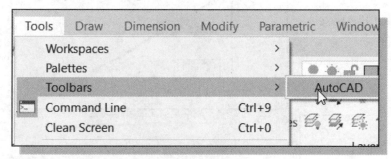

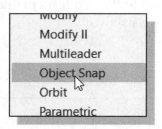

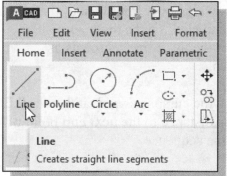

3. Activate the **Line** command in the *Draw* toolbar.

4. In the *command prompt* area, the message "*_line Specify first point:*" is displayed. Select **Snap From** in the *Object Snap* toolbar.

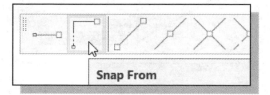

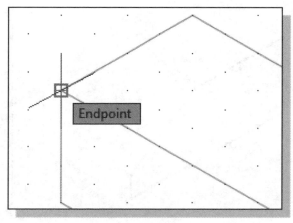

5. Select the **top left corner** of the base box as the reference point as shown.

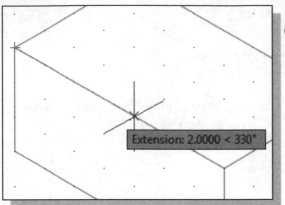

6. Move the cursor along the top edge of the front face and select the location at 2 units at 330 degrees (**2.0<330**) as shown.

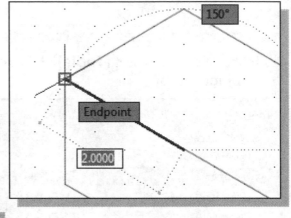

7. Snap to the **top left corner** of the base box as shown.

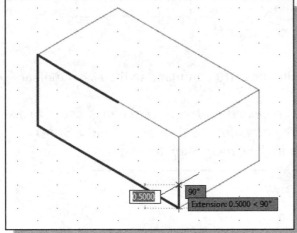

8. Follow along the front face of the base box and click on the three corners to create the three line segments as shown.

9. Move the cursor upward, and enter **0.5** as the distance to the next end point of the line sequence.

10. Right-click once to bring up the *option* list, and select **Close** to complete the front face of the design as shown.

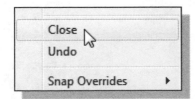

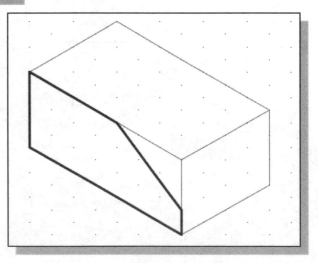

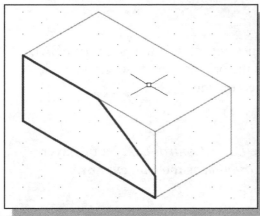

11. Hit the [**F5**] key or [**Ctrl+E**] once and notice the isoplane is now set to the *Top plane* as shown.

12. In the *Status Bar area*, switch on **Object Snap tracking** as shown.

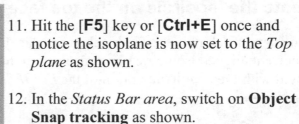

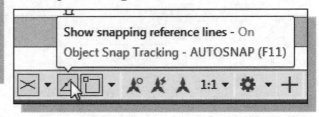

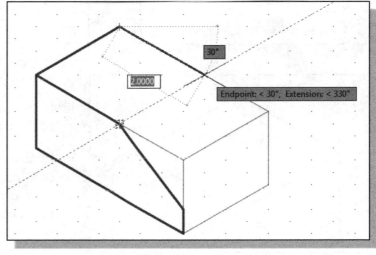

13. Activate the **Line** command in the *Draw* toolbar.

14. Snap to the **top left corner** of the base box and create the two line segment as shown. (Hint: Use the *AutoSnap* option to locate the corresponding corner.)

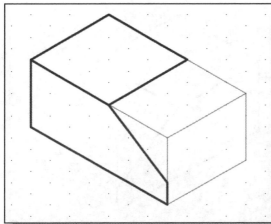

15. Complete the outlines of the top face of the design as shown.

16. Hit the [**Esc**] key once to end the line sequence.

Create the Isocircle on the top face

In an isometric drawing, cylindrical or circular shapes appear as ellipses. In AutoCAD, the current aligned isometric plane also determines the orientation of isometric circles created with the **Isocircle** option of the *ELLIPSE* command.

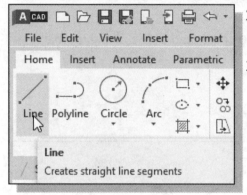

1. In the *Layer Control* box, set the layer *Construction* as the *Current Layer*.

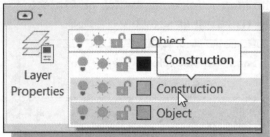

2. Activate the **Line** command in the *Draw* toolbar.

3. In the *command prompt* area, the message "*_line Specify first point:*" is displayed. Select **Midpoint** in the *Object Snap* toolbar.

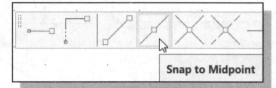

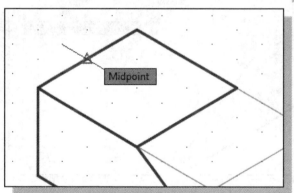

4. Select the midpoint of the top left edge of the top face of the design as shown.

5. Repeat the above step and create the two construction lines as shown. Note the intersection of the two lines identifies the center point of the circle.

6. In the *Layer Control* box, set the layer *Object* as the *Current Layer*.

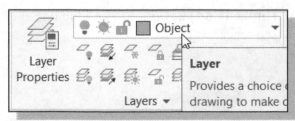

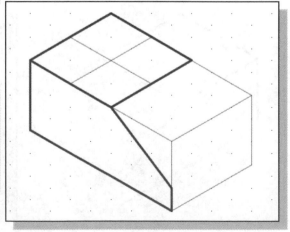

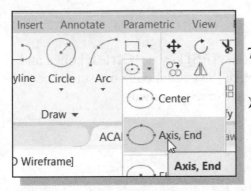

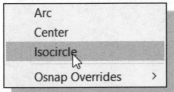

7. In the *Draw* toolbar, activate the **Axis, End Ellipse** command.

➤ In AutoCAD, the isometric circles are created with the **Isocircle** option of the *ELLIPSE* command (available with *Axis, End* or *Elliptical Arc* options.)

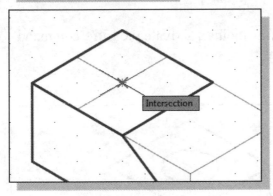

8. Right-click once to bring up the *option* list, and select **Isocircle** to create an isometric circle in an isometric view.

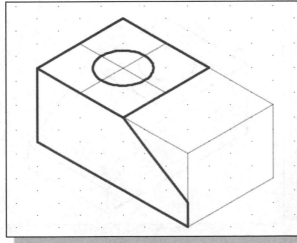

9. Select the intersection of the two construction lines we created in the previous section.

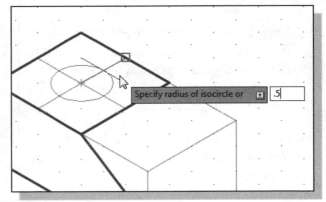

10. Enter **0.5** as the radius of the isocircle.

➤ In AutoCAD, the current aligned isometric top plane also determines the orientation of isometric circles created with the **Isocircle** option of the *ELLIPSE* command.

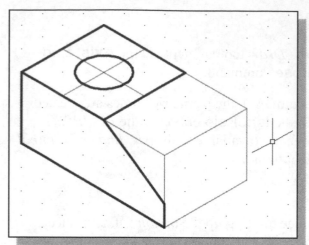

11. Hit the [**F5**] key or [**Ctrl+E**] once and notice the isoplane is now set to the *right plane* as shown.

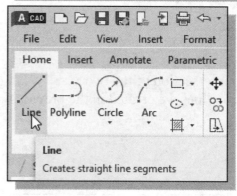

12. In the *Draw* toolbar, activate the **Line** command as shown.

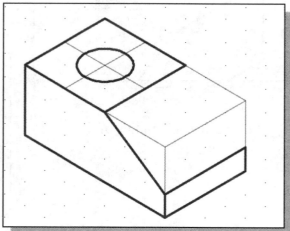

13. Snap to the **lower right corner** of the front face of the base box and create the two line segment as shown. (Hint: Use the *AutoSnap* option to locate the corresponding corner.)

14. On your own, complete the *Isometric drawing* for the *Angle support design* as shown.

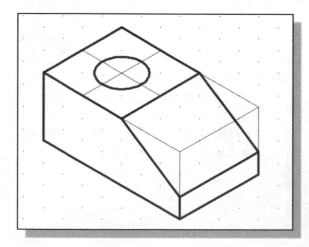

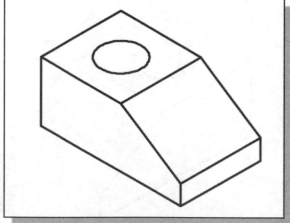

Using the Editing Tools in an Isometric drawing

In AutoCAD, the commonly used editing tools, such as **Move**, **Copy**, **Trim**, **Extend**, and **Erase**, can also be used to aid the construction of an isometric drawing. In the following section, the *Tube Anchor* design will be created using some of the AutoCAD editing tools.

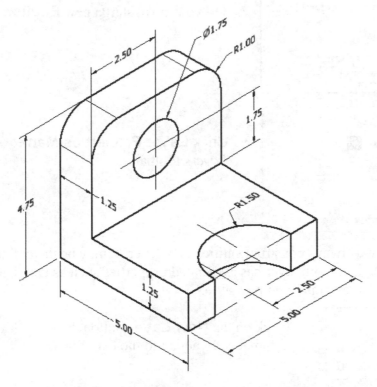

The Tube Anchor Design –Modeling Strategy

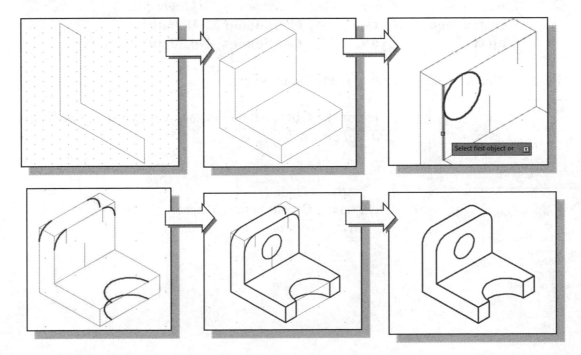

Start a New Drawing and Layers Setup

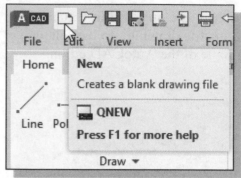

1. In the *Quick Access toolbar*, select **[New]**.

2. On your own, start a new **English Units** drawing.

3. Click **Layer Properties Manager** in the *Layers* toolbar.

❖ In AutoCAD, we always construct entities on a layer. It may be the default layer or a layer that we create. Each layer has associated properties such as the visibility setting, color, linetype, lineweight, and plot style.

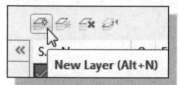

4. Click on the **New Layer** button. Notice a layer is automatically added to the list of layers.

5. Create **two new layers** with the following settings:

Layer	Color	Linetype	Lineweight
Construction	**Grey**	**Continuous**	**Default**
Object	**Cyan**	**Continuous**	**0.30mm**

6. Highlight the layer *Construction* in the list of layers.

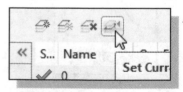

7. Click on the **Current** icon to set layer *Construction* as the *Current Layer*.

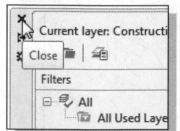

8. Click on the **Close** button to accept the settings and exit the *Layer Properties Manager* dialog box.

Create a Base Box of the Design

We will first use construction geometry to define the outer edges of the design.

1. In the *Status Bar* area, reset the option buttons so that only the *Grid Display, Dynamic Input, Ortho, Isodraft, Object Snap Tracking,* and *Object Snap* options are switched **ON**.

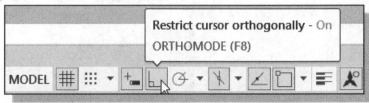

2. In the *Status Bar* area, set the *Isoplane* option to the *Left plane* as shown.

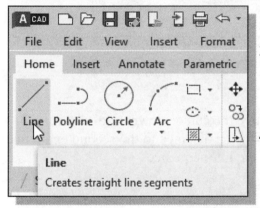

3. Activate the **Line** command in the *Draw* toolbar. Note that since the **Ortho** mode is switched on, the cursor is now aligned to the axes directions of the *left isoplane.*

4. Place the first endpoint of the line near the lower center of the graphics area.

5. Move the cursor downward, and enter **1.25** as the distance to the second end point of the line sequence.

6. Move the cursor toward the left side, and align the cursor in the 150 degrees direction. Enter **5.0** as the distance to the third end point of the line sequence.

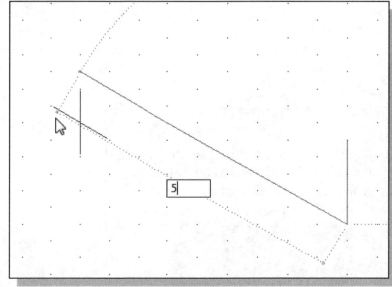

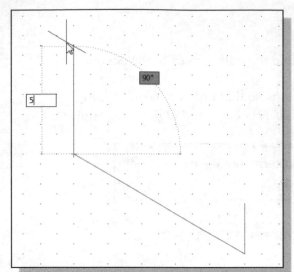

7. Move the cursor upward, and enter **5.0** as the distance to the next end point of the line sequence.

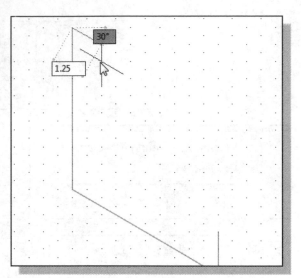

8. Move the cursor toward the left side, and align the cursor in the -30 degrees direction. Enter **1.25** as the distance to the next end point of the line sequence.

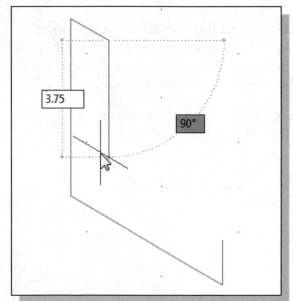

9. Move the cursor downward, and enter **3.75** as the distance to the second end point of the line sequence.

10. Right-click once to bring up the *option* list, and select **Close** to complete the front face of the design as shown.

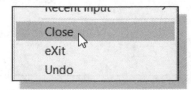

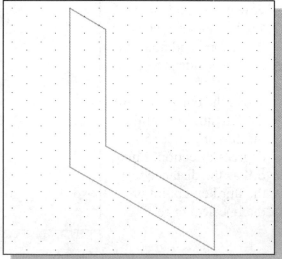

11. Hit the [**F5**] key or [**Ctrl+E**] once and notice the isoplane is now set to the *Top plane* as shown.

12. In the modify toolbar, activate the **Copy** command as shown.

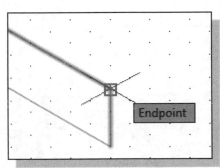

13. Select the four line segments as shown.

14. Right-click once to accept the selection and proceed with the **Copy** command.

15. Select **one of the corners** as the copy reference point.

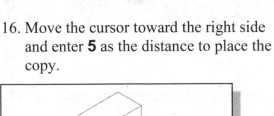

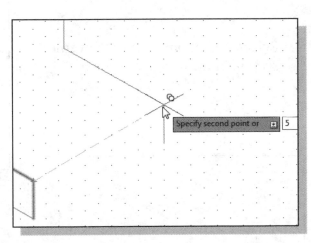

16. Move the cursor toward the right side and enter **5** as the distance to place the copy.

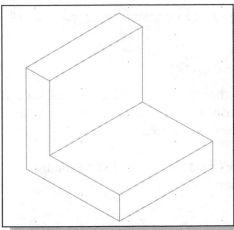

17. On your own, create additional lines to connect the corners of the front and back sides as shown.

Locate the Centers for the Isocircles

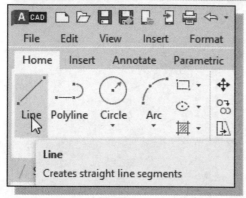

1. Hit the [**F5**] key or [**Ctrl+E**] once and notice the isoplane is now set to the *Right plane* as shown.

2. Activate the **Line** command in the *Draw* toolbar.

• We will first locate the center of the diameter 1.75 center circle of the design.

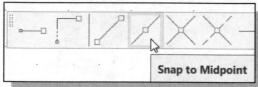

3. Activate the **Snap to Midpoint** option as shown.

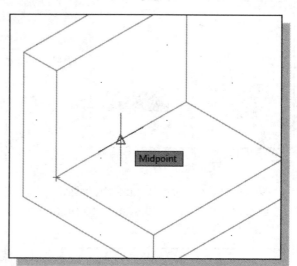

4. Select the bottom edge of the vertical section of the design as shown.

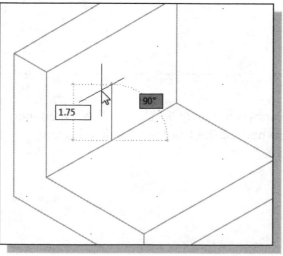

5. Move the cursor upward and enter **1.75** as the distance to the center point of the diameter 1.75 circle.

6. Hit the [**Esc**] key once to end the line command.

- We will also need to locate the center locations for the two rounded corners of the vertical section of the design.

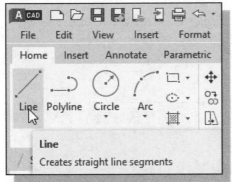

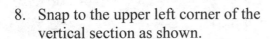

7. Activate the **Line** command in the *Draw* toolbar.

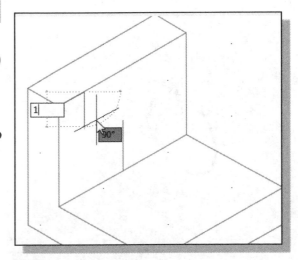

8. Snap to the upper left corner of the vertical section as shown.

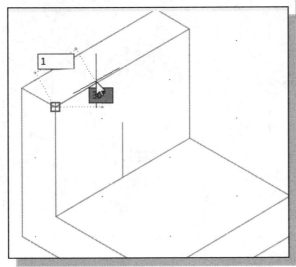

9. Move the cursor along the top edge and enter **1.0** as the distance as shown.

10. Move the cursor downward and enter **1.0** as the distance to locate the center point for the rounded corner as shown.

11. On your own, create two line segments to locate the center location of the rounded corner on **the other side**.

Create the Isocircles

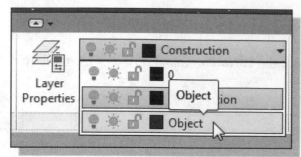

1. In the *Layer Control* box, set the layer *Object* as the *Current Layer*.

2. In the *Draw* toolbar, activate the **Axis, End Ellipse** command.

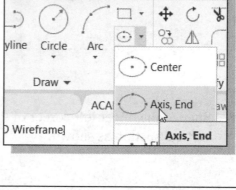

3. In the command prompt area, click Isocircle as shown.

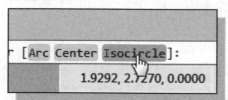

4. Snap to the bottom endpoint of the left line segment to place the center of the isocircle.

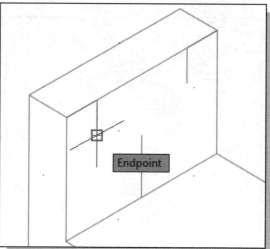

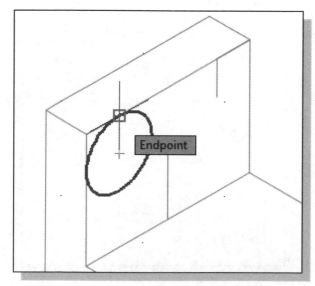

5. Snap to the top endpoint of the left line segment to create the isocircle representing the rounded corner.

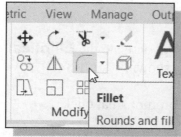

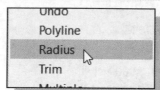

- Note that the *Fillet* command **does not** create the correct rounded corner in AutoCAD *Isometric views*; we will confirm this is the case.

6. In the *Modify* toolbar, activate the **Fillet** command.

7. Right-click once to bring up the option list and select the **Radius** option.

8. Enter **1.0** as the radius of the fillet.

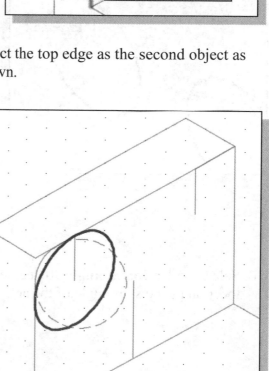

9. Select the left vertical edge as the first object as shown.

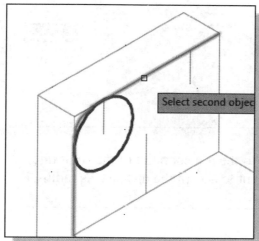

10. Select the top edge as the second object as shown.

11. On your own, create a circle on top of the arc created with the fillet command to confirm the circle is aligned to the screen, not the isometric axes.

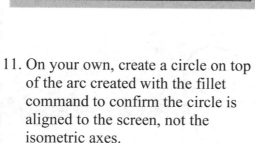

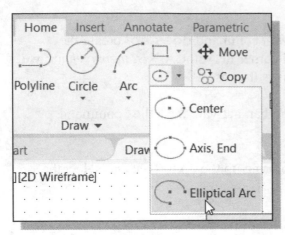

12. In the *Draw* toolbar, activate the **Elliptical Arc** command as shown.

13. In the command prompt area, click Isocircle as shown.

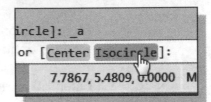

14. Snap to the bottom endpoint of the right line segment to place the center of the isocircle.

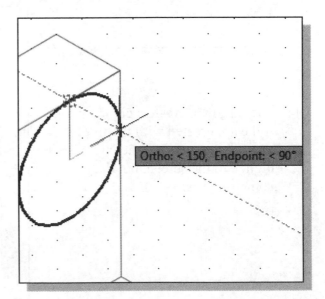

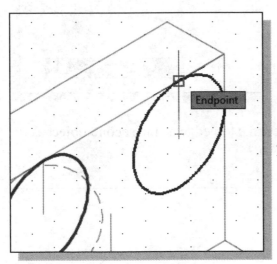

15. Snap to the top endpoint of the right line segment to accept the distance as radius.

16. Select the **Endpoint angle** 90 degree location as the start angle of the arc.

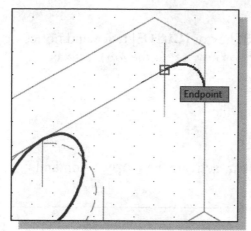

17. Snap to the top endpoint of the right line segment to accept the location as the end angle of the arc.

18. In the *Modify* toolbar, activate the **Trim** command as shown.

19. In the command prompt area, click **cuTting edges** as shown.

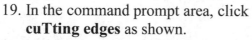

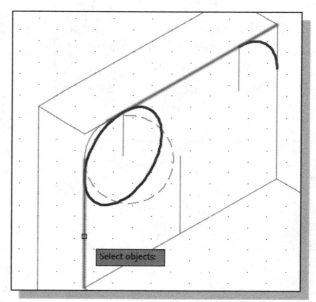

20. Select the top and the left edges as the trimming boundary as shown.

21. Click once with the right-mouse-button to accept the selection.

22. Click on the lower side of the isocircle to remove the lower portion as shown.

23. On your own, exit the trim command with the option list as shown.

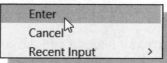

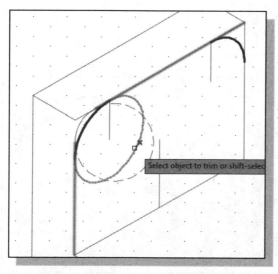

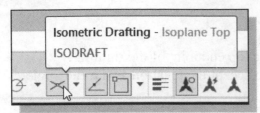

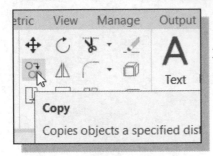

24. Hit the [**F5**] key or [**Ctrl+E**] twice and notice the isoplane is now set to the *top plane* as shown.

25. In the modify toolbar, activate the **Copy** command.

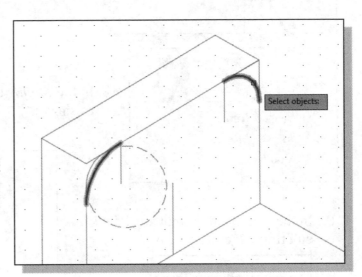

26. Select the two elliptical arcs as shown.

27. Right-mouse-click once to accept the selection and proceed with the **Copy** command.

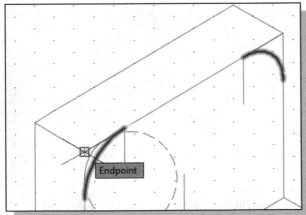

28. Select the **Front left corner** as the copy reference point.

29. Select the **back left corner** to place the two copies of arcs.

30. Hit the [**Esc**] key once to end the copy command.

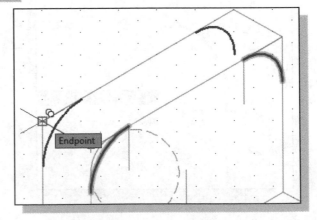

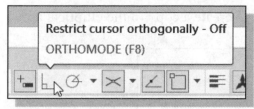

31. In the status bar area, turn off the **Ortho** mode as shown.

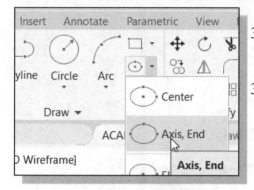

32. In the *Draw* toolbar, activate the **Axis, End Ellipse** command.

33. In the command prompt area, click Isocircle as shown.

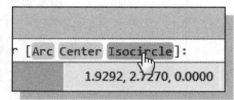

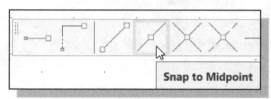

34. Activate the **Snap to Midpoint** option as shown.

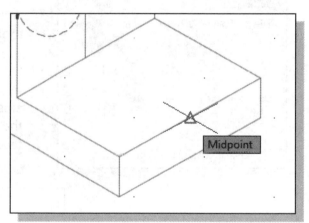

35. Snap to the midpoint of the top edge of the horizontal section of the design to place the center of the isocircle.

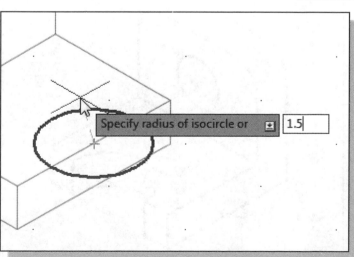

36. Enter **1.5** as the radius and create the isocircle.

37. On your own, use the **trim** and **copy** commands to create a copy of the elliptical arc to the bottom surface as shown.

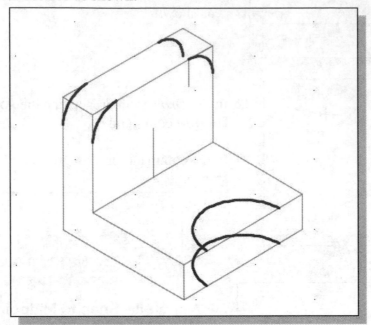

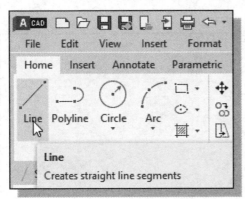

38. Activate the **Line** command in the *Draw* toolbar.

39. On your own, create the outline of the design by creating line segments connecting to the arcs as shown in the below figure. Also add the diameter 1.75 center hole on the vertical face as shown.

- Note that we need to identify the tangent edge for the top corner of the design.

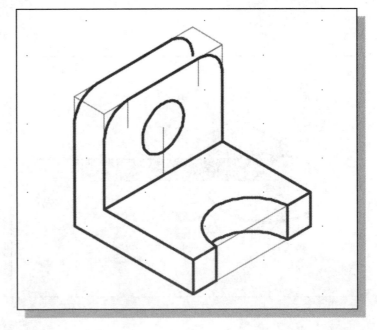

Using the Parametric tools to complete the drawing

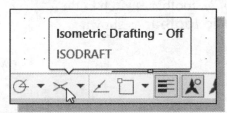

1. In the *Status Bar* area, turn off the *Isodraft*, *Object Snap Tracking* and *Object Snap* options as shown.

2. In the *modify toolbar*, activate the **Copy** command.

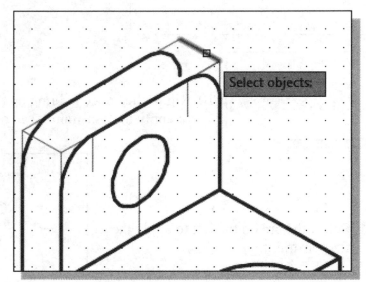

3. Select the **right edge** on the top face as shown.

4. Right-click once to accept the selection and proceed with the **Copy** command.

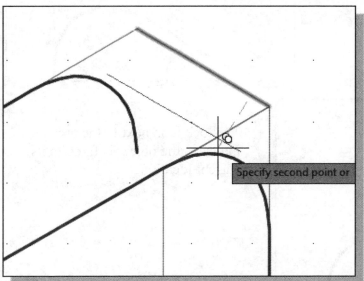

5. Select near the **right corner** as the copy reference point.

6. Place the line next to the two elliptical arcs; we will intentionally leave a gap.

7. Hit the [**Esc**] key once to exit the copy command.

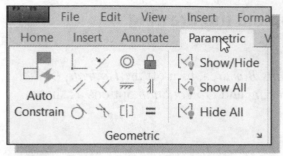

8. In the *Ribbon* toolbar, switch to the **Parametric toolbar** as shown.

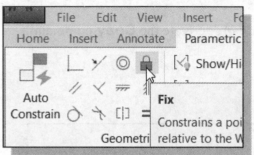

9. In the *Geometric* toolbar, activate the **Fix** constraint by clicking on the **Lock** icon as shown.

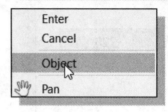

10. Right-click once to bring up the *option* list and select **Object** to apply the constraint to an object.

11. Select the arc on the left to apply the fix constraint.

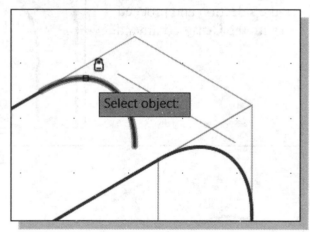

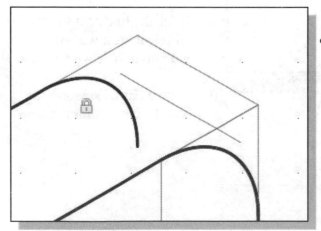

• The lock icon next to the arc indicates the object is fixed to its current location.

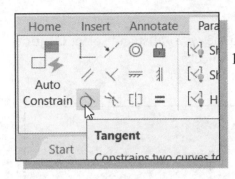

12. In the *Geometric* toolbar, activate the **Tangent** constraint by clicking on the icon as shown.

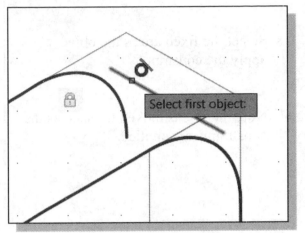

13. Select the inclined line as the first object to apply the tangent constraint.

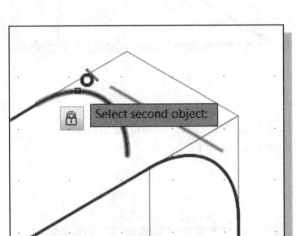

14. Select the **fixed arc** as the second object to apply the tangent constraint.

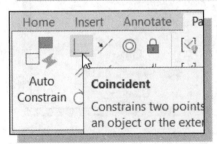

15. Click on the **Coincident** constraint as shown.

16. Click on the left end point of the line as shown.

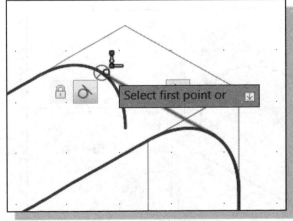

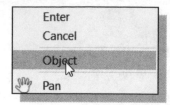

17. Right-click once to bring up the *option* list and select **Object** to apply the constraint to an object.

18. Select the fixed arc as the object to apply the constraint.

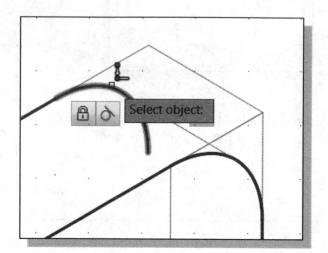

- Note the line is moved in place as the constraints are applied.

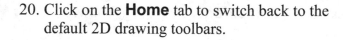
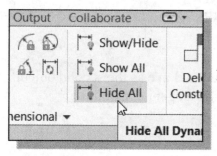

19. Click on the **Hide All** icon to remove all the constraints on the screen.

20. Click on the **Home** tab to switch back to the default 2D drawing toolbars.

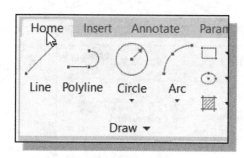

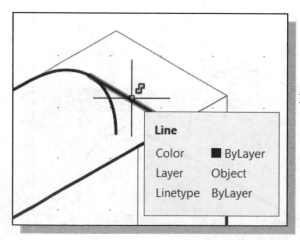

21. On your own, move the tangent line to the **Object layer**.

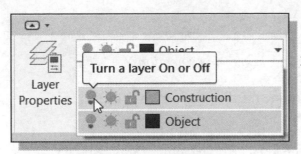

22. In the *Layer control* box, turn off the Construction layer as shown.

23. In the *Modify* toolbar, activate the **Trim** command as shown.

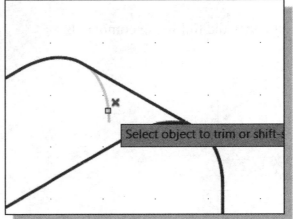

24. Trim the lower portion of the arc and complete the isometric drawing as shown.

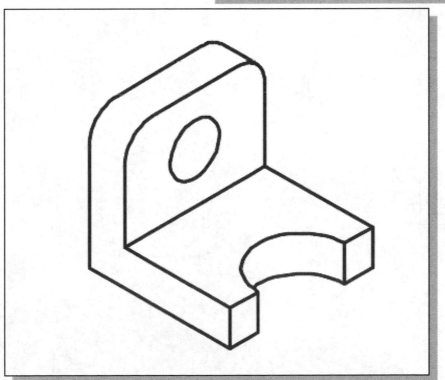

Review Questions:

1. What is the difference between *Isometric* and *Trimetric projections*?

2. In an isometric drawing, how do we determine the directions of the ellipses in representing cylindrical or circular shapes?

3. Describe some of the advantages and disadvantages of using the AutoCAD **2D Isometric** drawings over creating 3D models.

4. Can the Fillet command be used to create accurate rounded corners in AutoCAD 2D isometric drawings?

5. What are the axes' directions when the AutoCAD **Isoplane** is set to **Left plane**?

6. Identify the following commands:

(a)

(b)

(c)

(d)

Exercises:

1. Dimensions are in inches. Thickness: Base 0.5 & Boss height 0.25
    The two holes, diameter 1.00, are through holes.

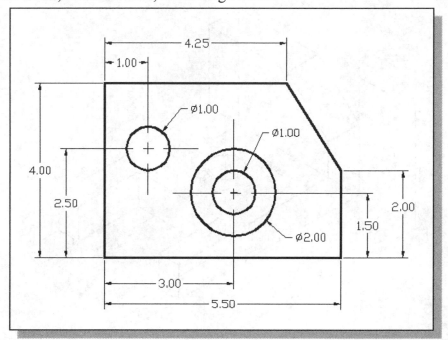

2. Dimensions are in inches.

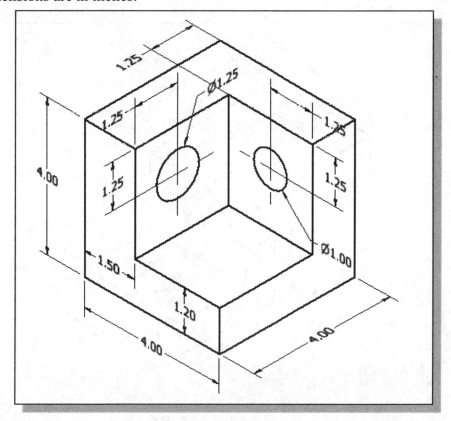

3.   Dimensions are in inches.

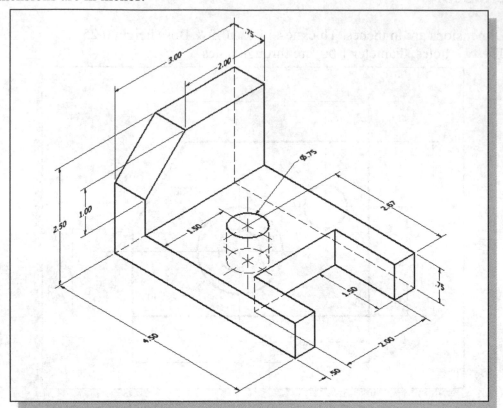

4.   Dimensions are in inches.

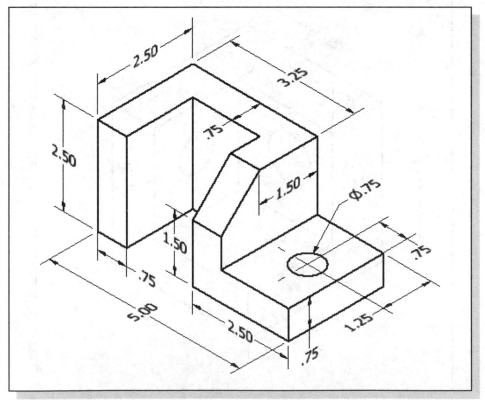

Chapter 7
Basic Dimensioning and Notes

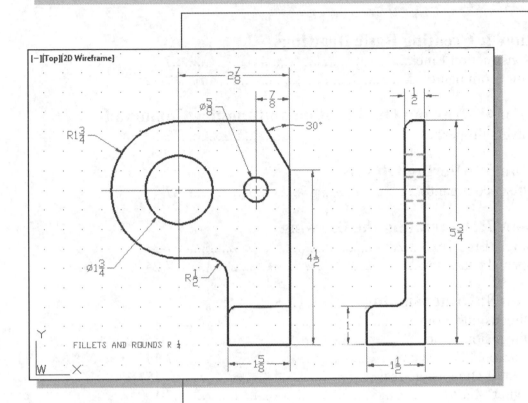

Learning Objectives

- ◆ **Understand dimensioning nomenclature and basics**
- ◆ **Display and use the Dimension toolbar**
- ◆ **Use the AutoCAD Dimension Style Manager**
- ◆ **Create Center Marks**
- ◆ **Add Linear and Angular Dimensions**
- ◆ **Use the TEXT command**
- ◆ **Create SPECIAL CHARACTERS in Notes**

AutoCAD Certified User Examination Objectives Coverage

This table shows the pages on which the objectives of the Certified User Examination are covered in Chapter 7.

Certified User Reference Guide

Introduction

In order to manufacture a finalized design, the complete *shape* and *size* description must be shown on the drawings of the design. Thus far, the use of AutoCAD 2023 to define the shape of designs has been illustrated. In this chapter, the procedures to convey the *size* definitions of designs using **AutoCAD 2023** are discussed. The *tools of size description* are known as *dimensions* and *notes*.

Considerable experience and judgment are required for accurate size description. Detail drawings should contain only those dimensions that are necessary to make the design. Dimensions for the same feature of the design should be given only once in the same drawing. Nothing should be left to chance or guesswork on a drawing. Drawings should be dimensioned to avoid any possibility of questions. Dimensions should be carefully positioned, preferably near the profile of the feature being dimensioned. The designer and the CAD operator should be as familiar as possible with materials, methods of manufacturing, and shop processes.

Traditionally, detailing a drawing is the biggest bottleneck of the design process; and when doing board drafting, dimensioning is one of the most time consuming and tedious tasks. Today, most CAD systems provide what is known as an **auto-dimensioning feature**, where the CAD system automatically creates the extension lines, dimensional lines, arrowheads, and dimension values. Most CAD systems also provide an **associative dimensioning feature** so that the system automatically updates the dimensions when the drawing is modified.

The Bracket Design

Starting Up AutoCAD 2023

1. Select the **AutoCAD 2023** option on the *Program* menu or select the **AutoCAD 2023** icon on the *Desktop*.

2. In the *Startup* window, select **Start from Scratch**, and select the **Imperial (feet and inches)** as the drawing units.

3. In the *Menu Bar*, select:

   ### [Tools] → [Drafting Settings]

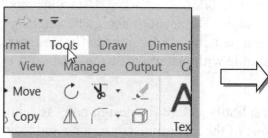

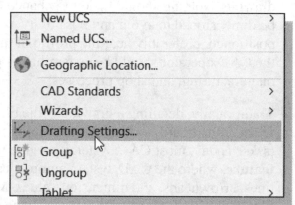

4. In the *Drafting Settings* dialog box, select the **SNAP and GRID** tab if it is not the page on top.

5. Change *Grid Spacing* to **1.0** for both X and Y directions.

6. Also adjust the *Snap Spacing* to **0.5** for both X and Y directions.

7. Pick **OK** to exit the *Drafting Settings* dialog box.

8. In the *Status Bar* area, reset the option buttons so that only *SNAP Mode* and *GRID Display* are switched **ON**.

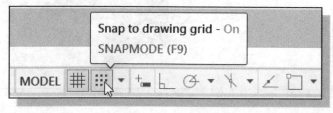

Layers Setup

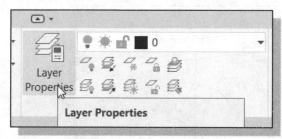

1. Pick *Layer Properties Manager* in the *Object Properties* toolbar.

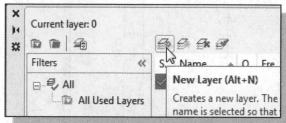

2. In the *Layer Properties Manager* dialog box, click on the **New** button (or the key combination [**Alt+N**]) to create a new layer.

3. Create **layers** with the following settings:

Layer	Color	Linetype	Lineweight
Construction	Gray(9)	Continuous	Default
ObjectLines	Blue	Continuous	0.6mm
HiddenLines	Cyan	Hidden	0.3mm
CenterLines	Red	Center	Default
Dimensions	Magenta	Continuous	Default
SectionLines	White	Continuous	Default
CuttingPlaneLines	Dark Gray(8)	Phantom	0.6mm
Title_Block	Green	Continuous	1.2mm
Viewport	White	Continuous	Default

4. Highlight the layer *Construction* in the list of layers.

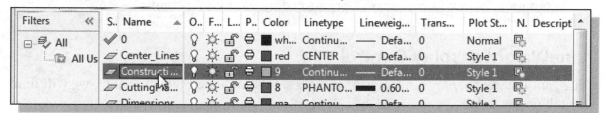

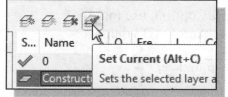

5. Click on the **Current** button to set layer *Construction* as the *Current Layer*.

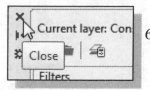

6. Click on the **Close** button to accept the settings and exit the *Layer Properties Manager* dialog box.

The Bracket Design

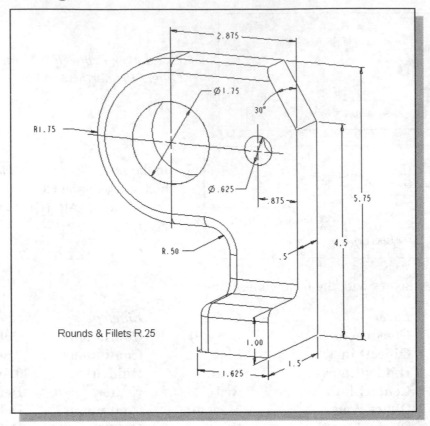

> Before going through the tutorial, make a rough sketch of a multiview drawing of the part. How many 2D views will be necessary to fully describe the part? Based on your knowledge of **AutoCAD 2023** so far, how would you arrange and construct these 2D views? Take a few minutes to consider these questions and do preliminary planning by sketching on a piece of paper. You are also encouraged to construct the orthographic views on your own prior to going through the tutorial.

LineWeight Display Control

The *LineWeight Display* control can be accessed through the Status bar area; we will first switch on the icon.

1. To show the icon for the AutoCAD *LineWeight Display* option, use the *Customization option* at the bottom right corner of the AutoCAD window.

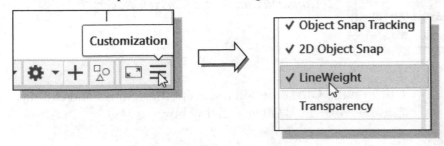

Drawing Construction Lines

We will place the construction lines on the *Construction* layer so that the layer can later be frozen or turned off.

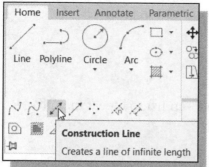

1. Select the **Construction Line** icon in the *Draw* toolbar. In the command prompt area, the message "*_xline Specify a point or [Hor/Ver/Ang/Bisect/Offset]:*" is displayed.

➢ To orient construction lines, we generally specify two points, although other orientation options are also available.

2. Place the first point at world coordinate (**3,2**) on the screen.

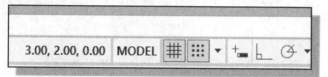

3. Pick a location above the last point to create a **vertical line**.

4. Move the cursor toward the right of the first point and pick a location to create a **horizontal line**.

5. In the *Status Bar* area, turn **OFF** the *SNAP* option.

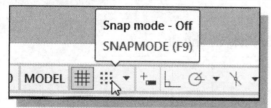

Using the Offset Command

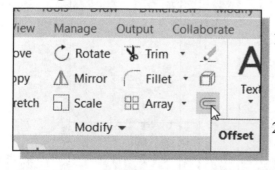

1. Select the **Offset** icon in the *Modify* toolbar. In the command prompt area, the message "*Specify offset distance or [Through]:*" is displayed.

2. In the command prompt area, enter **2.875** [**ENTER**].

3. In the *command prompt area*, the message "*Select object to offset or <exit>:*" is displayed. Pick the **vertical line** on the screen.

4. AutoCAD next asks us to identify the direction of the offset. Pick a location that is to the **right** of the vertical line.

5. Inside the *Drawing Area*, right-click once and choose **Enter** to end the Offset command.

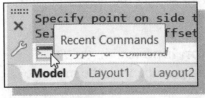

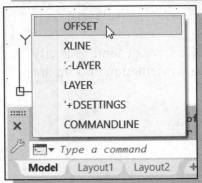

6. In the *command prompt area*, click on the small icon to access the list of recent commands.

7. Select **Offset** in the pop-up list to repeat the Offset command.

8. In the *command prompt area*, enter **5.75 [ENTER]**.

9. In the *command prompt area*, the message "*Select object to offset or <exit>:*" is displayed. Pick the **horizontal line** on the screen.

10. AutoCAD next asks us to identify the direction of the offset. Pick a location that is **above** the horizontal line.

11. Inside the *Drawing Area*, right-click once and choose **Enter** to end the Offset command.

12. Repeat the **Offset** command and create the lines as shown.

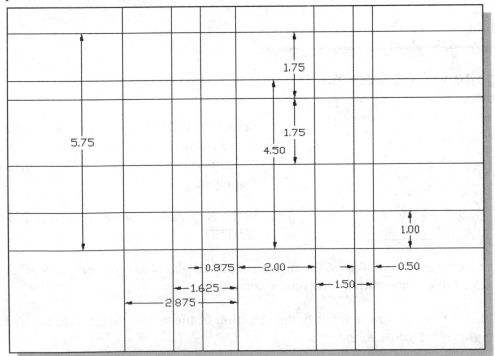

Set Layer ObjectLines as the Current Layer

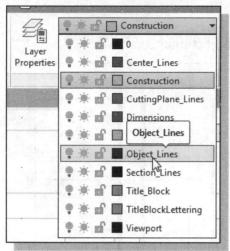

1. On the *Object Properties* toolbar, choose the **Layer Control** box with the left-mouse-button.

2. Move the cursor over the name of layer **ObjectLines** and the tool tip "*ObjectLines*" appears.

3. **Left-click once** and layer *ObjectLines* is set as the *Current Layer*.

4. In the *Status Bar* area, turn **ON** the *Object SNAP, Object TRACKing*, and *Line Weight Display* options.

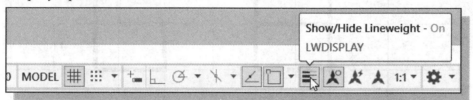

Creating Object Lines

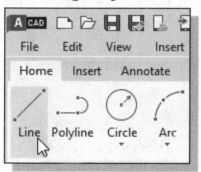

1. Select the **Line** command icon in the *Draw* toolbar. In the command prompt area, the message "*Line Specify first point:*" is displayed.

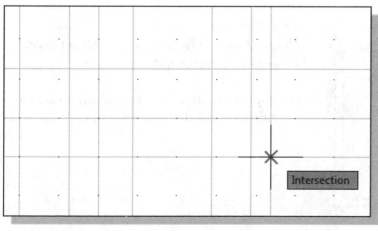

2. Move the cursor to the intersection of the construction lines at the lower right corner of the current sketch, and notice the visual aid automatically displayed at the intersection.

3. Create the object lines as shown below. (Hint: Use the *relative coordinate entry method* and the **Trim** command to construct the **30° line**.)

4. Use the Arc and Circle commands to complete the object lines as shown.

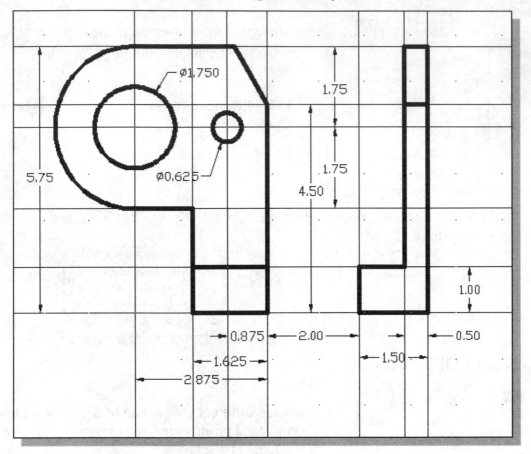

Creating Hidden Lines

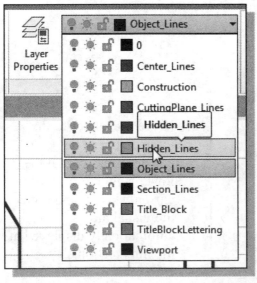

1. On the *Object Properties* toolbar, choose the *Layer Control* box with the left-mouse-button.

2. Move the cursor over the name of layer **HiddenLines**, **left-click once**, and set layer *HiddenLines* as the *Current Layer*.

3. On your own, create the **five hidden lines** in the side view as shown on the next page.

Creating Center Lines

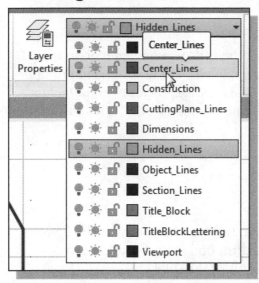

1. On the *Object Properties* toolbar, choose the *Layer Control* box to display the *Layer control list*.

2. Move the cursor over the name of layer *CenterLines* and **left-click once** to set the layer as the *Current Layer*.

3. On your own, create the center line in the side view as shown in the below figure. (The center lines in the front-view will be added using the *Center Mark* option.)

Turn Off the Construction Lines

1. On the *Layers* toolbar, choose the icon next to the *Layer Control* box with the left-mouse-button.

2. Move the cursor over the *lightbulb* icon for layer *ConstructionLines*, **left-click once**, and notice the icon color is changed to a gray tone color, representing the layer (layer *ConstructionLines*) is turned *OFF*.

3. On your own, set layer **ObjectLines** as the *Current Layer*.

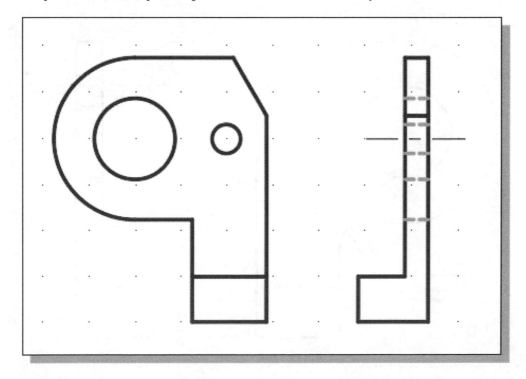

Using the Fillet Command

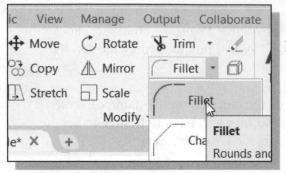

1. Select the **Fillet** command icon in the *Modify* toolbar. In the command prompt area, the message "*Select first object or [Polyline/Radius/Trim]:*" is displayed.

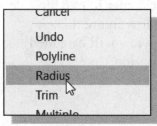

2. Inside the *Drawing Area*, right-click to activate the option menu and select the **Radius** option with the left-mouse-button to specify the radius of the fillet.

3. In the *command prompt area*, the message "*Specify fillet radius:*" is displayed.

   Specify fillet radius: **0.5** [**ENTER**].

4. Pick the **bottom horizontal line** of the front view as the first object to fillet.

5. Pick the **adjacent vertical line** connected to the arc to create a rounded corner as shown.

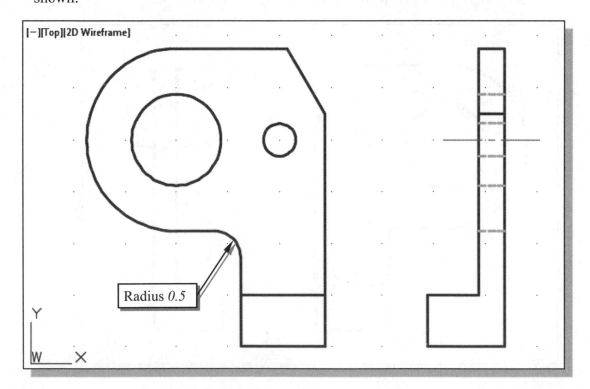

6. Repeat the **Fillet** command and create the four rounded corners (radius **0.25**) as shown.

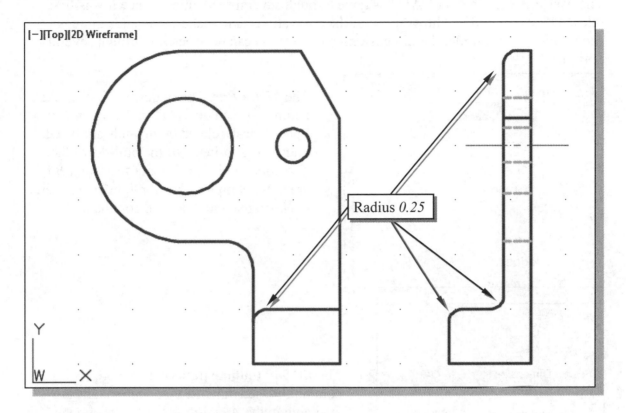

Radius *0.25*

Saving the Completed CAD Design

1. In the *Application menu*, select:
[Save As] → [Drawing]

2. In the *Save Drawing As* dialog box, select the folder in which you want to store the CAD file and enter **Bracket** in the *File name* box.

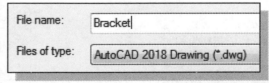

File name:	Bracket
Files of type:	AutoCAD 2018 Drawing (*.dwg)

3. Click **Save** in the *Save Drawing As* dialog box to accept the selections and save the file.

Accessing the Dimensioning Commands

The user interface in AutoCAD has gone through several renovations since it was first released back in 1982. The purpose of the renovations is to make commands more accessible. For example, the dimensioning commands can be accessed through several options:

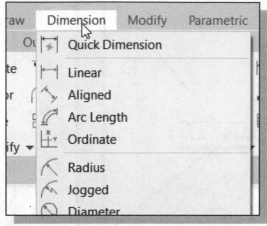

➢ The ***Menu Bar***: The majority of AutoCAD commands can be found in the *Menu Bar*. Specific task related commands are listed under the sub-items of the pull-down list. The commands listed under the *Menu Bar* are more complete, but it might take several clicks to reach the desired command.

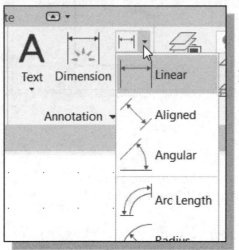

➢ The ***Ribbon* toolbar panels**: The more commonly used dimensioning commands are available in the *Annotation* toolbar as shown.

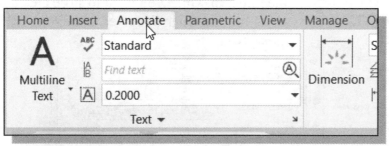

➢ The ***Dimensions* toolbar** under the **Annotate tab**: This panel contains a more complete set of dimensioning commands.

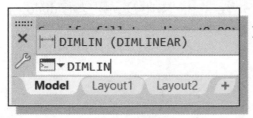

➢ The ***command prompt***: We can also type the command at the command prompt area. This option is always available, with or without displaying the toolbars.

The Dimension Toolbar

The *Dimension* toolbar offers the most flexible option to access the different dimensioning commands. The toolbar contains a more complete list than the *Ribbon* toolbars and this toolbar can be repositioned anywhere on the screen.

1. Move the cursor to the *Menu Bar* area and select:
    [Tools] → [Toolbars] → [AutoCAD] → [Dimension]

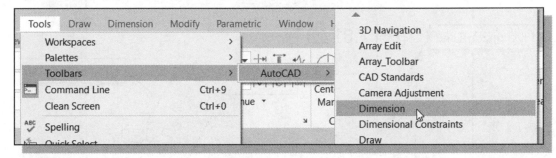

2. Move the cursor over the icons in the *Dimension* toolbar and read the brief description of each icon.

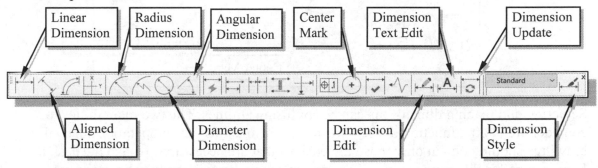

Using Dimension Style Manager

The appearance of the dimensions is controlled by *dimension variables*, which we can set using the *Dimension Style Manager* dialog box.

1. In the *Dimension* toolbar, pick **Dimension Style**. The *Dimension Style Manager* dialog box appears on the screen.

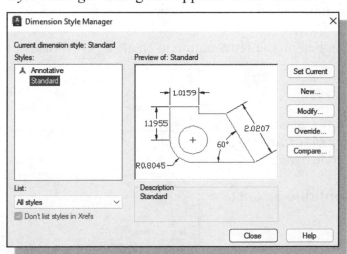

Dimensions Nomenclature and Basics

As it was stated in *Chapter 1*, the rule for creating CAD designs and drawings is that they should be created **full size** using real-world units. The importance of this practice is evident when we begin applying dimensions to the geometry. The features that we specify for dimensioning are measured and displayed automatically.

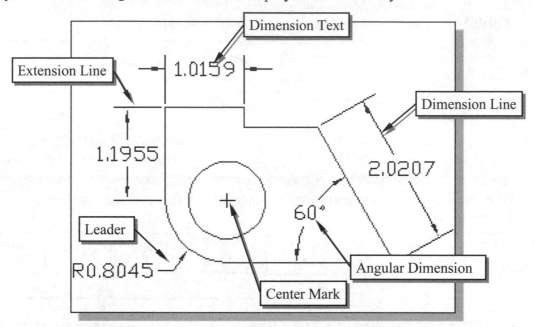

Selecting and placing dimensions can be confusing at times. The two main things to consider are (1) the function of the part and (2) the manufacturing operations. Detail drawings should contain only those dimensions that are necessary to make the design. Dimensions for the same feature of the design should be given only once in the same drawing. Nothing should be left to chance or guesswork on a drawing. Drawings should be dimensioned to avoid any possibility of questions. Dimensions should be carefully positioned, preferably near the profile of the feature being dimensioned.

Notice in the *Dimension Style Manager* dialog box, the AutoCAD default style name is *Standard*. We can create our own dimension style to fit the specific type of design we are working on, such as mechanical or architectual.

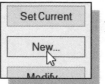

1. Click on the **New** button to create a new dimension style.

2. In the *Create New Dimension Style* dialog box, enter **Mechanical** as the dimension style name.

3. Click on the **Continue** button to proceed.

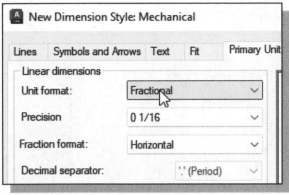

4. Click on the **Primary Units** tab.

5. Select *Fractional* as the *Unit format* under the **Primary Units** tab.

❖ On your own, examine the different options available; most of the settings are self-explanatory.

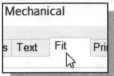

6. Select the **Fit** tab and notice the two options under *Scale for Dimension Features*.

❖ We can manually adjust the dimension scale factor or let AutoCAD automatically adjust the scale factor. For example, our current drawing will fit on A-size paper, and therefore we will use the scale factor of 1; this will assure the height of text size will be 1/8 of an inch on plotted drawing. If we change the Dimension Scale to 2.0 then the plotted text size will be ¼ of an inch on plotted drawing. In the next chapter, we will go over the procedure on using the AutoCAD layout mode, which allows us to plot the same design with different drawing sizes.

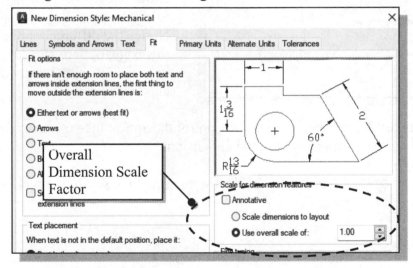

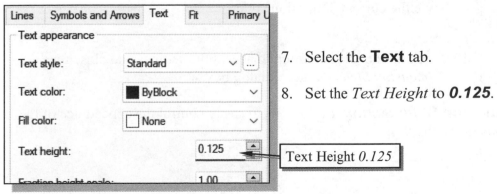

7. Select the **Text** tab.

8. Set the *Text Height* to **0.125**.

9. Select the **Lines** tab and set *Extend beyond dim lines* to **0.125**.

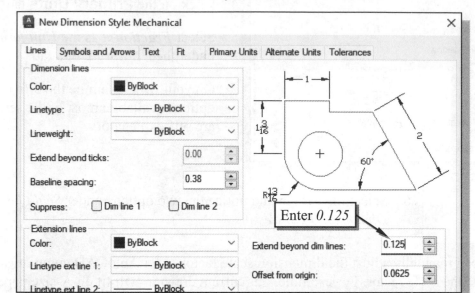

10. Select the **Symbols and Arrows** tab and set *Arrow size* and *Center Marks* to **0.125**. Also set the *Center Marks* type to **Line**.

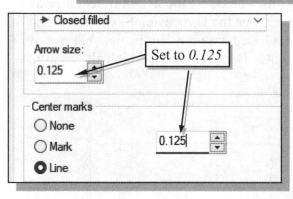

❖ Notice the different options available on this page, options that let us turn off one or both extension lines, dimension lines, and arrowheads.

➤ The **Center Mark** option is used to control the appearance of center marks and centerlines for diameter and radial dimensions.

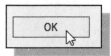

11. Click on the **OK** button to accept the settings and close the dialog box.

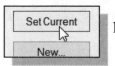

12. Pick the **Set Current** button to make the *Mechanical* dimension style the current dimension style.

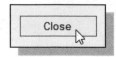

13. Click on the **Close** button to accept the settings and close the *Dimension Style Manager* dialog box.

➤ The **Dimension Style Manager** allows us to easily control the appearance of the dimensions in the drawing.

Using the Center Mark Command

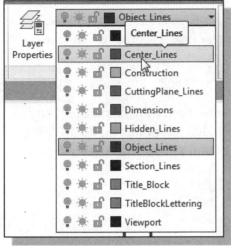

1. On the *Object Properties* toolbar, choose the *Layer Control* box with the left-mouse button.

2. Move the cursor over the name of layer *CenterLines*, **left-click once**, and set layer *CenterLines* as the *Current Layer*.

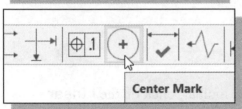

3. In the *Dimension* toolbar, click on the **Center Mark** icon.

4. Pick the **radius 1.75 arc** in the front view and notice AutoCAD automatically places two centerlines through the center of the arc.

5. Repeat the **Center Mark** command and pick the **small circle** to place the centerlines as shown. Also create a center line in the side view as shown.

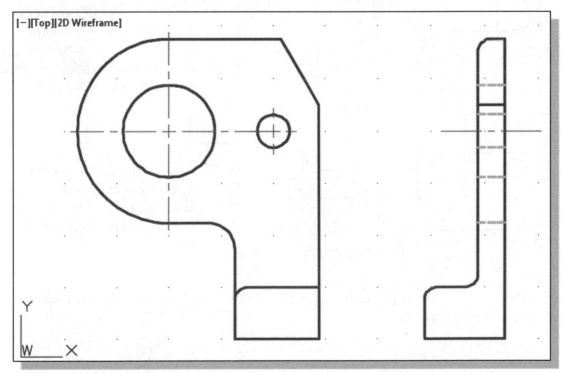

Adding Linear Dimensions

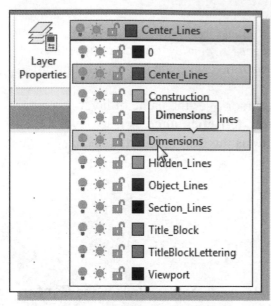

1. On the *Object Properties* toolbar, choose the **Layer Control** box with the left-mouse-button.

2. Move the cursor over the name of layer **Dimensions**, **left-click once**, and set layer *Dimensions* as the *Current Layer*.

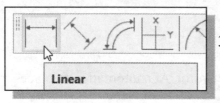

3. In the *Dimension* toolbar, click on the **Linear Dimension** icon.

➤ The Linear Dimension command measures and annotates a feature with a horizontal or vertical dimension.

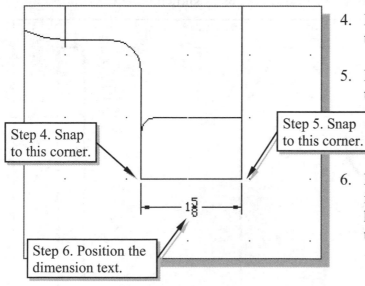

Step 4. Snap to this corner.

Step 5. Snap to this corner.

Step 6. Position the dimension text.

4. Pick the **lower left corner** of the front view of the part.

5. Pick the **lower right corner** of the front view of the part.

6. Pick a point that is about 0.5 inch below the bottom horizontal line of the front view to place the dimension text.

❖ Adding dimensions is this easy with AutoCAD's auto-dimensioning and associative-dimensioning features.

7. Repeat the **Linear Dimension** command and add the necessary linear dimensions as shown.

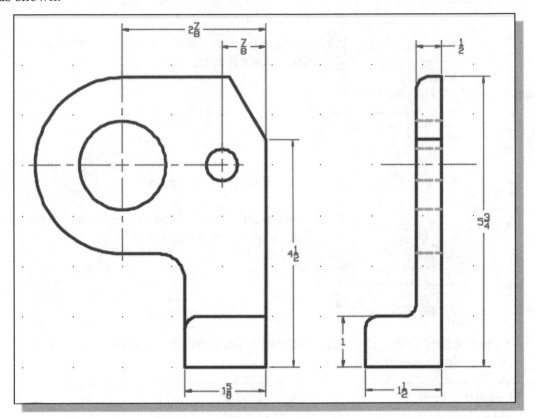

Adding an Angular Dimension

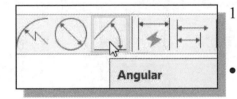

1. In the *Dimension* toolbar, click on the **Angular Dimension** icon.

- The Angular Dimension command measures and annotates a feature with an angle dimension.

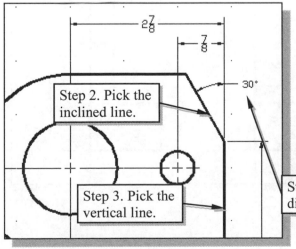

Step 2. Pick the inclined line.

Step 3. Pick the vertical line.

Step 4. Position the dimension text.

2. Pick the **inclined line** of the part in the front view.

3. Pick the **right vertical line** of the part in the front view.

4. Pick a point inside the desired quadrant and place the dimension text as shown.

Adding Radius and Diameter Dimensions

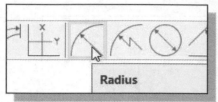

1. In the *Dimension* toolbar, click on the **Radius Dimension** icon.

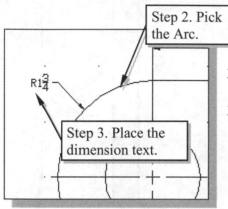

2. Pick the **large arc** in the front view.

3. Pick a point toward the left of the arc to place the dimension text.

4. Use the Radius Dimension and Diameter Dimension commands to add the necessary dimensions as shown.

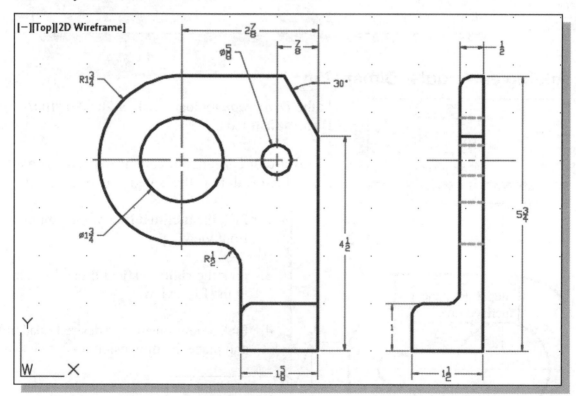

Using the Multiline Text Command

AutoCAD provides two options to create notes. For simple entries, we can use the **Single Line Text** command. For longer entries with internal formatting, we can use the **Multiline Text** command. The procedures for both of these commands are self-explanatory. The **Single Line Text** command, also known as the **Text** command, can be used to enter several lines of text that can be rotated and resized. The text we are typing is displayed on the screen. Each line of text is treated as a separate object in AutoCAD. To end a line and begin another, press the **[ENTER]** key after entering characters. To end the Text command, press the **[ENTER]** key without entering any characters.

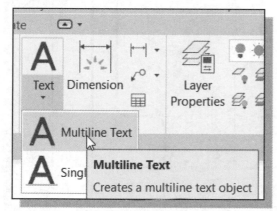

1. In the *Ribbon* toolbar panel, select:

   [Multiline Text]

2. In the command prompt area, the message "*Specify start point of text or [Justify/Style]:*" is displayed. Pick a location near the world coordinate (*1,1.5*).

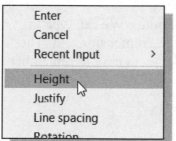

3. Right-click once to bring up the option list and choose **Height**. In the command prompt area, the message "*Specify Height:*" is displayed. Enter *0.125* as the text height.

4. Click a position toward the right and create a rectangle representing roughly the area for the notes.

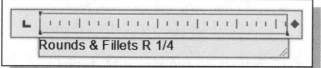

5. In the Edit Box, enter ***Rounds & Fillets R 1/4***.

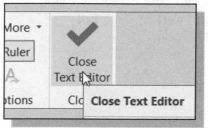

6. Note in the *Ribbon toolbar area* the different command options are available to adjust the multiline text entered. Click **[Close Text Editor]** once to end the command.

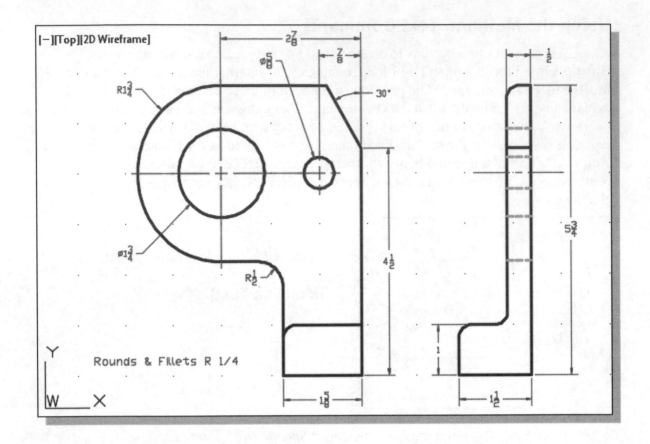

Adding Special Characters

We can add special text characters to the dimensioning text and notes. We can type in special characters during any text command and when entering the dimension text. The most common special characters have been given letters to make them easy to remember.

Code	Character	Symbol
%%C	Diameter symbol	Ø
%%D	Degree symbol	°
%%P	Plus/Minus sign	±

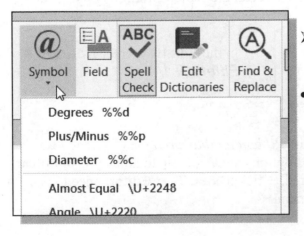

➢ On your own, create notes containing some of the special characters listed.

• Note the complete list of special characters can be accessed through the [**Multiline Text**] → [**Symbol**] dashboard.

Saving the Design

In the *Standard Toolbar* area, select the **Save** icon.

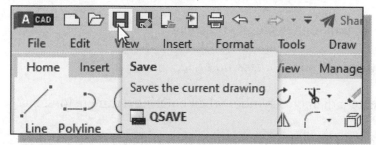

A Special Note on Layers Containing Dimensions

AutoCAD creates several hidden **BLOCKS** when we create associative dimensions and we will take a more in depth look at *blocks* in *Chapter 12*. AutoCAD treats **blocks** as a special type of object called a *named object*. Each kind of *named object* has a *symbol table* or a *dictionary*, and each table or dictionary can store multiple *named objects*. For example, if we create five dimension styles, our drawing's dimension style *symbol table* will have five dimension style records. In general, we do not work with *symbol tables* or *dictionaries* directly.

When we create dimensions in AutoCAD, most of the hidden blocks are placed in the same layer where the dimension was first defined. Some of the definitions are placed in the *DEFPOINTS* layer. When moving dimensions from one layer to another, AutoCAD does not move these definitions. When deleting layers, we cannot delete the current layer, *layer 0*, xref-dependent layers, or a layer that contains visible and/or invisible objects. Layers referenced by block definitions, along with the *DEFPOINTS* layer, cannot be deleted even if they do not contain visible objects.

To delete layers with hidden blocks, first use the **Purge** command **[Application →Drawing Utilities → Purge → Blocks]** to remove the invisible blocks. (We will have to remove all visible objects prior to using this command.) The empty layer can now be *deleted* or *purged*.

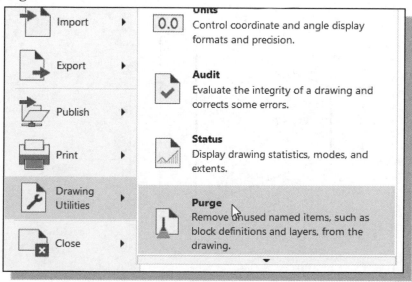

Review Questions: (Time: 25 minutes)

1. Why are dimensions and notes important to a technical drawing?

2. List and describe some of the general-dimensioning practices.

3. Describe the procedure in setting up a new *Dimension Style*.

4. What is the special way to create a diameter symbol when entering a dimension text?

5. What is the text code to create the **Diameter** symbol (Ø) in AutoCAD?

6. Which command can be used to delete layers with hidden blocks? Can we delete *Layer 0*?

7. Which quick-key is used to display and hide the *AutoCAD Text Window*?

8. What are the length and angle of the inclined line, highlighted in the figure below, in the front view of the *LBracket* design?

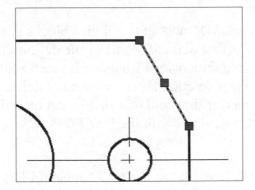

9. Construct the following drawing and measure the angle α.

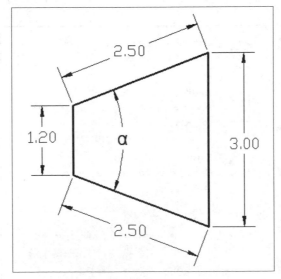

Exercises: (Time: 210 minutes)

1. **Shaft Guide** (Dimensions are in inches. .25 drill hole to the front part of the design)

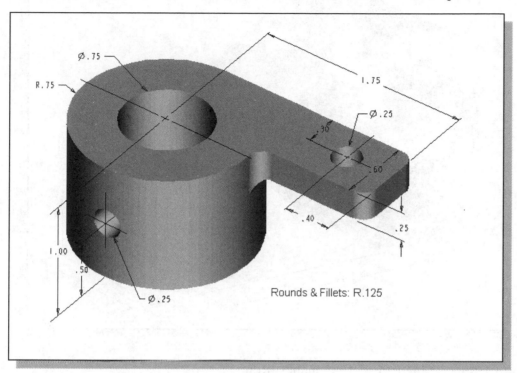

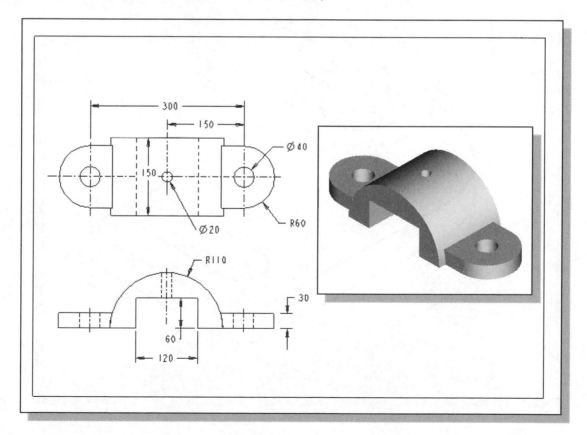

2. **Fixture Cap** (Dimensions are in millimeters.)

3. Cylinder Support (Dimensions are in inches.)

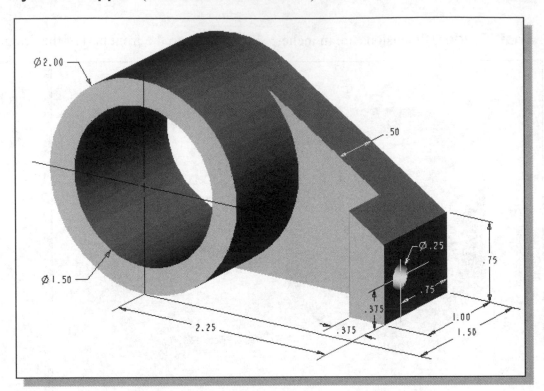

4. Swivel Base (Dimensions are in inches. Rounds: R 0.25)

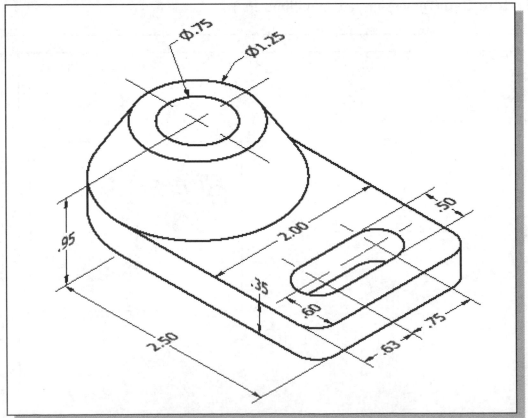

5. Pivot Holder (Dimensions are in inches.)

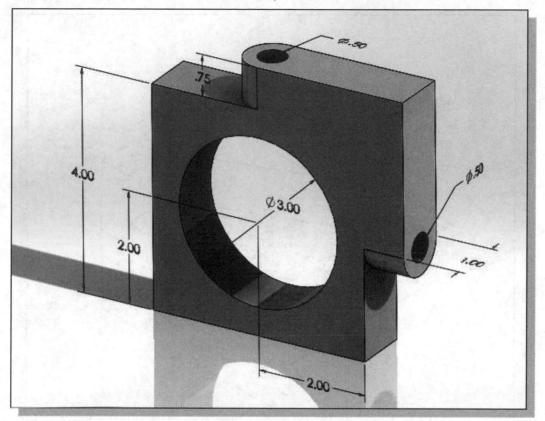

6. Shaft Guide (Dimensions are in inches.)

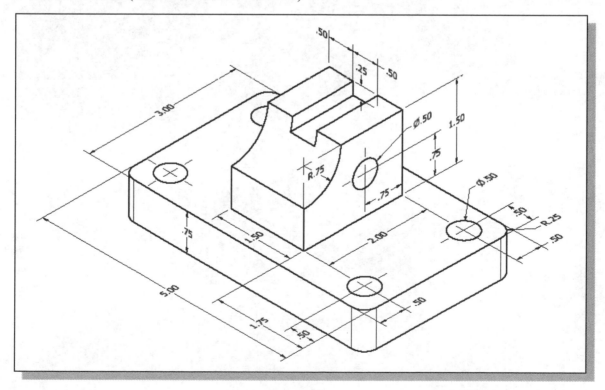

7. Indexing Stop (Dimensions are in millimeters.)

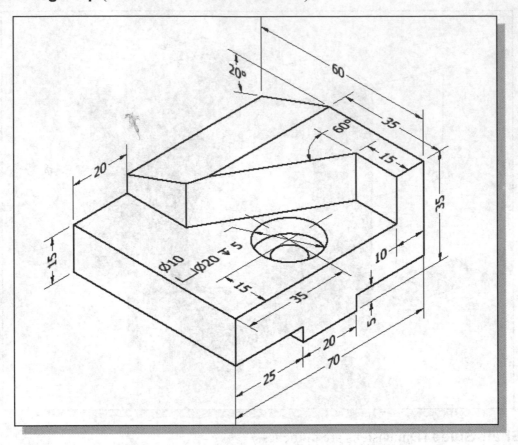

Chapter 8
Templates and Plotting

Learning Objectives

♦ **Set up the AutoCAD Plot Style option**
♦ **Create a Template file**
♦ **Use the Mirror Command**
♦ **Create Multiple Copies of objects**
♦ **Set up Layouts in Paper Space**
♦ **Create Viewports in Paper Space**
♦ **Use the Properties Command**
♦ **Adjust the Text Scale for Plotting**

AutoCAD Certified User Examination Objectives Coverage

This table shows the pages on which the objectives of the Certified User Examination are covered in Chapter 8.

Introduction

One of the main advantages of using CAD systems is that we can easily reuse information that is already in the system. For example, many of the system settings, such as setting up layers, colors, linetypes and grids, are typically performed in all AutoCAD files. In **AutoCAD 2023**, we can set up **template files** to eliminate these repetitive steps and make our work much more efficient. Using template files also helps us maintain consistent design and drafting standards. In this chapter, we will illustrate the procedure to set up template files that can contain specific plotting settings, system units, environment settings and other drafting standard settings.

We can also reuse any of the geometry information that is already in the system. For example, we can easily create multiple identical copies of geometry with the **Array** command, or create mirror images of objects using the **Mirror** command. In this chapter, we will examine the use of these more advanced construction features and techniques in **AutoCAD 2023**.

Also in this chapter, we will demonstrate the printing/plotting procedure to create a hardcopy of our design. **AutoCAD 2023** provides plotting features that are very easy to use. The **AutoCAD 2023** plotting features include WYSIWYG (What You See Is What You Get) layouts, onscreen lineweights, plot style tables, device-accurate paper sizes, and creating custom paper sizes.

The Geneva Cam Design

Starting Up AutoCAD 2023

1. Select the **AutoCAD 2023** option on the *Program* menu or select the **AutoCAD 2023** icon on the *Desktop*.

2. In the *Startup* window, select **Start from Scratch**, as shown in the figure below.

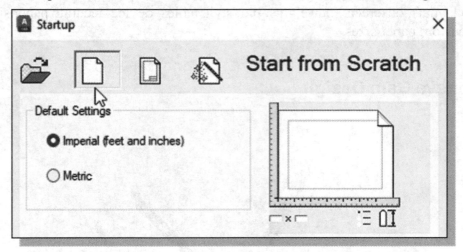

3. In the *Default Settings* section, pick **Imperial (feet and inches)** as the drawing units.

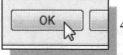   4. Pick **OK** in the startup dialog box to accept the selected settings.

❖ Note the units setting of *Drawing1* is set to the *English units* system. One of the important rules for creating CAD designs and drawings is that they should be created **full size**. The importance of this practice is evident when we are ready to create hardcopies of the design, or transfer the designs electronically to manufacturing equipment, such as a CNC machine. Internally, CAD systems do not distinguish whether the one unit of measurement is one inch or one millimeter. **AutoCAD 2023** provides several options to control the units settings.

Setting up the Plot Style Mode

Using **AutoCAD 2023** *plot styles* and *plot style tables* allows us to control the way drawings look at plot time. We can reassign object properties, such as color, linetype, and lineweight, and plot the same drawing differently. The default **AutoCAD 2023** *Plot Style Mode* is set to use the *Color-Dependent plot style*, which controls the plotting of objects based on the object colors and is the *traditional* method of adjusting the plotted hardcopy in AutoCAD. The other plot method in **AutoCAD** is to use the *Named plot style table* that works independently of color. In this chapter, we will learn to plot with the new AutoCAD *Named plot style*, which provides a very flexible and fast way to control the plotting of our designs.

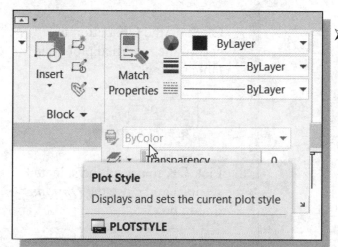

➢ Notice the *Plot Style* box, the first box in the *Object Properties* toolbar, displays the setting of **ByColor** and is grayed out. This indicates the plot style is set to the default *Color-Dependent* plot style, and therefore the object color is used to control plotting of the drawing.

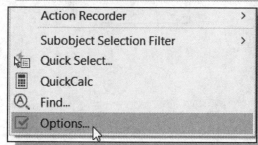

1. Inside the graphics area, **right-click** to bring up the option menu.

2. Select **Options** as shown in the figure.

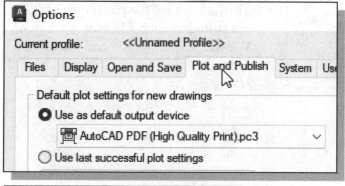

3. In the *Options* dialog box, select the **Plot and Publish** tab if it is not the page on top.

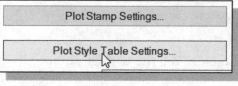

4. Click on the **Plot Style Table Settings** button as shown.

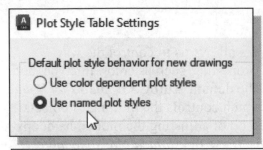

5. Switch *ON* the **Use named plot styles** as shown.

6. In the *Default plot style table*, select *acad.stb* from the list of plot style tables.

7. Click on the **OK** button to accept the modified plot style table settings.

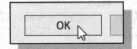

8. Pick **OK** to accept the selected settings and close the *Options* dialog box.

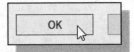

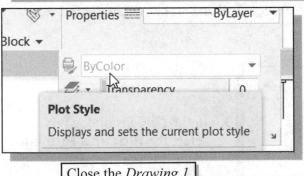

❖ Note the *Plot Style* box in the *Object Properties* toolbar is still grayed out. This is because the plotting settings are stored in each file as it is created. We will close this file and start a new file to have the new settings take effect.

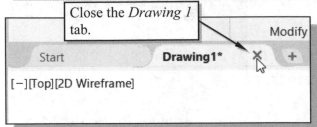

9. Click on the **[X]** button of the *Drawing 1 tab*, to close the file.

10. In the AutoCAD *Warning* dialog box, select **NO** to close the file without saving.

Starting a New File

1. Click the **New** icon in the *Quick Access* toolbar.

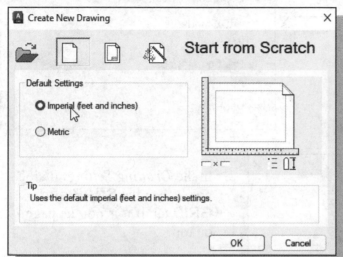

2. In the AutoCAD *Create New Drawing* dialog box, select the **Start from Scratch** option as shown.

3. Pick **Imperial (Feet and Inches)** as the drawing units.

4. Pick **OK** in the startup dialog box to accept the selected settings.

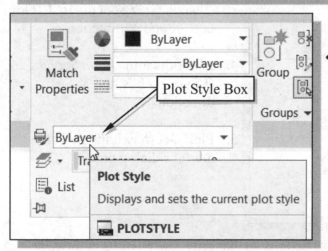

❖ Notice the *Plot Style* box in the *Object Properties* toolbar now displays the setting of *ByLayer* and is no longer grayed out. This indicates we are now using the *Named plot style*, with different options available to control plotting of the design.

➢ We will demonstrate the use of the ***Named plot style*** to create a hardcopy of the *Geneva Cam* design. The steps described in the above sections are required to set up the settings for the named plot style to be used in new drawings, and it should be done prior to creating the design. We can also convert an existing drawing to use named plot styles; it will require installing the ***AutoCAD Migration*** application and using the **Convertpstyles** command to perform the conversion. Note that after the conversion, any color-dependent plot style tables attached to layouts in the drawing are removed.

Grid and Snap Intervals Setup

1. In the *Menu Bar*, select:

   [Tools] → [Drafting Settings]

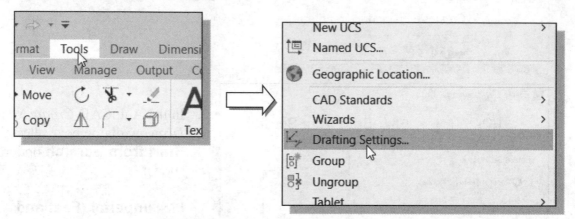

2. In the *Drafting Settings* dialog box, select the **SNAP and GRID** tab if it is not the page on top.

3. Change *Grid Spacing* to **1.0** for both X and Y directions.

4. Also adjust the *Snap Spacing* to **0.5** for both X and Y directions.

5. Pick **OK** to exit the *Drafting Settings* dialog box.

6. In the *Status Bar* area, reset the option buttons so that only *SNAP Mode* and *GRID Display* are switched **ON**.

SNAP & GRID ON

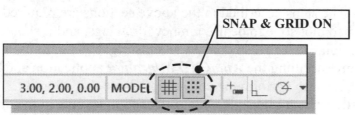

Layers Setup

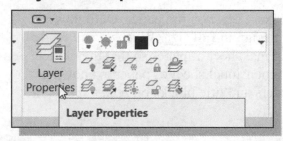

1. Pick *Layer Properties Manager* in the *Object Properties* toolbar.

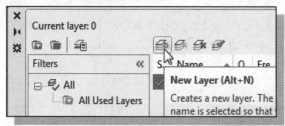

2. In the *Layer Properties Manager* dialog box, click on the **New** button (or the key combination [**Alt+N**]) to create a new layer.

3. Create **layers** with the following settings:

Layer	Color	Linetype	Lineweight	PlotStyle
CenterLines	Red	Center	Default	Normal
Construction	Gray	Continuous	Default	Normal
CuttingPlaneLines	Dark Gray	Phantom	0.6mm	Normal
Dimensions	Magenta	Continuous	Default	Normal
HiddenLines	Cyan	Hidden	0.3mm	Normal
ObjectLines	Blue	Continuous	0.6mm	Normal
SectionLines	White	Continuous	Default	Normal
Title_Block	Green	Continuous	1.2mm	Normal
TitleBlockLettering	Blue	Continuous	Default	Normal
Viewport	White	Continuous	Default	Normal

➤ Using the *Normal PlotStyle* enables plotting of *lineweights* defined in the specific layer. Note that the **Lineweight** settings are set for proper printing of different linetypes and some may appear thicker on screen.

4. Highlight the layer *Construction* in the list of layers.

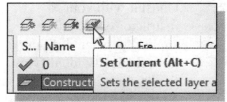

5. Click on the **Current** button to set layer *Construction* as the *Current Layer*.

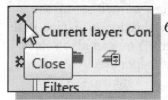

6. Click on the **Close** button to accept the settings and exit the *Layer Properties Manager* dialog box.

Adding Borders and Title Block in the Layout

AutoCAD 2023 allows us to create plots to any exact scale on the paper. Until now, we have been working in *model space* to create our design in **full size**. When we are ready to plot, we can arrange our design on a two-dimensional sheet of paper so that the plotted hardcopy is exactly what we wanted. This two-dimensional sheet of paper is known as the *paper space* in AutoCAD. We can place borders and title blocks on *paper space*, the objects that are less critical to our design.

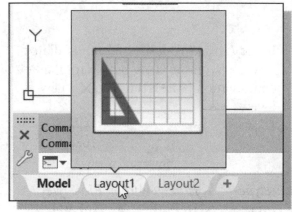

1. Click the **Layout1** icon to switch to the two-dimensional paper space.

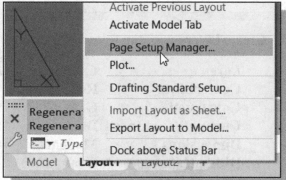

2. To adjust any *Page Setup* options, right-click once on the tab and select **Page Setup Manager**.

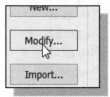

3. Choose the default layout in the *Page Setup Manager* and click **Modify**.

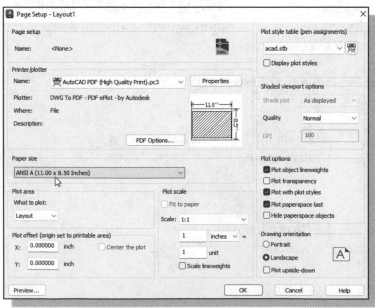

4. In the *Page Setup* dialog box, select a plotter or printer that is available to plot/print your design. Consult with your instructor or technical support personnel if you have difficulty identifying the hardware.

❖ Here we will set up the plotting for an A-size PDF plot. You can also select A4 paper size if you need to do a metric size plot.

➢ Also notice the *Printable area* listed in the *Paper size and paper units* section is typically smaller than the actual paper size, which is due to the limitations of the hardware.

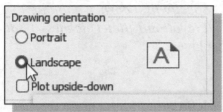

5. Set the *Paper size* to ANSI A (8.5″ by 11″) or equivalent and the *Drawing orientation* is set to **Landscape**.

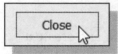

6. Click on the **OK** button and **Close** button to accept the settings and exit the *Page Setup* dialog boxes.

❖ In the Drawing Area, a rectangular outline on a gray background indicates the paper size. The dashed lines displayed within the paper indicate the *printable area*.

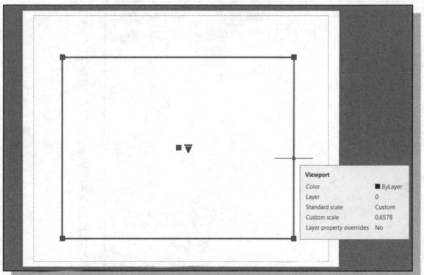

7. Select the rectangular viewport inside the dashed lines.

8. Hit the [**Delete**] key once to remove it.

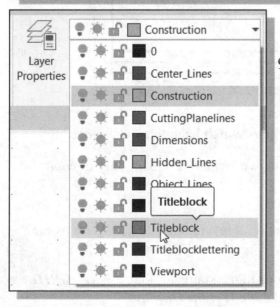

9. In the *Object Properties* toolbar area, select the **Layer Control** box and set layer **Title_Block** as the *Current Layer*.

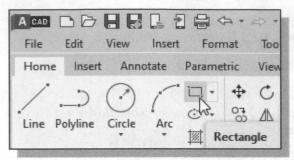

10. Select the **Rectangle** icon in the *Draw* toolbar. In the command prompt area, the message "*Specify first corner point or [Chamfer/Elevation/Fillet/Thickness/Width]:*" is displayed.

11. Pick a location that is on the inside and near the lower left corner of the dashed rectangle.

12. In the command prompt area, use the *relative coordinate entry method* and create a 10.25″ × 7.75″ rectangle.

13. Complete the title block as shown.

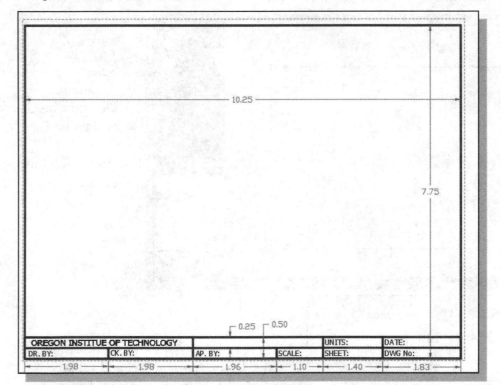

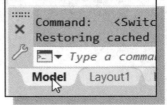

14. Click the **Model** tab to switch back to *model space*.

➢ Notice the title block we created is shown only in *paper space*.

15. On your own, select the ***Layer Control*** box and set layer *Construction Lines* as the *Current Layer*.

16. On your own, switch ***ON*** the following options in the status toolbar: ***SNAP***, ***GRID***, ***Object Snap***, and ***Object Snap Tracking*** and ***Line Weight Display***.

Create a Template File

The heart of any CAD system is the ability to reuse information that is already in the system. In the preceding sections, we spent a lot of time setting up system variables, such as layers, colors, linetypes and plotting settings. We will make a **template file** containing all of the settings and the title block we have created so far.

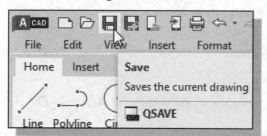

1. In the *Quick Access* toolbar, click the **Save** icon.

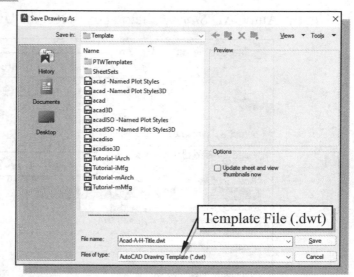

2. In the *Save Drawing As* dialog box, save to the default AutoCAD template folder (or select the folder in which you want to store the *template file).* Enter **Acad-A-H-Title** in the *File name* box.

3. Pick **Save** in the *Save Drawing As* dialog box to close the dialog box.

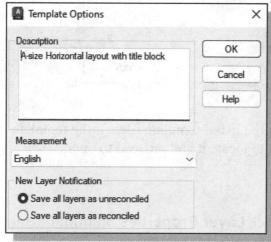

4. In the *Template Description* dialog box, enter **A-size Horizontal layout with title block** in the *Description* box.

5. Pick **OK** to close the dialog box and save the template file.

➢ The only difference between an AutoCAD template file and a regular AutoCAD drawing file is the filename extension, (.dwt) versus (.dwg). We can convert any AutoCAD drawing into an AutoCAD template file by simply changing the filename extension to *.dwt.* It is recommended that you keep a second copy of any template files on a separate disk as a backup.

Exit AutoCAD 2023

To demonstrate the effects of using a template file, we will exit and restart **AutoCAD 2023**.

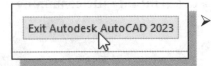

> Click on the *Application Bar* and select **Exit AutoCAD** from the pull-down menu as shown.

Starting Up AutoCAD 2023

1. Select the **AutoCAD 2023** option on the *Program* menu or select the **AutoCAD 2023** icon on the *Desktop*.

2. In the AutoCAD *Startup* dialog box, select the **Use a Template** option.

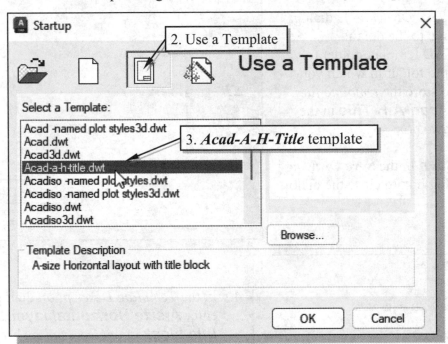

3. Select the *Acad-A-H-Title* template file from the list of template files. If the template file is not listed, click on the **Browse** button to locate it and proceed to open a new drawing file.

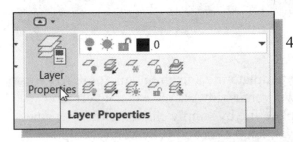

4. Pick **Layer Properties Manager** in the *Object Properties* toolbar.

5. Examine the layer property settings in the *Layer Properties Manager* dialog box.

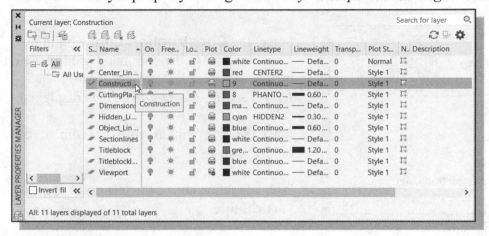

6. Confirm layer ***Construction*** is set as the *Current Layer*.

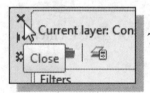

7. Click on the **Close** button to exit the *Layer Properties Manager* dialog box.

The Geneva Cam Drawing

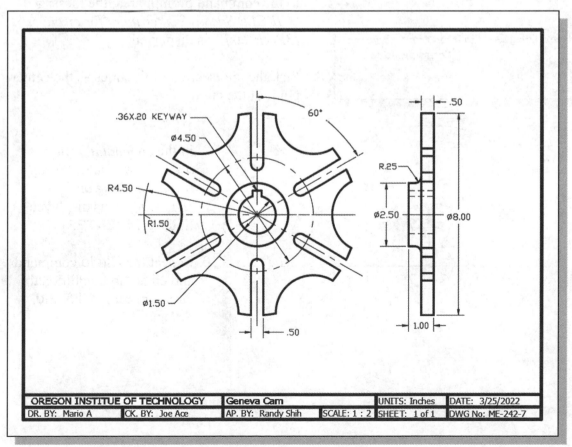

Drawing Construction Lines

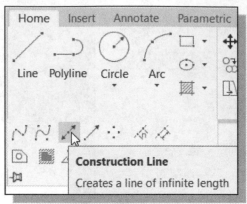

1. Select the **Construction Line** icon in the *Draw* toolbar. In the command prompt area, the message "*_xline Specify a point or [Hor/Ver/Ang/Bisect/Offset]:*" is displayed.

2. Place the first point at world coordinate (**5,4.5**) on the screen.

3. Pick a location above the last point to create a **vertical line**.

4. Move the cursor toward the right of the first point, and then pick a location to create a **horizontal line**.

5. Next, to create a construction line that is rotated 30 degrees from horizontal, enter **@2<30** [ENTER].

6. Inside the *Drawing Area*, **right-click** to end the Construction Line command.

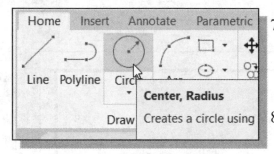

7. Select the **Circle** icon in the *Draw* toolbar. In the command prompt area, the message "*CIRCLE Specify center point for circle or [3P/2P/Ttr]:*" is displayed.

8. Pick the intersection of the lines as the center point of the circle.

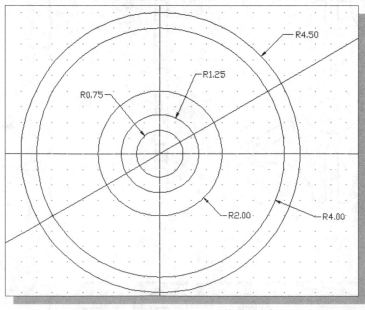

9. In the *command prompt area*, the message "*Specify radius of circle or [Diameter]:*" is displayed. Enter **0.75** [ENTER].

10. Repeat the Circle command and create four additional circles of radii **1.25**, **2.0**, **4.0** and **4.5** as shown.

Creating Object Lines

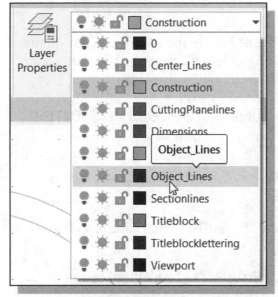

1. On the *Object Properties* toolbar, choose the ***Layer Control*** box with the left-mouse-button.

2. Move the cursor over the name of layer ***Object*Lines**; the tool tip "*ObjectLines*" appears.

3. **Left-click once** and layer *ObjectLines* is set as the *Current Layer*.

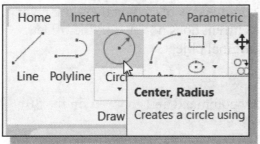

4. Select the **Circle** icon in the *Draw* toolbar. In the command prompt area, the message "*CIRCLE Specify center point for circle or [3P/2P/Ttr]:*" is displayed.

5. Move the cursor to the center of the circles, then left-click once to select the intersection as the center of the new circle.

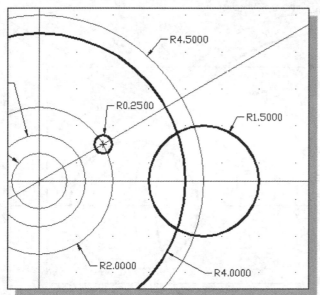

6. In the *command prompt area*, the message "*Specify radius of circle or [Diameter]:*" is displayed. Pick the **right intersection** of the *horizontal line* and the *radius 4.0 circle*.

7. Repeat the Circle command and pick the **right intersection** of the *horizontal line* and the *radius 4.5 circle* as the center point of the circle.

8. In the *command prompt area*, the message "*Specify radius of circle or [Diameter]:*" is displayed. Enter **1.5 [ENTER]**.

9. Repeat the **Circle** command and create a circle of radius **0.25** centered at the intersection of the inclined line and the radius **2.0** circle as shown.

Using the Offset Command

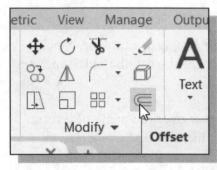

1. Select the **Offset** icon in the *Modify* toolbar. In the command prompt area, the message "*Specify offset distance or [Through]:*" is displayed.

2. In the command prompt area, enter **0.25** [ENTER].

3. In the *command prompt area*, the message "*Select object to offset or <exit>:*" is displayed. Pick the **inclined line** on the screen.

4. AutoCAD next asks us to identify the direction of the offset. Pick a location that is **below** the inclined line.

5. Inside the *Drawing Area*, **right-click** and pick **Enter** to end the **Offset** command.

❖ Notice that the new line created by the **Offset** command is placed on the same layer as the line we selected to offset. Which layer is current does not matter; the offset object will always be on the same layer as the original object.

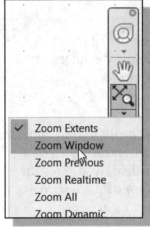

6. Use the **Zoom Window** command and zoom-in on the 30 degrees region as shown.

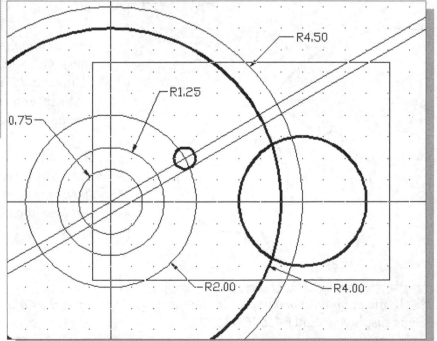

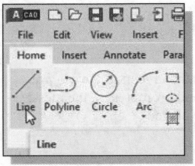

7. Select the **Line** command icon in the *Draw* toolbar. In the command prompt area, the message *"Line Specify first point:"* is displayed.

8. Move the cursor to the **intersection** of the *small circle* and the lower *inclined line* and notice the visual aid that automatically displays at the intersection. **Left-click** once to select the point.

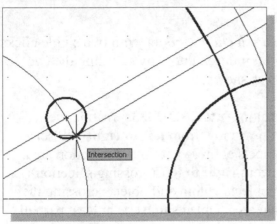 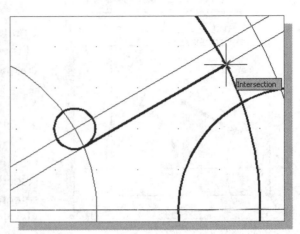

9. Pick the next **intersection point**, toward the right side, along the inclined line.

➢ On your own, use the **Trim** and **Erase** commands to remove the unwanted portions of the objects until your drawing contains only the objects shown below.

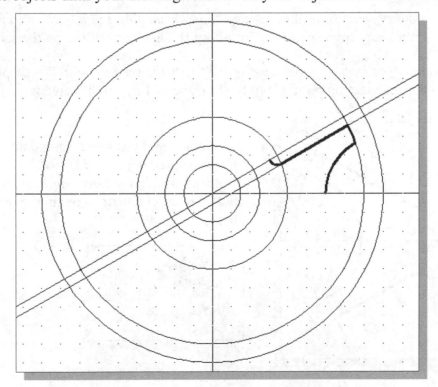

Using the Mirror Command

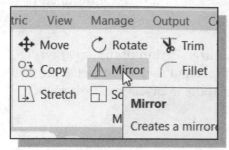

1. Select the **Mirror** command icon in the *Modify* toolbar. In the command prompt area, the message "*Select objects:*" is displayed.

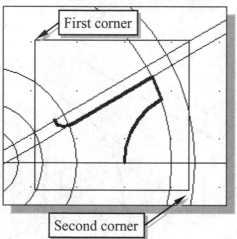

First corner

Second corner

2. Create a *selection window,* with two single clicks of the left mouse button, by selecting the two corners as shown.

➢ Note that in **AutoCAD 2023**, creating the selection window from **left to right** will select only objects entirely within the selection area. Going from **right to left** (crossing selection) selects objects within and objects crossing the selection area. Objects must be at least partially visible to be selected.

3. Inside the *Drawing Area*, **right-click** to accept the selection and continue with the Mirror command.

4. In the *command prompt area*, the message "*Specify the first point of the mirror line:*" is displayed. Pick any **intersection point along the horizontal line** on the screen.

5. In the *command prompt area*, the message "*Specify the second point:*" is displayed. Pick any other intersection point along the **horizontal line** on the screen.

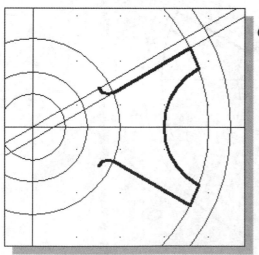

6. In the *command prompt area*, the message "*Delete source objects? [Yes/No] <N>:*" is displayed. Inside the *Drawing Area*, **right-click** and select **Enter** to retain the original objects.

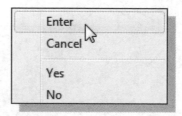

Using the Array Command

We can make multiple copies of objects in polar, rectangular or path arrays (patterns). For polar arrays, we control the number of copies of the object and whether the copies are rotated. For rectangular arrays, we control the number of rows and columns and the distance between them.

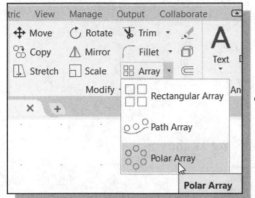

1. Select the **Polar Array** command icon in the *Modify* toolbar. In the command prompt area, the message "*Select objects:*" is displayed.

* Note the three different Array commands available in AutoCAD.

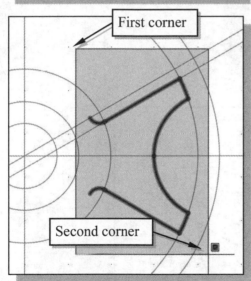

First corner

Second corner

2. Using two single clicks of the left mouse button to create a *selection* **window**, enclose the objects we mirrored and the mirrored copies as shown.

3. In the command prompt area, the message "*Select Objects:*" is displayed. Inside the Drawing Area, **right-click** to end the selection.

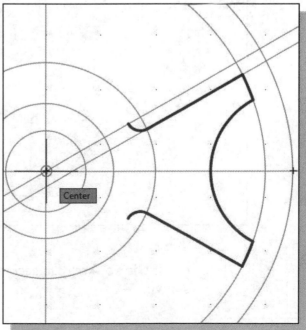

Center

4. In the command prompt area, the message "*Specify the center point of array:*" is displayed. Pick the **Center point location** of the design as shown.

5. In the *Item* toolbar area, set the total number of *Items* as **6** as shown.

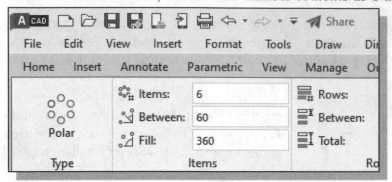

6. Set the angle to *Fill* to **360** as shown.

7. On your own, confirm the array *Properties* options are as shown.

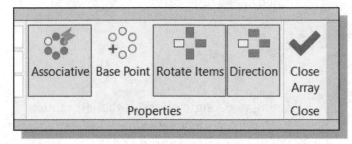

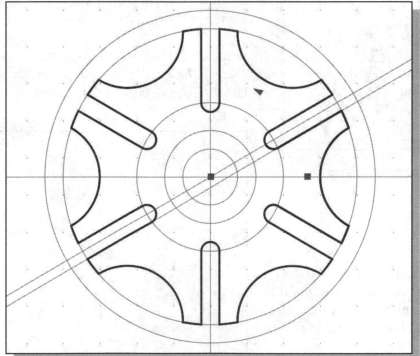

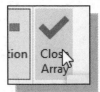

8. Click **Close Array** to accept the settings and create the array as shown.

9. On your own, construct the **two inner circles** and the *0.36 × 0.20* keyway. (Hint: First create parallel lines at 0.95 and 0.18 distance of the horizontal and vertical construction lines.)

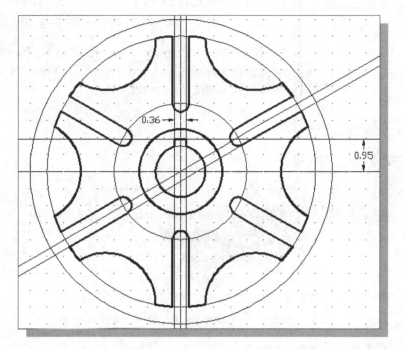

➢ On your own, complete the two views with dimensions. (In the *Dimension Style Manager*, set options under the **Fit** tab to control the appearance of radius and diameter dimensions.)

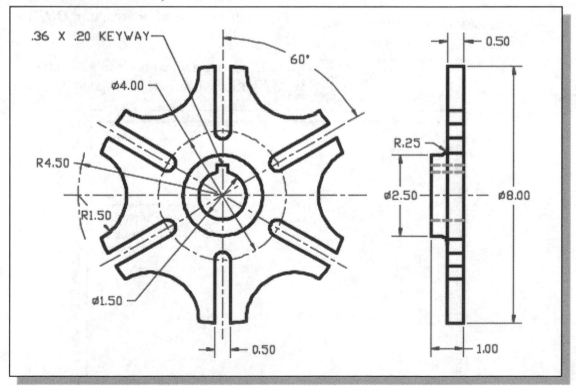

Creating a Viewport Inside the Title Block

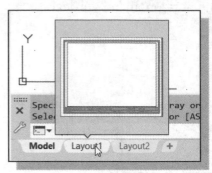

1. Click the **Layout1** tab to switch to the two-dimensional paper space containing the title block.

2. If a view is displayed inside the title block, use the **Erase** command and delete the view by selecting any edge of the **viewport**.

3. Set the *Viewport* layer as the *Current Layer*.

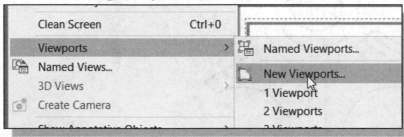

4. In the *Pull-down menu*, select **[View]** → **[Viewports]** → **[New Viewport]**

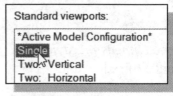

5. Select **Single** in the *Standard viewports* list.

6. Click **OK** to proceed with the creation of the new viewport.

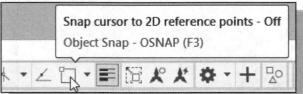

7. In the *Status Bar* area, turn **OFF** the *OSNAP* option.

8. Create a viewport inside the title block area as shown.

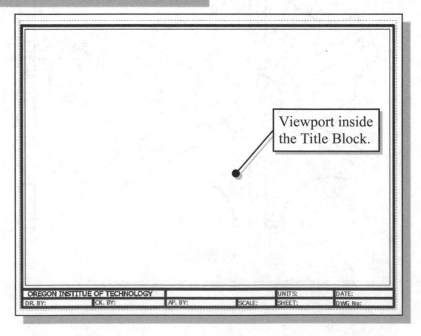

Viewport inside the Title Block.

Viewport Properties

1. Pre-select the **viewport** by left-clicking once on any edge of the viewport.

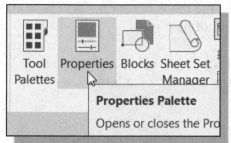

2. Switch to the **View** tab in the *Ribbon toolbars*, and select the **Properties Palette** icon.

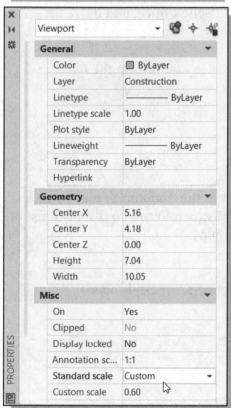

3. In the *Properties* dialog box, scroll down to the bottom of the list. Notice the current scale is set to *Custom, 0.60*. (The number on your screen might be different.)

4. **Left-click** the *Standard scale* box and notice an arrowhead appears.

5. Click on the **arrowhead button** and a list of standard scales is displayed. Use the scroll bar to look at the list.

6. Set the ***Custom scale*** to **0.5** as shown. This will set the plotting scale factor to half scale.

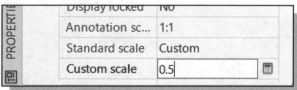

7. Click on the **[X]** button to exit the *Properties* dialog box.

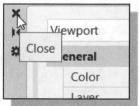

Hide the Viewport Borders

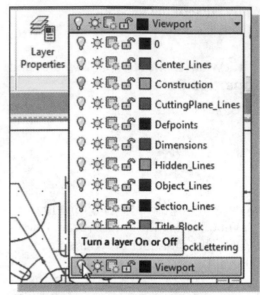

❖ We will **turn off** the *viewport borders* so that the lines will not be plotted.

1. With the viewport pre-selected, choose the *Layer Control* box with the left-mouse-button.

2. Move the cursor over the name of layer *Viewport*, **left-click once**, and move the viewport to layer *Viewport*.

3. Turn *OFF* layer *Viewport* in the *Layer Control* box.

Adjusting the Dimension Scale

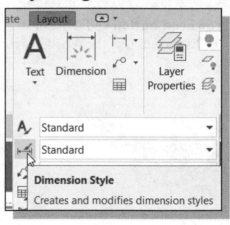

1. Move the cursor to the *Annotation* toolbar area and **left-click** the **down arrow** to display a list of toolbar menu groups.

2. Click on the **Dimension Style** icon.

3. In the *Dimension Style Manager* dialog box, select **Modify**.

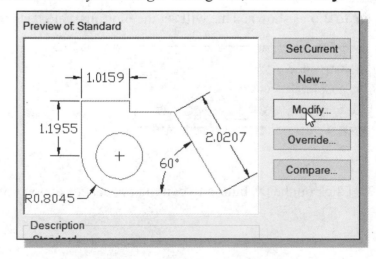

4. Under the **Fit** tab, choose the *Text* option in the *Fit options* section and set the overall scale for the dimension features to **2.0** as shown. (Why 2.0? What is the scale of the viewport in Layout1?)

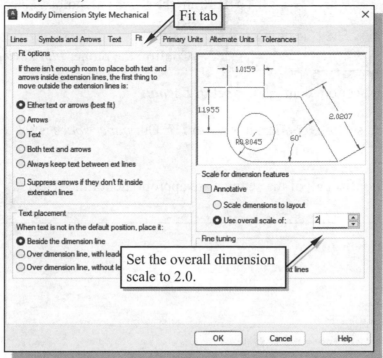

5. Click on the **Primary Units** tab, and set the Precision to two digits after the decimal point.

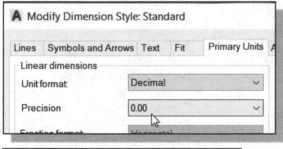

6. Set the *Zero suppression* of linear dimension to **Leading** as shown.

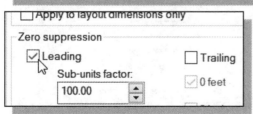

7. Click on the **OK** button to close the *Modify Dimension Style* dialog box.

8. Click on the **Close** button to close the *Dimension Style Manager* dialog box.

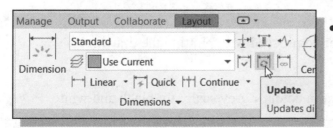

• Note the **Dimension Update** option, located under the **Annotate** tab, is available to apply the current dimension style to specific dimensions.

Plot/Print the Drawing – Color or Grey Scale Prints

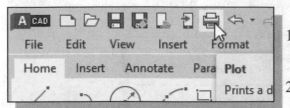

1. In the *Standard* toolbar, select the **Plot** icon.

2. Confirm the proper ***plot device*** is selected.

3. In the *Plot area* section, confirm it is set to ***Layout***.

4. In the *Plot scale* section, confirm it is set to ***1:1***. Our *paper space* is set to the correct paper size.

5. On your own, confirm all of the settings are appropriate for the selected printer/plotter.

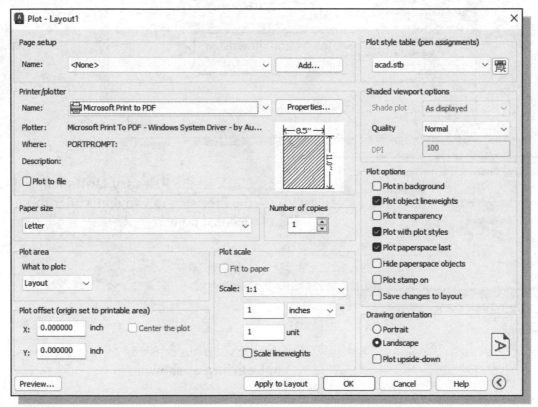

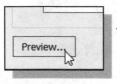

❖ Note the **Preview** option is also available.

6. Click on the **OK** button to proceed with plotting the drawing.

*   Note the current layout is set to print in color or grey scale; for black and white prints refer to the settings on page 10-24.

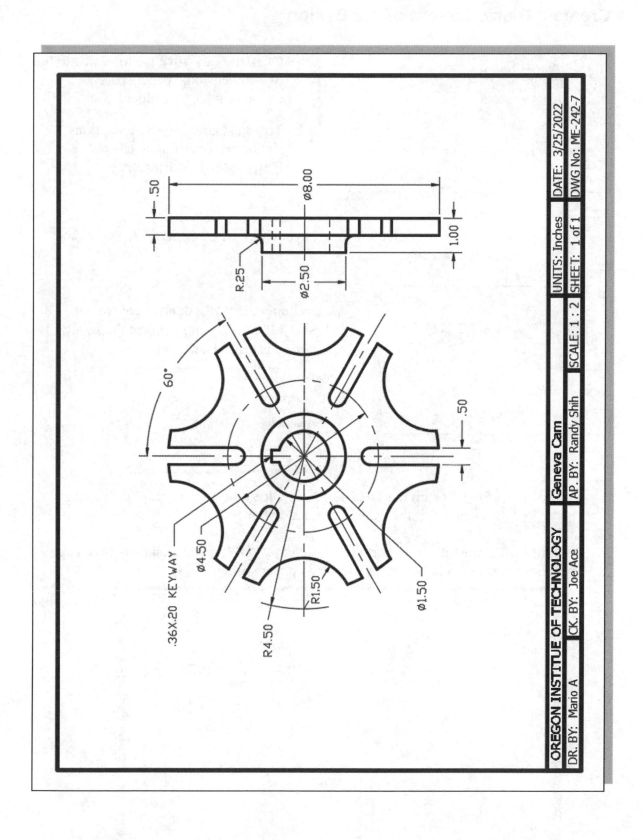

.50
Ø8.00
1.00
R.25
Ø2.50

60°
.50
.36X.20 KEYWAY
Ø4.50
R1.50
R4.50
Ø1.50

OREGON INSTITUTE OF TECHNOLOGY
Geneva Cam
UNITS: Inches
DATE: 3/25/2022
DWG No: ME-242-7
AP. BY: Randy Shih
SCALE: 1 : 2
SHEET: 1 of 1
DR. BY: Mario A
CK. BY: Joe Ace

Create a B size Layout of the Design

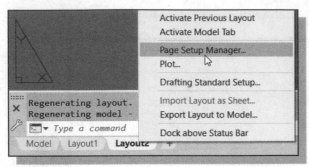

1. Click the **Layout2** tab to switch to the two-dimensional paper space containing the title block.

2. To adjust any *Page Setup* options, right-click once on the tab and select **Page Setup Manager**.

3. Choose the default layout, Layout2, in the *Page Setup Manager* and click **Modify**.

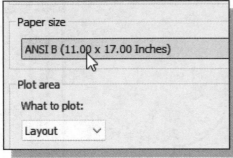

4. Set the *Paper size* to **Ledger** or equivalent (ANSI B 11″ by 17″) and confirm the *Drawing orientation* is set to **Landscape**.

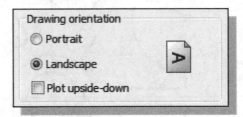

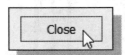

5. Click on the **OK** button and **Close** button to accept the settings and exit the *Page Setup* dialog boxes.

6. On your own, adjust the viewport to fit inside the *11X17* page as shown. (See page 8-25 on how to set up and adjust a viewport in paper space.)

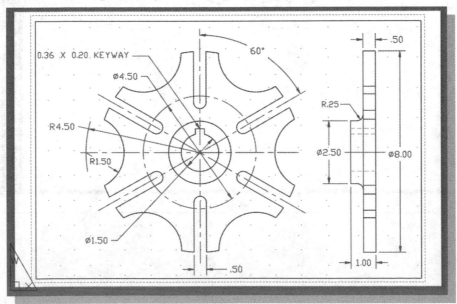

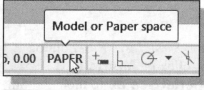

7. Switch to the **Model Space** by clicking on the status tool bar as shown.

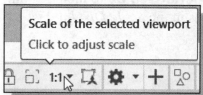

8. In the bottom of the main window, notice the scale of the current viewport is displayed as shown.

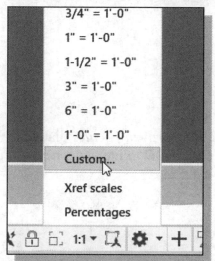

9. Click on the scale of the viewport to display a list of commonly used scales. Note the default *English scale format* is set to architectural style, for example 6" = 1' is really 1/2 scale.

10. Click on **Custom** to bring up the *Edit Drawing Scale* dialog box.

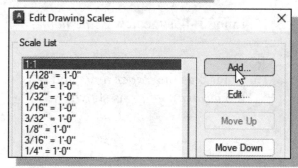

11. In the *Edit Drawing Scale* dialog box, note that we can add more scales to the list and/or edit any of the scales in the list.

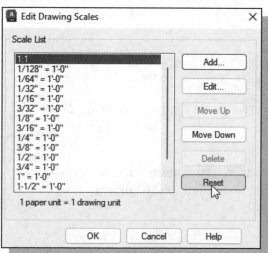

12. Instead of using the architectural scales, we will reset the list to include the basic engineering format. Click **Reset** to enter the reset scale list mode.

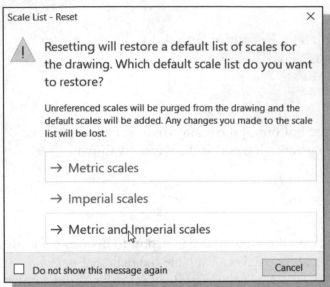

13. In the *Scale List –Reset* dialog box, click on **Metric and Imperial scales** to include both types in the *Scale Display List*.

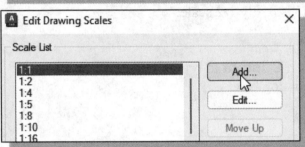

14. In the *Edit Drawing Scales* dialog box, click **ADD** to include a new drawing scale.

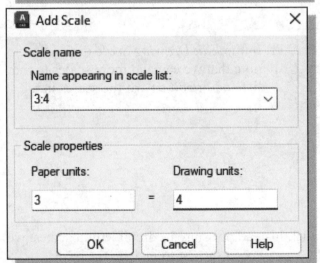

15. Enter **3:4** as the new *Scale name*.

16. Enter **3** as the *Paper units* and **4** as the *Drawing units* as shown.

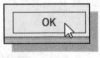

17. Click **OK** to proceed with the creation of the new *Drawing Scale*.

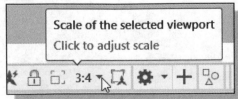

18. In the bottom of the main window, notice the new scale is applied to the current viewport as shown.

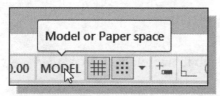

19. Switch back to the *Paper Space* by clicking on the **Model Icon** in the status toolbar.

20. On your own, add some text near the bottom of the current layout; set the text height to 0.125.

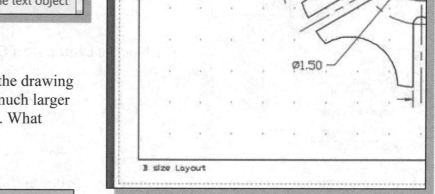

- The text height of the drawing dimensions seem much larger than 1/8 of an inch. What happened?

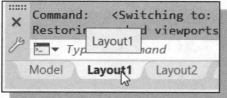

21. Click the **Layout1** tab to switch to the A size layout we created.

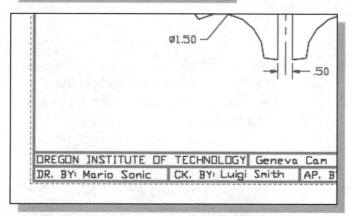

- Note the text height of the drawing dimensions matches the 1/8 of an inch text height in the title block.

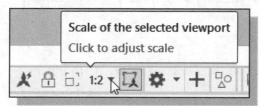

22. On your own, switch to the *Model Space* by clicking on the **Model Icon** in the status toolbar, and notice the scale of the selected viewport in *Layout1* is set to **1/2** scale.

- The text height of the drawing dimensions was set based on the viewport scale in Layout1; refer to page 8-25 and page 8-27 on the set up of the two scales.

Adjust the Dimension Scale for the B size Print

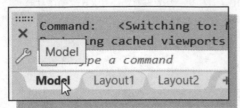

1. Click the **Model** tab to switch back to *Model Space*.

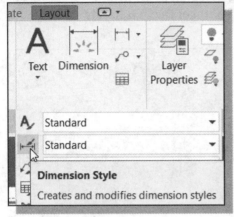

2. Move the cursor to the *Annotation* toolbar area and **left-click** the **down arrow** to display a list of toolbar menu groups.

3. Click on the **Dimension Style** icon.

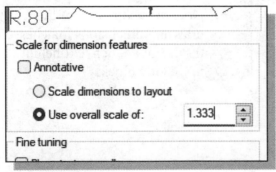

4. In the *Dimension Style Manager* dialog box, select **Modify**.

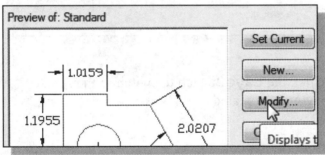

5. Under the **Fit** tab, choose the *Text* option in the *Fit options* section and set the overall scale for the dimension features to **1.333** as shown.

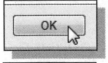

6. Click on the **OK** button to close the *Modify Dimension Style* dialog box.

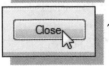

7. Click on the **Close** button to close the *Dimension Style Manager* dialog box.

- Note the text height of the drawing dimensions has been adjusted to the scale of **4/3**; this number came from the inverse of the viewport scale of **3/4**.

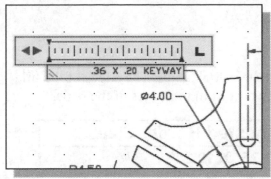

8. Double click on the local note for the keyway to enter the **Text Editor** mode as shown.

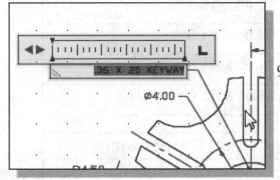

9. Select all of the text in the *Text Edit* box as shown.

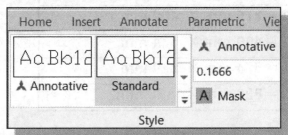

10. On your own, adjust the text height to **0.166** as shown. (Hit the **Enter** key once to make the change.)

   • Note the text height is adjusted based on the viewport scale **0.125 X 1.333 = 0.1666**.

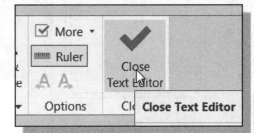

11. Click **Close Text Editor** to accept the adjustment and exit the Text Editor.

12. On your own, print a copy of the B size layout and confirm the text height is 1/8 of an inch.

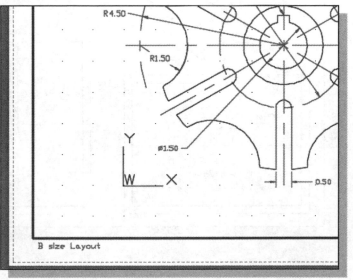

Additional Title Blocks

Drawing Paper and Border Sizes

The standard drawing paper sizes are as shown in the below tables. The edges of the title block border are generally 0.5 ~ 1 inches or 10~20 mm from the edges of the paper.

American National Standard	Suggested Border Size
A – 8.5″ X 11.0″	A – 7.75″ X 10.25″
B – 11.0″ X 17.0″	B – 10.0″ X 16.0″
C – 17.0″ X 22.0″	C – 16.0″ X 21.0″
D – 22.0″ X 34.0″	D – 21.0″ X 33.0″
E – 34.0″ X 44.0″	E – 33.0″ X 43.0″

International Standard	Suggested Border Size
A4 – 210 mm X 297 mm	A4 – 190 mm X 276 mm
A3 – 297 mm X 420 mm	A3 – 275 mm X 400 mm
A2 – 420 mm X 594 mm	A2 – 400 mm X 574 mm
A1 – 594 mm X 841 mm	A1 – 574 mm X 820 mm
A0 – 841 mm X 1189 mm	A0 – 820 mm X 1168 mm

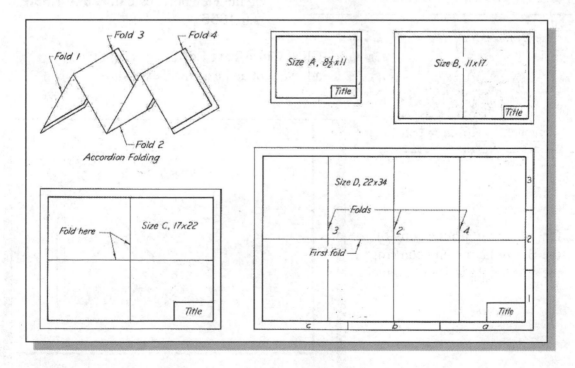

- **English Title Block** (For A size paper, dimensions are in inches)

OREGON INSTITUE OF TECHNOLOGY				UNITS:	DATE:
DR. BY:	CK. BY:	AP. BY:	SCALE:	SHEET:	DWG No:

.25 .50

|← 1.98 →|← 1.98 →|← 1.96 →|← 1.10 →|← 1.40 →|← 1.83 →|

- **Metric Title Block** (For A4 size paper, dimensions are in mm.)

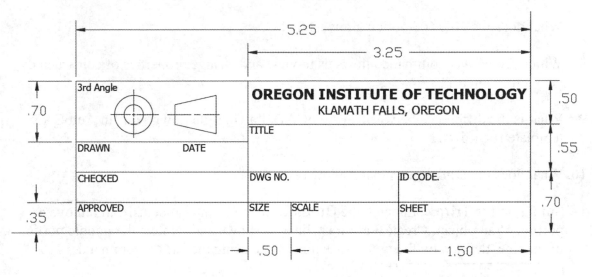

OREGON INSTITUE OF TECHNOLOGY				UNITS:	DATE:
DR. BY:	CK. BY:	AP. BY:	SCALE:	SHEET:	DWG No:

6 12

|← 54 →|← 54 →|← 52 →|← 30 →|← 35 →|← 49 →|

- **English Title Block** (For B size or larger paper, dimensions are in inches.)

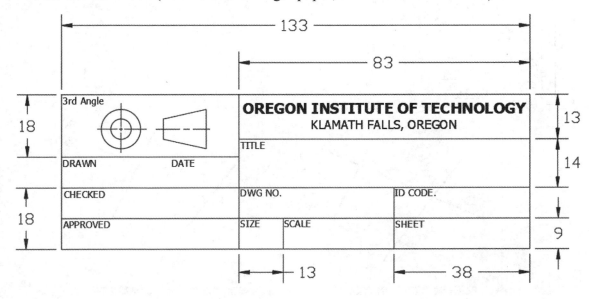

- **Metric Title Block** (For A3 size or larger paper, dimensions are in mm.)

Review Questions: (Time: 25 minutes)

1. List and describe three advantages of using *template files*.

2. Describe the items that were included in the *Acad-A-H-Title* template file.

3. List and describe two methods of creating copies of existing objects in **AutoCAD 2023**.

4. Describe the procedure in determining the scale factor for plotting an **AutoCAD 2023** layout.

5. What is the difference between the AutoCAD *Model Space* and the *Paper Space*?

6. Why should we use the AutoCAD *Paper Space*?

7. What does the *ORTHO* option allow us to do?

8. Which AutoCAD command allows us to view and change properties of constructed geometric objects?

9. What is the difference between an AutoCAD drawing file (.dwg) and an AutoCAD template file (.dwt)?

10. What does the **Divide** command allow us to do?

11. Similar to the **Trim** command, the **Break** command can also be used to remove a portion of an object. Create 3 arbitrary lines as shown and remove the middle portion of the line, defined by the two intersecting lines, using the **Break** command.

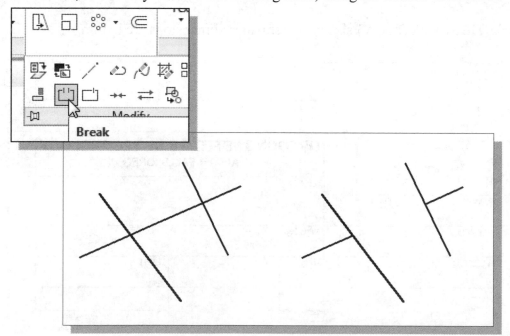

Exercises:

(Unless otherwise specified, dimensions are in inches.) (Time: 150 minutes)

1. Ratchet Plate (thickness: 0.125 inch)

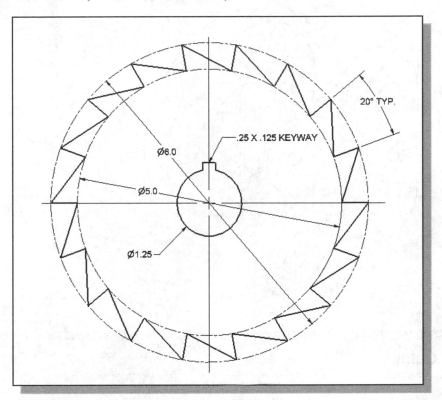

2. Switch Base

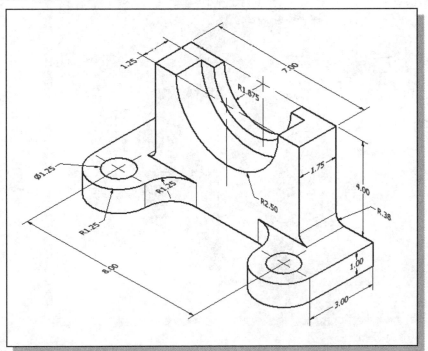

3. Auxiliary Support

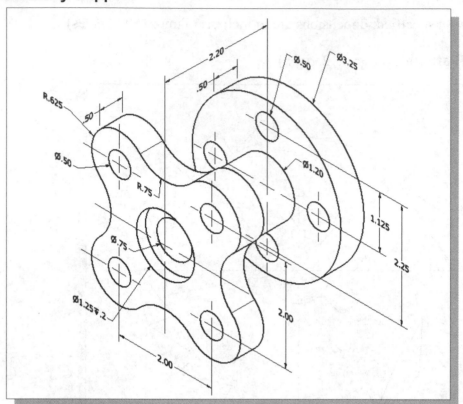

4. Indexing Guide

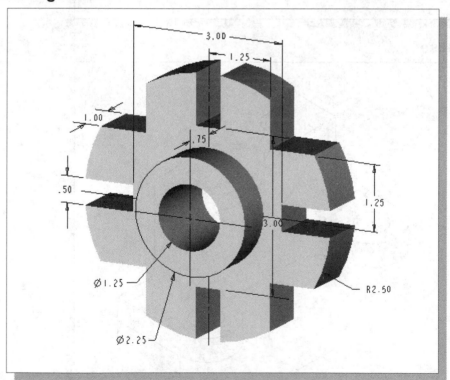

5. Coupling Base

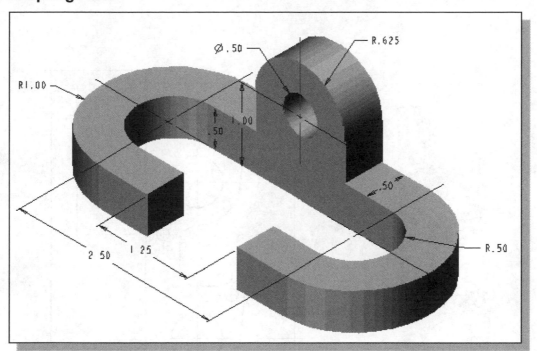

6. Anchor Base

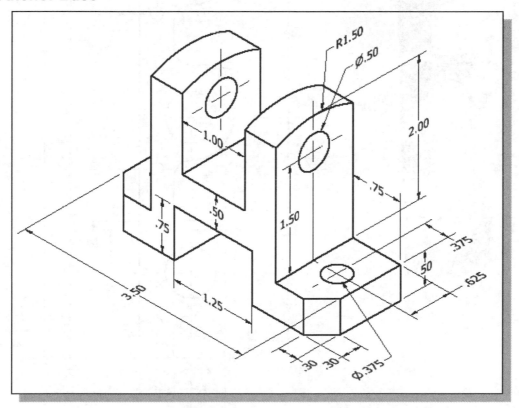

7. Intake Flange

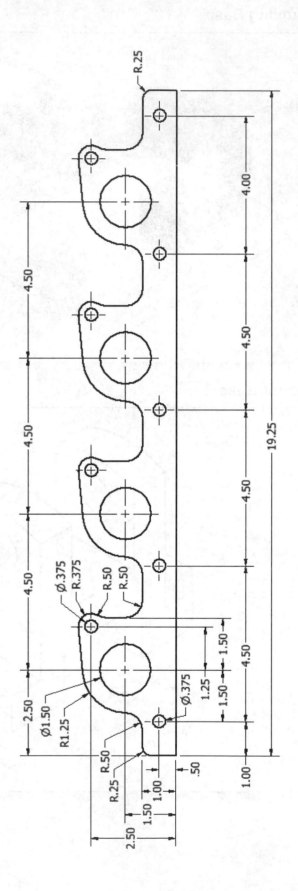

Chapter 9
Parametric Drawing Tools

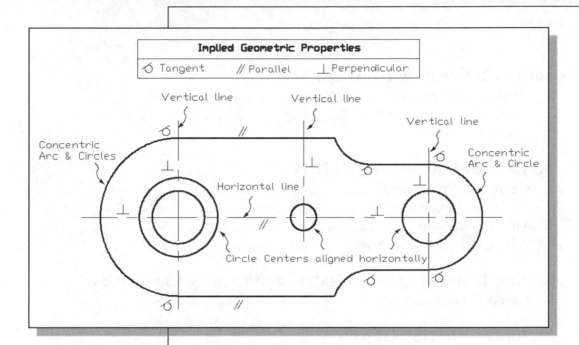

Implied Geometric Properties		
⌀ Tangent	// Parallel	⊥ Perpendicular

Vertical line

Vertical line

Vertical line

Concentric Arc & Circles

Concentric Arc & Circle

Horizontal line

Circle Centers aligned horizontally

Learning Objectives

- ◆ **Understand the Basics of Parametric Modeling**
- ◆ **Apply the Geometric Constraints Manually and Automatically**
- ◆ **Use the Dimensional Constraints command**
- ◆ **Control Geometry through the use of Constraints**
- ◆ **Show and Hide the Applied Constraints**

AutoCAD Certified User Examination Objectives Coverage

This table shows the pages on which the objectives of the Certified User Examination are covered in Chapter 9.

Certified User Reference Guide

Introduction

One of the most important advancements in CAD/CAE technology was the invention of *parametric modeling* tools in the late 1980s. The introduction of parametric technology revolutionized the CAD industry by allowing users to use CAD software as true design tools. The parametric modeling approach has elevated the traditional CAD technology to the level of a very powerful design tool. Parametric modeling techniques can be used to automate the design and revision procedures; this is done through the use of parametric features. Parametric features control the model geometry by the use of design variables. In 2009, a new set of *parametric drawing tools* was first introduced in **AutoCAD**. The word ***parametric*** means that the geometric definitions of the design, such as dimensions, can be varied at any time during the design process. The concept of parametric modeling makes the way CAD works more closely match the actual design-manufacturing process than the mathematics of a CAD program. By using the parametric drawing tools available in **AutoCAD 2023**, CAD drawings can now be updated more easily when the design is refined.

The main characteristic of *parametric modeling* involves the use of **Constraints**. *Constraints* are geometric rules and restrictions applied to 2D geometry.

There are two general types of constraints: **Geometric constraints** and **Dimensional constraints**.

Geometric constraints are used to control the geometric relationships of objects with respect to each other; for example, a line that is tangent to an arc, a line that is horizontal, or two lines that are collinear.

Dimensional constraints are used to control the *size* and *location* of geometric entities; for example, the distance between two parallel lines, the length of a line, the angle of two lines, or radius values of arcs.

When a design is created or changed, a drawing will be in one of three states:
  ➢ **Unconstrained:** No constraints are applied to the constructed geometry.
  ➢ **Under constrained:** Some constraints, but not all, are applied to the constructed geometry.
  ➢ **Fully constrained:** The necessary definitions of the design have been properly applied to the constructed geometry, which means all relevant geometric and dimensional constraints are present.

  ❖ Note that AutoCAD will prevent the user from applying any constraints that result in an *over-constrained* condition (having duplicating or conflicting definitions).

Generally speaking, during the initial stages of a design process, creating an under constrained, or even unconstrained, drawing can be very beneficial in helping a designer to determine the forms and shapes of the design. But as the design begins to reach the final stages, a fully constrained drawing is necessary eventually, as this will assure the manufacturability of the finalized design.

In this lesson, we will examine the use of the very powerful *parametric drawing tools* that are available in **AutoCAD 2023**. The parametric drawing tools can be used to (1) assist the construction of designs, especially when more complex geometric relations are present, and (2) help maintain the design intents and thus ease the tedious tasks involved in design modifications.

The concepts and procedures described in this chapter represent the basic techniques involved in the use of 3D parametric modeling software, such as the *Autodesk Inventor* software.

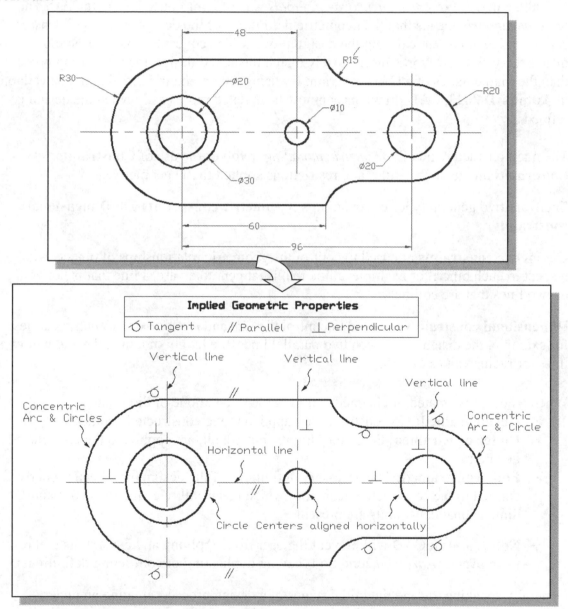

❖ Note the *parametric drawing tools* allow users to concentrate on the design itself. With the parametric tools, the users now have full control of the specific geometric properties, as well as the size and location definitions. This approach can be quite effective, as it can be used to supplement traditional geometric construction techniques.

Starting Up AutoCAD 2023

1. Select the AutoCAD 2023 option on the *Program* menu or select the AutoCAD 2023 icon on the *Desktop*. Once the program is loaded into the memory, the **AutoCAD 2023** drawing window will appear on the screen.

2. In the *Startup* window, select **Start from Scratch**, as shown in the figure below.

3. In the *Default Settings* section, pick **Metric** as the drawing units.

4. On your own, set the display to **no digits** after the decimal point and also both the **Snap and Grid** options to *10* for both X and Y directions.

Layers Setup

1. Pick **Layer Properties Manager** in the *Layers* toolbar.

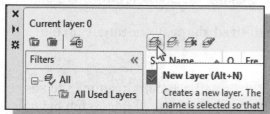

2. Click on the **New** icon to create new layers.

3. Create two **new** layers with the following settings:

Layer	Color	Linetype	Lineweight
Construction	White	Continuous	Default
Center	Red	CENTER	Default
Object	Blue	Continuous	0.3mm

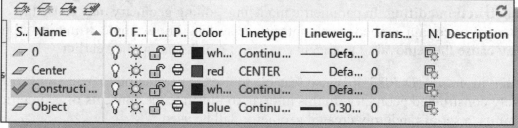

4. On your own, set layer *Construction* as the *Current Layer*.

5. In the *Status Bar* area, reset the option buttons so that only *GRID Display* is switched *ON*.

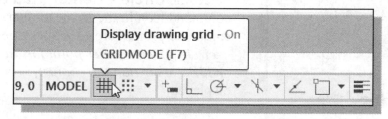

Creating Rough Sketches

Quite often during the early design stage, the shape of a design may not have any precise dimensions. Most conventional CAD systems require the user to input the precise lengths and locations of all geometric entities defining the design, which are not available during the early design stage. With *parametric modeling*, we can use the computer to elaborate and formulate the design idea further during the initial design stage. With Autodesk *parametric drawing tools*, we can use the computer as an electronic sketchpad to help us concentrate on the formulation of forms and shapes for the design. This approach is the main advantage of *parametric drawing* over conventional CAD drawing techniques.

As the name implies, a **rough sketch** is not precise at all. When sketching, we simply sketch the geometry so that it closely resembles the desired shape. Precise scale or lengths are not needed. AutoCAD provides us with many tools to assist us in finalizing sketches. For example, geometric entities such as horizontal and vertical lines can be set at any time. Here are some general guidelines for creating sketches in *AutoCAD 2023*:

- **Create a sketch that is proportional to the desired shape.** Concentrate on the shapes and forms of the design.

- **Keep the sketches simple.** Initially, leave out the small geometry features such as fillets, rounds and chamfers. Add those in after the major parts of the sketch have been established.

- **Exaggerate the geometric features of the desired shape.** For example, if the desired angle is 85 degrees, create an angle that is 50 or 60 degrees. It is not necessary to construct everything precisely.

- **Confirm all necessary constraints are maintained correctly throughout the construction/editing.** In parametric modeling, editing geometry may cause the removal of certain applied constraints by the system. For example, trimming a line may cause the removal of a tangent constraint which was applied earlier.

Note that in *AutoCAD 2023*, it is also feasible to use the parametric drawing tools on precisely constructed geometry. In this tutorial, to illustrate the concepts of geometric constraints, a set of randomly created geometry will be used.

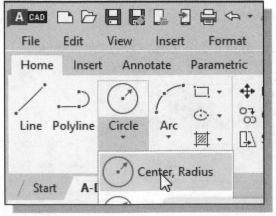

1. Click on and switch to the **Home** tab in the *Ribbon* tabs and panels area.

2. Select the **Circle – Center, Radius** command icon in the *Draw* toolbar. In the command prompt area, the message "*_circle Specify center point for the circle or [3P/2P/Ttr (tan tan radius)]:*" is displayed.

3. On your own, create two circles of arbitrary sizes near the center of the screen as shown.

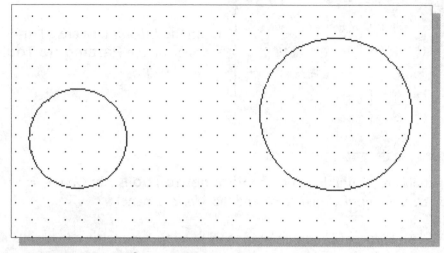

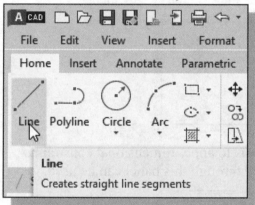

4. Select the **Line** command icon in the *Draw* toolbar. In the command prompt area, near the bottom of the AutoCAD drawing screen, the message "*Line Specify first point:*" is displayed. AutoCAD expects us to identify the starting location of a straight line.

5. On your own, create a line of arbitrary length just below the two circles as shown in the figure below.

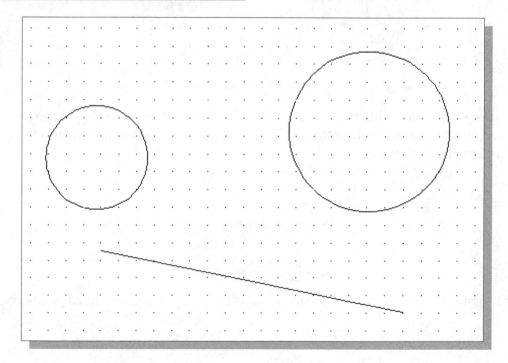

Parametric Drawing Tools

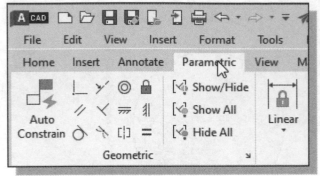

1. In the *Ribbon* tabs area, left-click once on the **Parametric Tools** tab as shown.

❖ Three tool panels are available under the **Parametric Tools** tab: *Geometric Constraints*, *Dimensional Constraints* and the *Manage* panel.

➤ *Geometric Constraints* panel contains the tools to apply geometric constraints either manually or automatically.

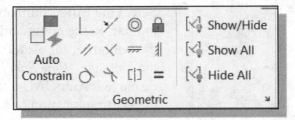

➤ *Dimensional Constraints* panel contains the tools to apply dimensional constraints manually. Note the dynamic dimensions applied through this panel can be used.

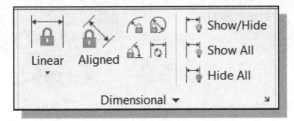

➤ The *Manage* panel contains two tools: **Delete Constraints** to manually remove unwanted constraints and **Parameters Manager** to allow parametric equations to be set up among dimensions.

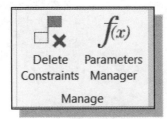

❖ We will first use the *Geometric Constraint* tools to control the constructed geometry.

Applying Geometric Constraints

In the *Geometric Constraints* panel, twelve types of constraints are available for 2D sketches.

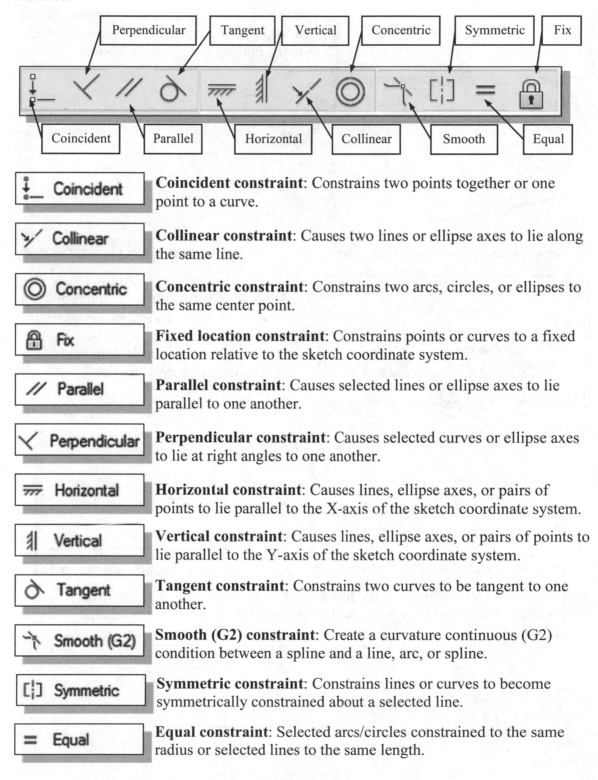

Coincident constraint: Constrains two points together or one point to a curve.

Collinear constraint: Causes two lines or ellipse axes to lie along the same line.

Concentric constraint: Constrains two arcs, circles, or ellipses to the same center point.

Fixed location constraint: Constrains points or curves to a fixed location relative to the sketch coordinate system.

Parallel constraint: Causes selected lines or ellipse axes to lie parallel to one another.

Perpendicular constraint: Causes selected curves or ellipse axes to lie at right angles to one another.

Horizontal constraint: Causes lines, ellipse axes, or pairs of points to lie parallel to the X-axis of the sketch coordinate system.

Vertical constraint: Causes lines, ellipse axes, or pairs of points to lie parallel to the Y-axis of the sketch coordinate system.

Tangent constraint: Constrains two curves to be tangent to one another.

Smooth (G2) constraint: Create a curvature continuous (G2) condition between a spline and a line, arc, or spline.

Symmetric constraint: Constrains lines or curves to become symmetrically constrained about a selected line.

Equal constraint: Selected arcs/circles constrained to the same radius or selected lines to the same length.

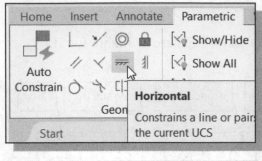

1. Select the **Horizontal** constraint by clicking on the icon as shown in the figure.

2. Click on the **line** with the left-mouse-button to apply the constraint. The line is adjusted to the horizontal position as shown.

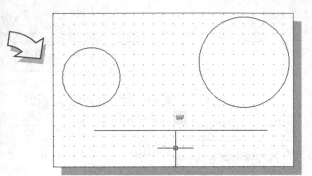

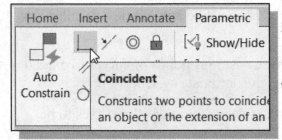

3. Select the **Coincident** constraint by clicking on the icon as shown in the figure.

4. Click on the **larger circle** with the left-mouse-button to apply the constraint.

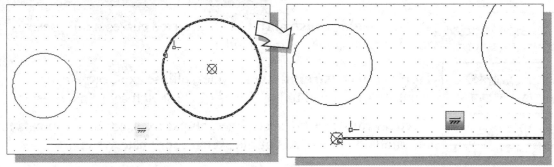

5. Click on the left endpoint of the **line** with the left-mouse-button to apply the constraint. Note that may move the line or the circle to maintain the applied constraint.

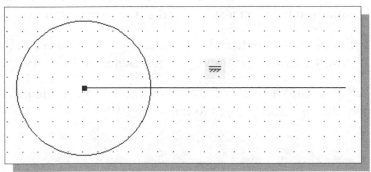

❖ Note that since both the circle and line can be moved, AutoCAD will move either geometry to satisfy the applied geometric property. We will take a different approach to have more controls in the editing of the geometry.

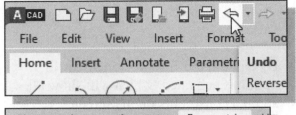

6. In the *Quick Access* toolbar, click the **Undo** button as shown.

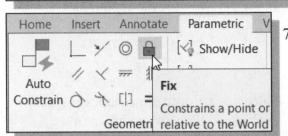

7. Select the **Fix** constraint by clicking on the icon as shown in the figure.

8. Click near the left endpoint of the **line** with the left-mouse-button to apply the constraint.

❖ Note the different constraint symbols next to the geometry.

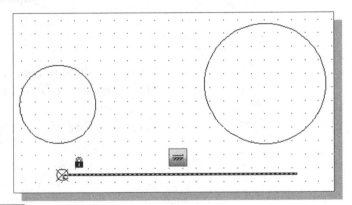

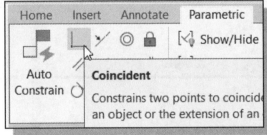

9. Select the **Coincident** constraint by clicking on the icon as shown in the figure.

10. Click on the **larger circle** with the left-mouse-button to apply the constraint.

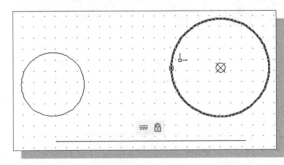

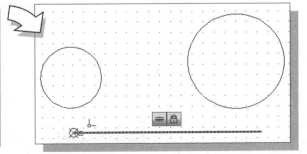

11. Click on the **left endpoint of the line** with the left-mouse-button to apply the constraint. Now the circle is moved and the alignment constraint is maintained.

Applying Dimensional Constraints

In the *Dimensional Constraints* panel, a set of dimensional constraints are available; we can apply the dimensional constraints, such as Linear, Radial, or Angular dimensions, to control geometry.

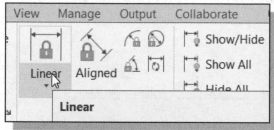

❖ By default, the **Show Constraints** option is activated, which will show all applied constraints on the screen.

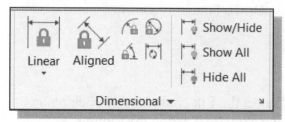

1. Select the **Linear** dimension constraint by clicking on the icon as shown in the figure.

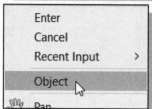

2. Inside the display area, right-click once to bring up the option menu and select **Object** as shown.

3. Select the **line** with a single click of the left-mouse-button.

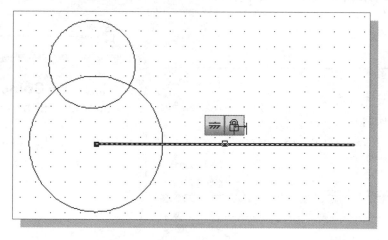

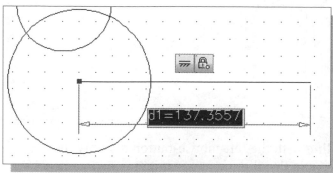

4. Place the dimension **below the line** with a single left-mouse-button click as shown.

❖ AutoCAD next expects us to edit the length of the line through the applied dimensional constraint.

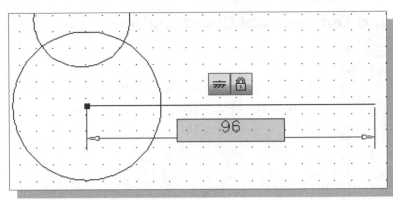

5. Enter **96** as the new length of the line.

❖ Note the length of the line is adjusted to the specified length.

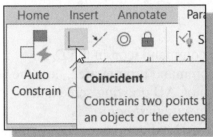

6. Select the **Coincident** constraint by clicking on the icon as shown in the figure.

7. On your own, align the **center of the smaller circle** to the **right endpoint of the line** as shown in the figure below.

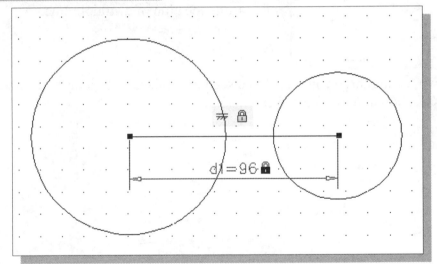

8. On your own, adjust the length of the line by **double-clicking on the dimension**. Experiment with different values and see how the geometry behaves as changes are made. Set the dimension back to **96** before continuing.

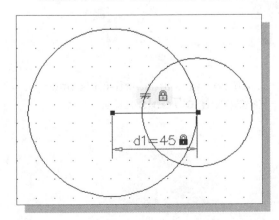

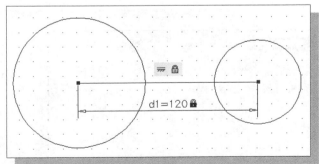

Additional Geometric and Dimensional Constructions

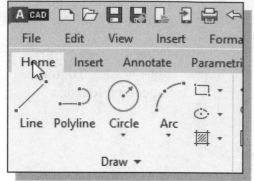

1. In the *Ribbon* tabs area, left-click once on the **Home** tab as shown.

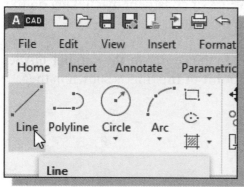

2. Select the **Line** command icon in the *Draw* toolbar. In the command prompt area, near the bottom of the AutoCAD drawing screen, the message "*Line Specify first point:*" is displayed. AutoCAD expects us to identify the starting location of a straight line.

3. On your own, create three connected line segments of arbitrary length just above the two circles as shown in the figure below.

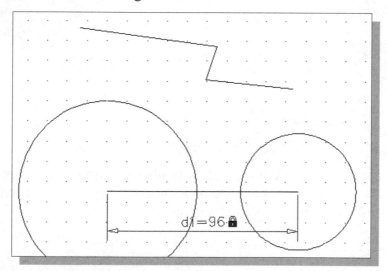

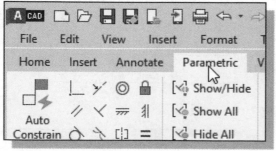

4. In the *Ribbon* tabs area, left-click once on the **Parametric Tools** tab as shown.

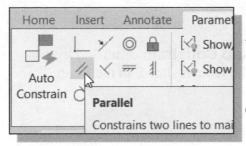

5. Select the **Parallel** constraint by clicking on the icon as shown in the figure.

6. Select the **line segment on the left** as shown.

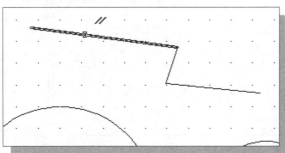

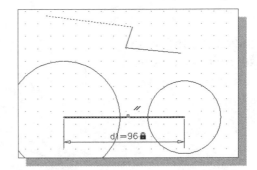

7. Select the **horizontal line** connected to the circle centers as shown.

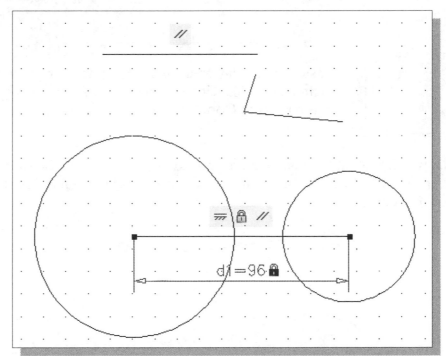

❖ Note that although the selected line segment now becomes a horizontal line, parallel to the first line created, the line segment is no longer connected to the two other lines. Although it is feasible to apply additional Coincident constraints, a better option is to use the **Auto Constrain** command.

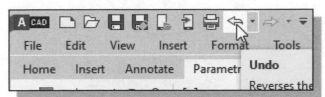

8. In the *Quick Access* toolbar, click the **Undo** button as shown.

9. Select the **Auto Constrain** command by clicking on the icon as shown in the figure.

10. Select the **three line segments** by using a selection window as shown.

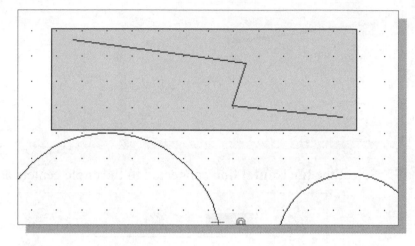

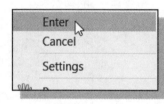

11. Inside the display area, click once with the right-mouse-button and select **Enter** to accept the selection and proceed with the Auto Constrain command.

❖   Note that two Coincident constraints are applied to the geometry. The Auto Constrain command will automatically apply constraints based on the existing geometric relations.

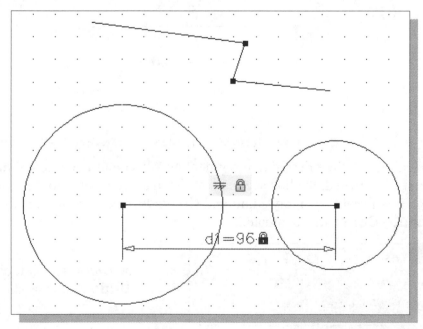

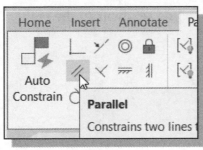

12. Select the **Parallel** constraint by clicking on the icon as shown in the figure.

13. Select the **line segment** on the left as shown.

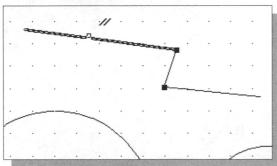

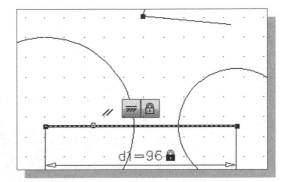

14. Select the **horizontal line** connected to the circle centers as shown in the figure above.

15. On your own, apply another Parallel constraint to the **right line segment** so that the three longer lines are parallel to each other as shown.

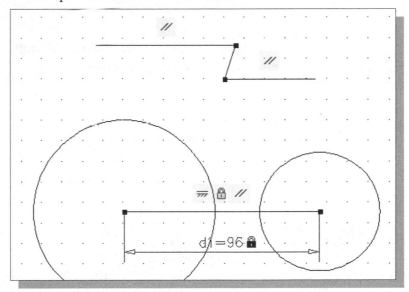

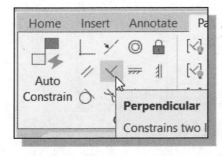

16. Select the **Perpendicular** constraint by clicking on the icon as shown in the figure.

17. On your own, apply the **Perpendicular** constraint to the **line segment to the right** and the short **line segment in the middle** as shown.

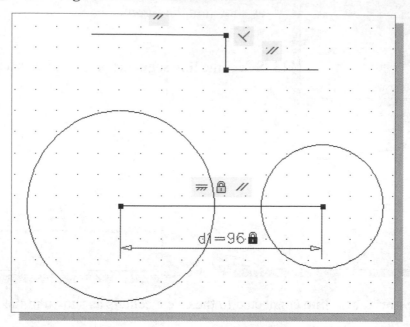

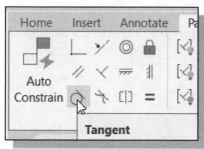

18. Select the **Tangent** constraint by clicking on the icon as shown in the figure.

19. On your own, select the **circle on the right** and the **line segment** just above it to apply the constraint.

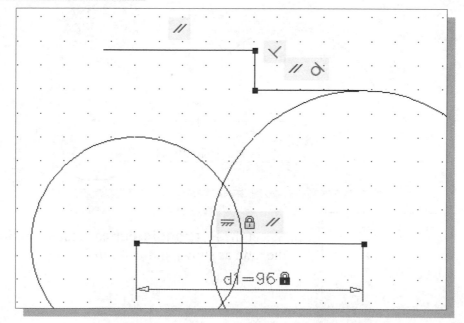

❖ Note the size of the circle is adjusted to satisfy the applied constraint.

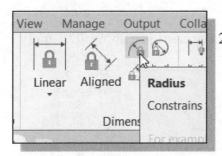

20. Select the **Radius Dimensional constraint** by clicking on the icon as shown in the figure.

21. Select the **circle on the right** and place the dimension to the right as shown.

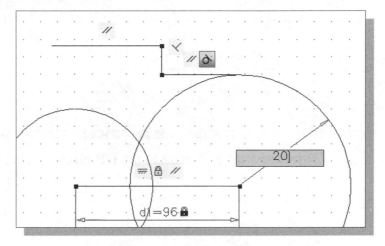

22. Enter **20** as the new radius of the arc and note the associated geometry is adjusted.

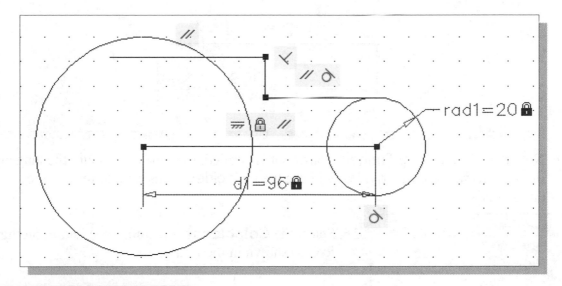

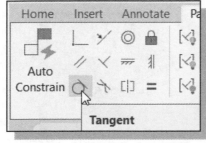

23. Select the **Tangent** constraint by clicking on the icon as shown in the figure.

24. On your own, select the **circle on the left** and the **line segment** just above it to apply the constraint.

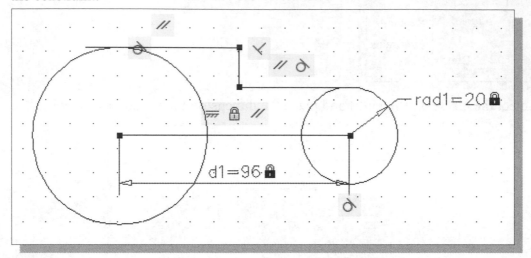

25. On your own, apply a **Dimensional** constraint to the **left circle** and set the radius to *30* as shown.

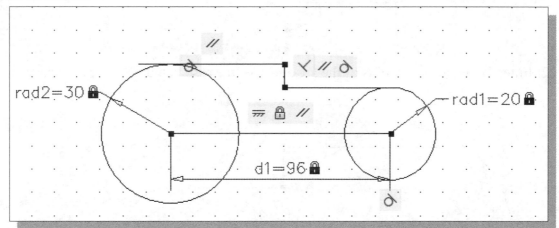

❖ Note that although the Tangent constraint is applied, that does not assure the tangent lines are trimmed correctly. We will use the Coincident constraint to restrict the endpoints.

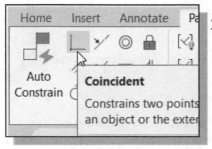

26. Select the **Coincident** constraint by clicking on the icon as shown in the figure.

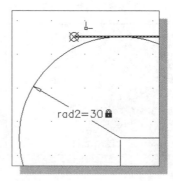

27. Select the **endpoint of the top horizontal line** as shown.

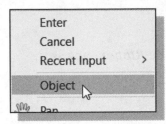

28. Inside the display area, right-click once to bring up the option menu and select **Object** as shown.

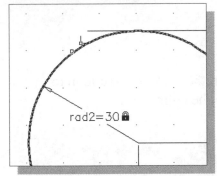

29. Select the **circle** on the left as shown.

❖ The Coincident constraint assures the endpoint of the line is always on the circle.

30. On your own, repeat the above steps to constrain the right endpoint of the right line to the circle. Your drawing should appear as shown in the figure below.

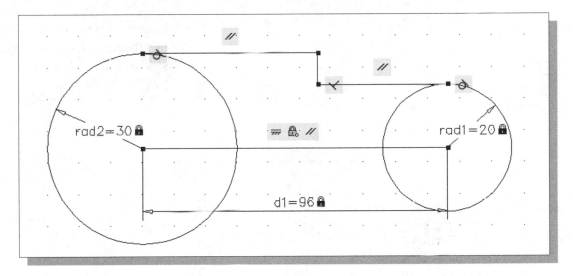

31. On your own, apply a **Linear** dimensional constraint to the **top horizontal line** and set its length to *60* as shown.

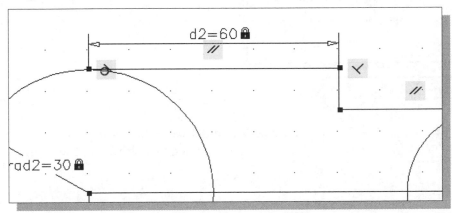

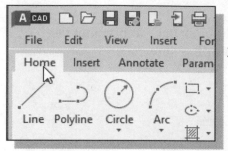

32. Click on the **Home** tab in the *Ribbon* tabs area.

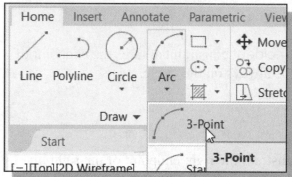

33. Click on the **3-Point arc** icon to activate the command.

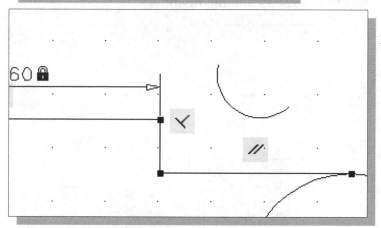

34. On your own, create an arc of arbitrary size to the right side of the top horizontal line as shown.

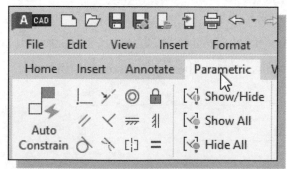

35. In the *Ribbon* tabs area, left-click once on the **Parametric Tools** tab as shown.

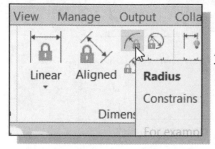

36. Select the **Radius** *dimensional constraint* by clicking on the icon as shown in the figure.

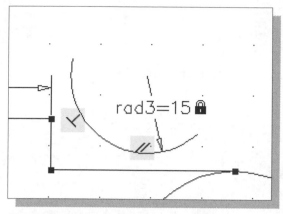

37. Select the **arc** by clicking once with the left-mouse-button.

38. Enter **15** as the new radius of the arc. Note the geometry adjustment is done instantly.

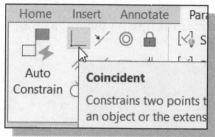

39. Select the **Coincident** constraint by clicking on the icon as shown in the figure.

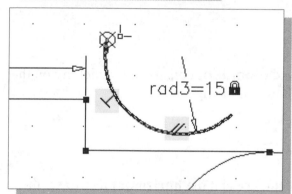

40. Select the **top endpoint of the arc** as shown.

41. Select the **right endpoint of the top horizontal line** as shown.

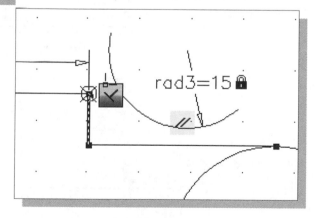

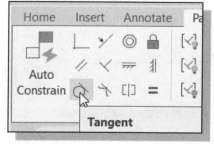

42. Select the **Tangent** constraint by clicking on the icon as shown in the figure.

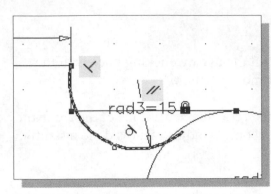

43. Select the **arc** as the first object.

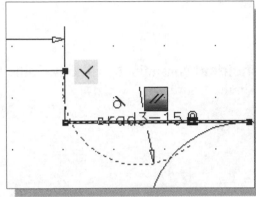

44. Select the **horizontal line** on the left as the second object.

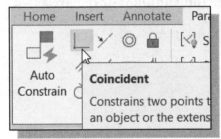

45. Select the **Coincident** constraint by clicking on the icon as shown in the figure.

46. On your own, select the **right endpoint of the arc** and the **horizontal line** to set the constraint.

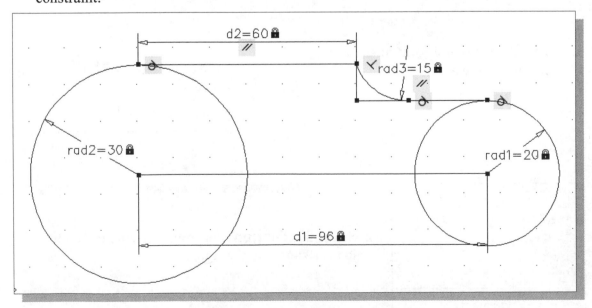

Using the Mirror Command

1. Select the **Mirror** command icon in the *Modify* toolbar. In the command prompt area, the message *"Select objects:"* is displayed.

2. Use a *selection window* to select the objects on the upper portion as shown.

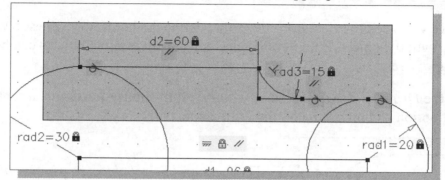

3. Inside the *Drawing Area*, right-click to accept the selection and continue with the Mirror command.

4. In the *command prompt area*, the message *"Specify the first point of the mirror line:"* is displayed. Pick the **center point of the circle on the left**. (Hint: Switch on the *Object Snap* option to assist the selection.)

5. In the *command prompt area*, the message *"Specify the second point:"* is displayed. Pick the **center point of the circle on the right**.

6. In the *command prompt area*, the message *"Delete source objects? [Yes/No] <No>:"* is displayed. Inside the *Drawing Area*, right-click and select **Enter** to retain the original objects.

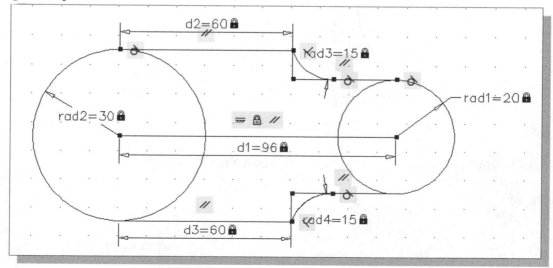

Using the Trim Command

The Trim command shortens an object so that it ends precisely at a selected boundary.

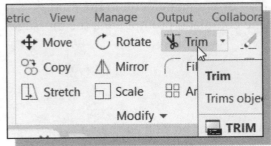

1. Select the **Trim** command icon in the *Modify* toolbar. In the command prompt area, the message *"Select boundary edges... Select objects:"* is displayed.

❖ First, we will need to select the objects that define the **boundary edges** to which we want to trim the object. If no item is selected, then all existing objects can be used as boundary edges.

2. Inside the *Drawing Area*, click once with the **right-mouse-button** to accept the default option and proceed with the Trim command.

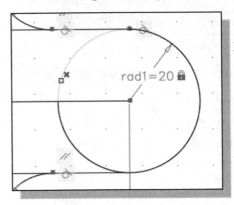

3. The message *"Select object to trim or shift-select object to extend or [Fence/Crossing/Project/Edge/eRase/Undo]:"* is displayed in the command prompt area. Pick the **inside top segment of the circle on the right** as shown.

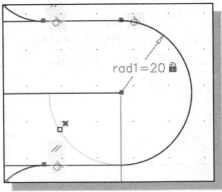

4. Pick the **left side of the arc** to trim the lower left portion of the arc.

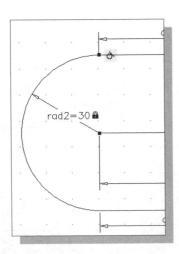

5. On your own, trim the other circle to remove the inside portion.

6. On your own, continue to trim the horizontal lines on the right, and the drawing should appear as shown in the figure below.

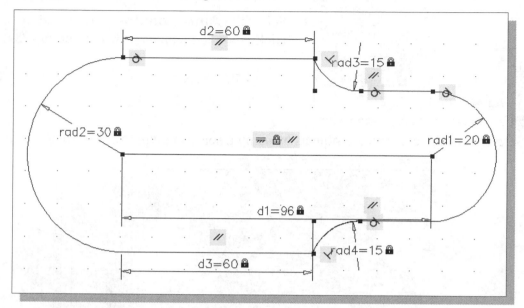

❖ Note that all constraints appear to be intact prior to exiting the Trim command.

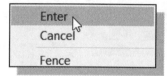

7. Inside the Drawing Area, right-click to activate the option menu and select **Enter** with the left-mouse-button to end the Trim command.

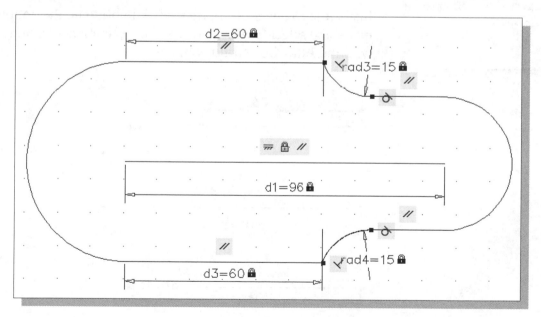

❖ Note that some of the constraints, both geometric and dimensional constraints, are removed as we exited the Trim command. The constraints are properties applied to the geometry, and the associated constraints will be removed when the geometry is edited/changed.

Using the Auto Constrain Command

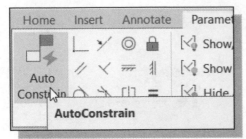

1. Select the **Auto Constrain** command by clicking on the icon as shown in the figure.

2. Select all of the constructed objects by using a selection window as shown.

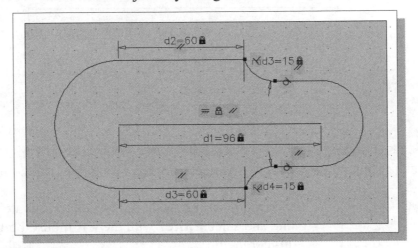

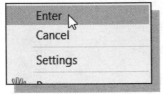

3. Inside the *Drawing Area*, right-click to bring up the option menu and select **Enter** to accept the selection and continue with the Auto Constrain command.

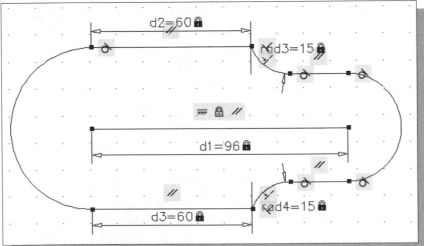

❖ Since the geometric entities were constructed precisely, AutoCAD is able to reapply the geometric constraints correctly; however, we will need to create the two radial dimensional constraints manually.

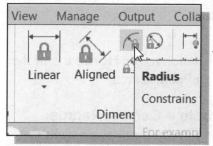

4. Select the **Radius** dimensional constraint by clicking on the icon as shown in the figure.

5. Select the **circle on the right** and place the dimension to the right as shown.

6. Hit the **Enter** key once to accept the default dimension.

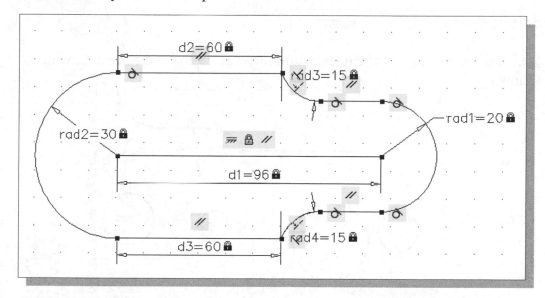

7. On your own, move the objects forming the outside loop to layer *Object* and switch *ON* the *Display Lineweight* option. Your drawing should appear as shown in the figure below.

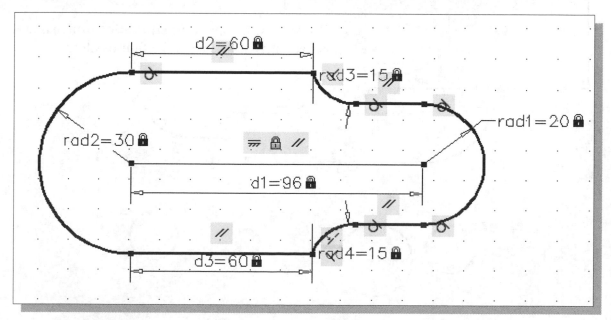

Creating and Constraining Additional Circles

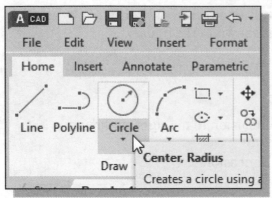

1. Click on and switch to the **Home** tab in the *Ribbon* tabs and panels area.

2. Select the **Circle – Center, Radius** command icon in the *Draw* toolbar. In the command prompt area, the message "*_circle Specify center point for the circle or [3P/2P/Ttr (tan tan radius)]:*" is displayed.

3. On your own, create four circles of arbitrary sizes below the current drawing as shown.

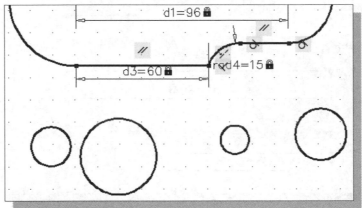

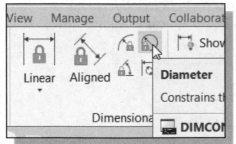

4. Select the **Diameter** dimensional constraint by clicking on the icon as shown in the figure.

5. On your own, apply four **Diameter dimensional constraints** as shown in the figure below.

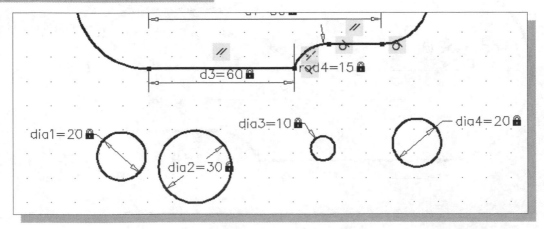

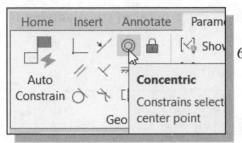

6. Select the **Concentric** constraint by clicking on the icon as shown in the figure.

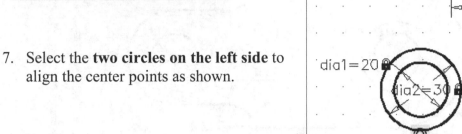

7. Select the **two circles on the left side** to align the center points as shown.

8. On your own, repeat the Concentric constraint command and align the three circles as shown in the figure below.

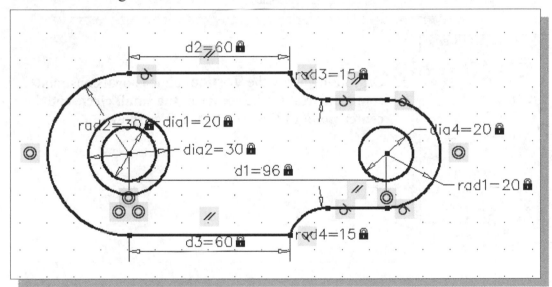

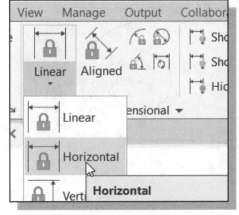

9. Click on the **triangle** symbol below the icon of the **Linear** dimensional constraint to show more options.

10. Select the **Horizontal** dimensional constraint by clicking on the icon as shown in the figure.

11. Select the **center point of the arc** on the left and the **center point of the diameter 10 circle**.

12. On your own, set the horizontal distance to **48** as shown.

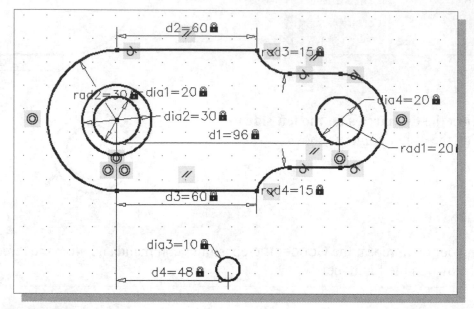

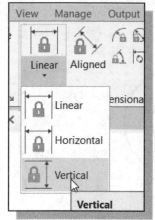

13. On your own, use the **Vertical** dimensional constraint command to align the **center of the small circle** and the **center point of the arc** on the left as shown.

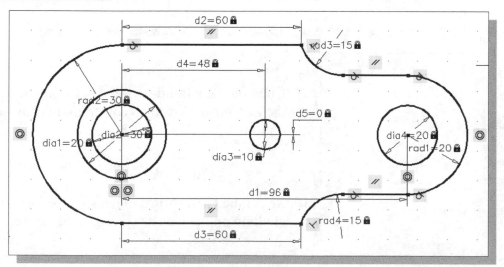

Control the Display of Constraints

1. Click on the **Hide All Dynamic Constraints** icon with the left-mouse-button.

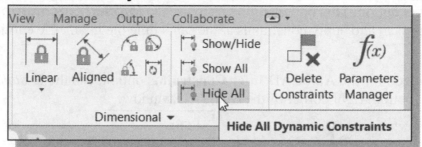

❖ Note all dimensional constraints are removed from the screen.

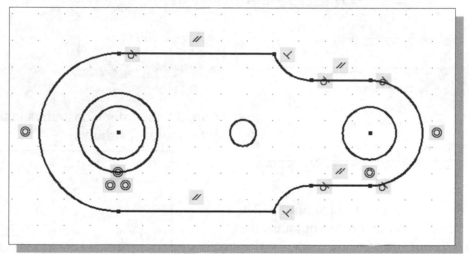

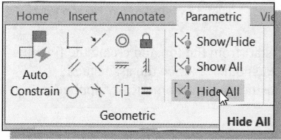

2. Click on the **Hide All Geometric Constraints** icon with the left-mouse-button.

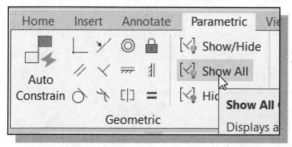

3. On your own, use the **Show** and **Show All** commands to display the applied geometric constraints.

•   Note that the procedures involved in using *parametric drawing tools* are very different than traditional drafting techniques. Although both approaches can be used to achieve the same results, the *parametric drawing* approach provides the more powerful functionalities for design revisions and modifications.

The Implicit Geometric Constraint Approach

In *AutoCAD 2023*, a parametric option is also available in the *Status toolbar* area, the **Infer Constraints** option. This option allows us to use the ***Implicit Geometric Constraint*** approach and apply constraints at the same time the geometric entities are constructed.

1. To show the icon for the AutoCAD **Infer Constraints** option, use the *Customization option* at the bottom right corner of the AutoCAD window.

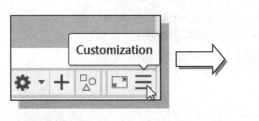
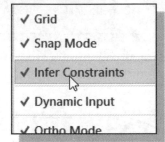

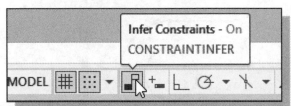

2. Turn **ON** the Infer Constraints in the status toolbar as shown.

3. Select the **Line** command icon in the *Draw* toolbar. In the command prompt area, the message "*Line Specify first point:*" is displayed.

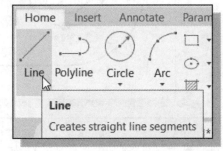

4. To the right side of the completed drawing, start from the top and create the five line segments that are perpendicular/parallel to each other. Notice the associated coincident constraints and perpendicular constraints are automatically applied.

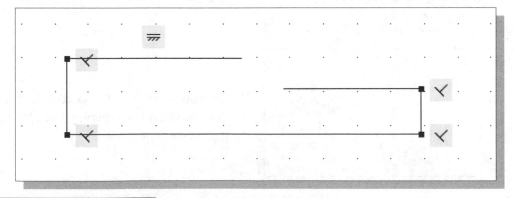

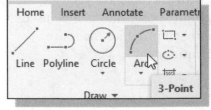

5. Click on the **3-Point arc** icon to activate the command.

6. Create a 3-point arc with the starting and ending points connected to the two horizontal lines as shown.

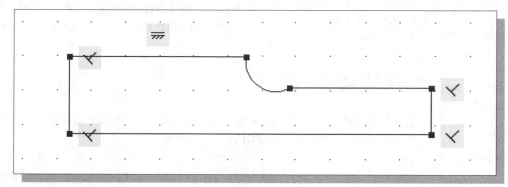

7. In the *Ribbon* tabs area, left-click once on the **Parametric Tools** tab as shown.

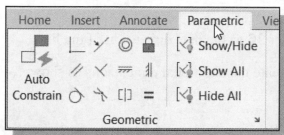

8. Select the **Tangent** constraint by clicking on the icon as shown in the figure.

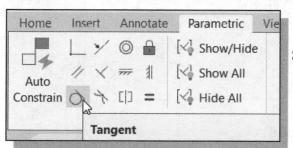

9. Select the **arc** as the first object.

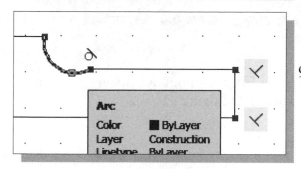

10. Select the **horizontal line** on the left as the second object.

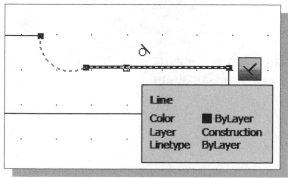

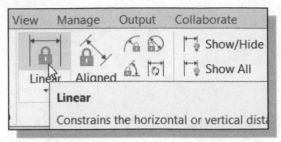

11. Select the **Linear** dimension constraint by clicking on the icon as shown in the figure.

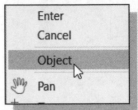

12. Inside the display area, right-click once to bring up the option menu and select **Object** as shown.

13. Select the **bottom horizontal line** with a single click of the left-mouse-button.

14. Place the dimension below the line and enter **96** as the new dimension of the line.

15. On your own, repeat the above steps and create the additional dimensions as shown.

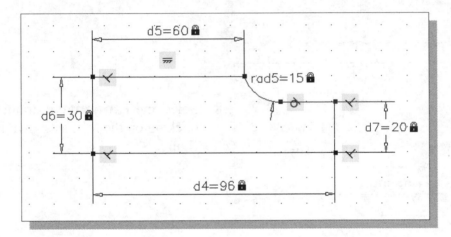

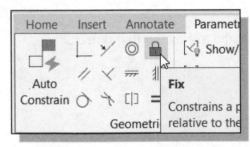

16. Select the **Fix** constraint by clicking on the icon as shown in the figure.

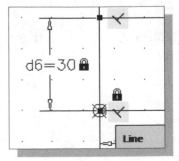

17. Click on the **bottom left corner** of the sketch with the left-mouse-button to apply the constraint.

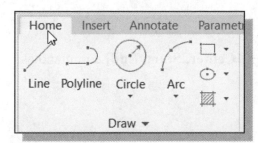

18. In the *Ribbon* tabs area, left-click once on the **Home** tab as shown.

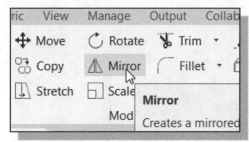

19. Select the **Mirror** command icon in the *Modify* toolbar. In the command prompt area, the message "*Select objects:*" is displayed.

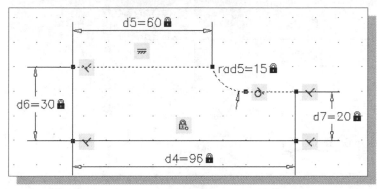

20. Select the three objects that are on top as shown.

21. Inside the *Drawing Area*, **right-click** to accept the selection and continue with the **Mirror** command.

22. In the command prompt area, the message "*Specify the first point of the mirror line:*" is displayed. Pick the **left endpoint of the bottom horizontal line**.

23. In the command prompt area, the message "*Specify the second point:*" is displayed. Pick the **right endpoint of the bottom horizontal line**.

24. In the command prompt area, the message "*Delete source objects? [Yes/No] <N>:*" is displayed. Hit the **[Enter]** key once to retain the original objects.

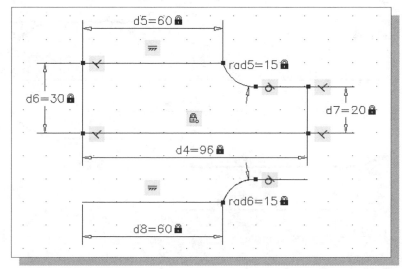

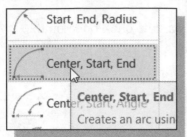

25. Activate the [Arc] → [Center, Start, End] command as shown.

26. Create the two arcs as shown in the figure below.

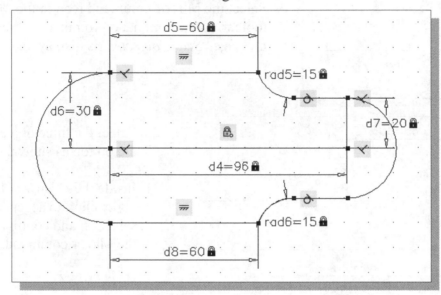

27. On your own, move the objects forming the outline of the design to the **Object** layer.

28. Switch on the **Line Weight** option in the status bar area.

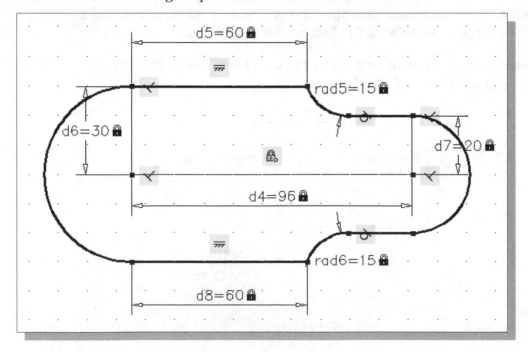

29. On your own, create the four additional circles.

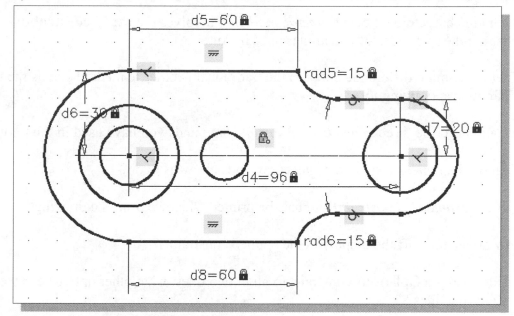

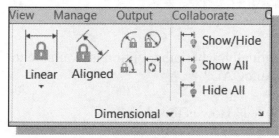

30. Complete the construction of the design by adding geometry/dimensional constraints to the circles.

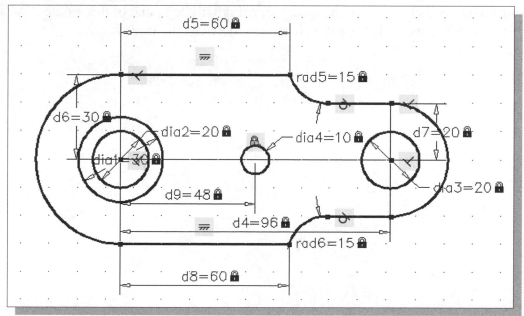

➤ Note that the implicit geometric constraint approach is the preferred method used in most 3D parametric modeling software, such as the *Autodesk Inventor* software.

Review Questions: (Time: 25 minutes)

1. The main characteristic of *parametric modeling* involves the use of **constraints**. What are the two types of constraints used in AutoCAD?

2. What is the main difference between the traditional geometric method versus the use of parametric drawing tools?

3. List and describe three *geometric constraint* commands you have used in the tutorial.

4. What do the *dimensional constraint* commands allow us to do?

5. Will the *geometric constraint* symbols be printed when we print the drawing?

6. How do we turn off the display of the applied constraints?

7. Besides using a Collinear constraint to align two lines, what other options can we use to align two lines?

8. When will AutoCAD automatically remove some of the applied constraints?

9. Besides applying *geometric constraints* individually, what other option is available to constrain precisely constructed geometry in AutoCAD?

10. Can the applied constraints be manually removed? How is this done?

11. When a design is created or changed, a drawing will be in one of three states. What are the three states? What are the differences between the three states?

Exercises: (Time: 200 minutes)

1. Indexing Base (Dimensions are in inches.)

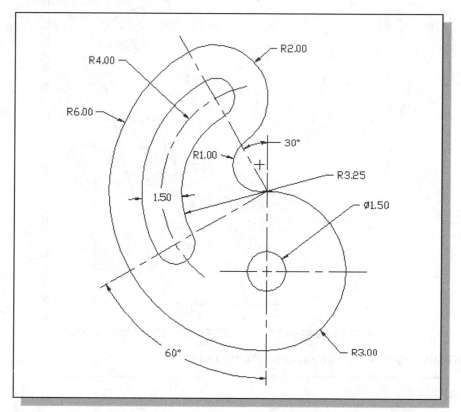

2. Positioning Spacer (Dimensions are in inches.)

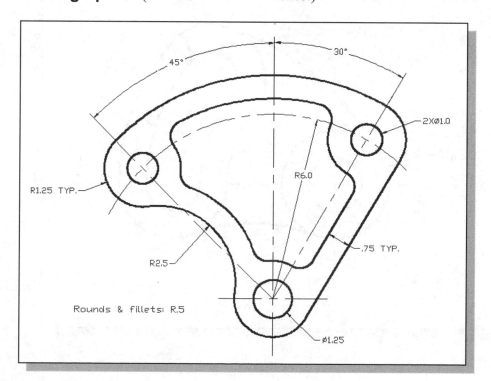

3. V-Slide Plate (Dimensions are in inches.)

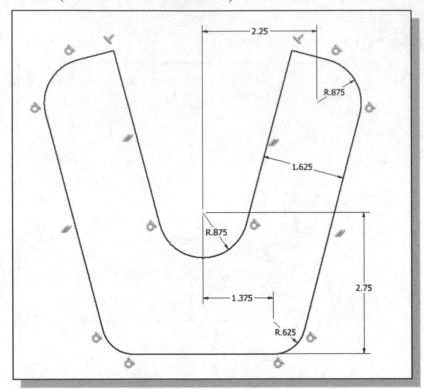

4. Adjustable Support (Dimensions are in inches.)

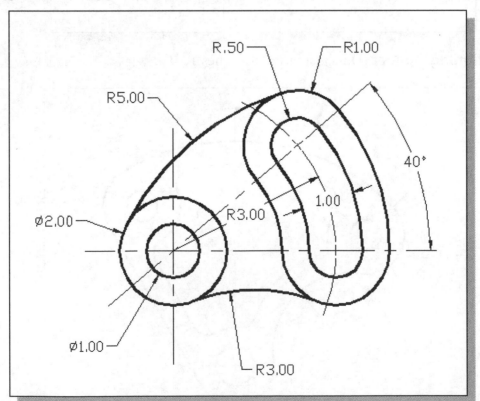

5. Sensor Mount (Dimensions are in inches.)

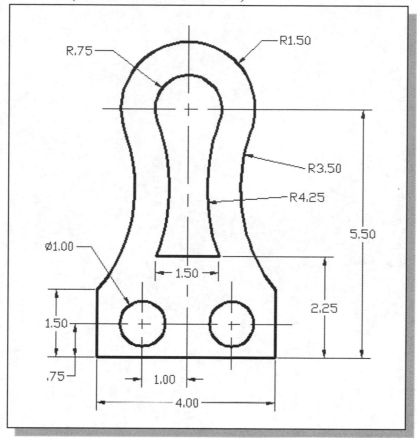

6. Flat Hook (Dimensions are in mm.)

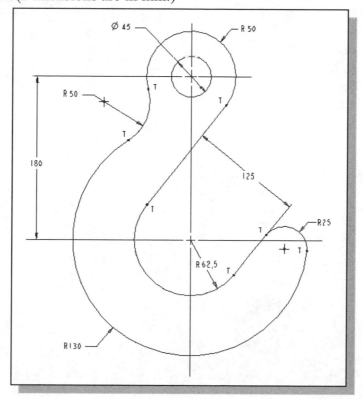

Notes:

Chapter 10
Auxiliary Views and Editing with GRIPS

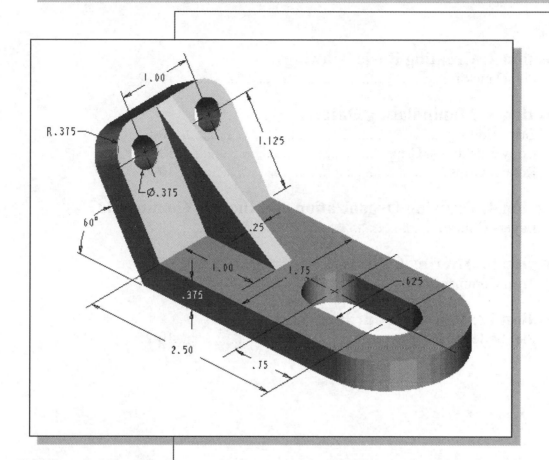

Learning Objectives

♦ **Use 2D Projection Method to Draw Auxiliary Views**
♦ **Create Rectangles**
♦ **Use the Basic GRIPS Editing Commands**
♦ **Create and Edit the Plot Style Table**
♦ **Set Up and Use the Polar Tracking Option**
♦ **Create Multiple Viewports in Paper Space**

AutoCAD Certified User Examination Objectives Coverage

This table shows the pages on which the objectives of the Certified User Examination are covered in Chapter 10.

Certified User Reference Guide

Introduction

An important rule concerning multiview drawings is to draw enough views to accurately describe the design. This usually requires two or three of the regular views, such as a front view, a top view and/or a side view. Many designs have features located on inclined surfaces that are not parallel to the regular planes of projection. To truly describe the feature, the true shape of the feature must be shown using an **auxiliary view**. An *auxiliary view* has a line of sight that is perpendicular to the inclined surface, as viewed looking directly at the inclined surface. An *auxiliary view* is a supplementary view that can be constructed from any of the regular views. This lesson will demonstrate the construction of an auxiliary view using various CAD techniques.

In this lesson, we will examine the use of the very powerful AutoCAD *GRIPS* feature. In **AutoCAD 2023**, a *GRIP* is a small square displayed on a **pre-selected object**. Grips are key control locations such as the endpoints and midpoints of lines and arcs. Different types of objects display different numbers of grips. Using grips, we can *stretch*, *move*, *mirror*, *scale*, *rotate*, and *copy* objects without entering commands or clicking toolbars. Grips reduce the keystrokes and object selection required in performing common editing commands. To edit with grips, we select the objects <u>before</u> issuing any commands. To remove a specific object from a selection set that displays grips, we hold down the **[SHIFT]** key as we select the object. To exit the *grips mode* and return to the *command prompt*, press the **[ESC]** key.

The V-Block Design

The V-Block Example

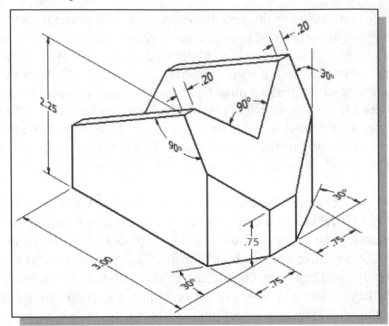

Before going through the tutorial, make a rough sketch of a multiview drawing of the part. How many 2D views will be necessary to fully describe the part? Based on your knowledge of **AutoCAD 2023** so far, how would you arrange and construct these 2D views? Take a few minutes to consider these questions and do preliminary planning by sketching on a piece of paper. You are also encouraged to construct the orthographic views on your own prior to going through the tutorial.

Starting Up AutoCAD 2023

1. Select the **AutoCAD 2023** option on the *Program* menu or select the **AutoCAD 2023** icon on the *Desktop*.

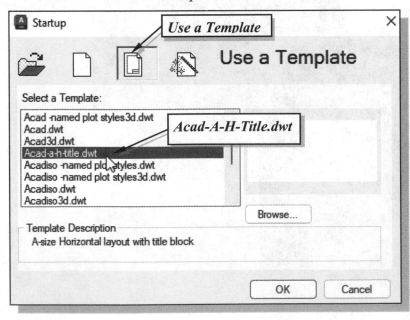

2. In the AutoCAD 2023 *Startup* dialog box, select the **Template** option.

3. Select the *Acad-A-H-Title* template file from the list of template files. If the template file is not listed, click on the **Browse** button to locate and proceed to open a new drawing file.

Setting up the Principal Views

1. In the *Layer Control* box, set layer **Construction** as the *Current Layer* if it is not the default layer.

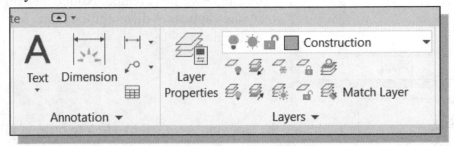

2. In the *Status Bar* area, switch *ON* the following options in the status toolbar: ***GRID Display***, ***Dynamic Input***, ***Object Snap Tracking***, ***Object Snap*** and ***Line Weight Display***.

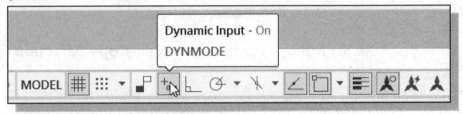

- We will first create construction geometry for the front view.

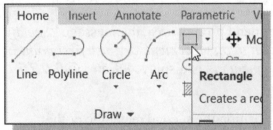

3. Select the **Rectangle** icon in the *Draw* toolbar. In the command prompt area, the message "*Specify first corner point or [Chamfer/Elevation/Fillet/Thickness/Width]:*" is displayed.

4. Place the first corner point of the rectangle near the lower left corner of the screen. Do not be overly concerned about the actual coordinates of the location; the CAD drawing space is a very flexible virtual space.

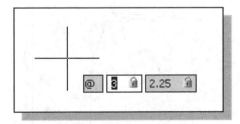

5. Create a 3" × 2.25" rectangle. Using the *Dynamic Input* option, enter **@3,2.25** [**ENTER**].

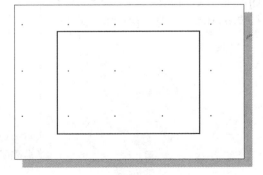

❖ The Rectangle command creates rectangles as *polyline* features, which means all segments of a rectangle are created as a single object.

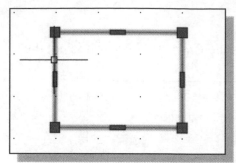

6. Next, we will use the *GRIPS* editing tools to make a copy of the rectangle. Pick any edge of the rectangle we just created. Notice that small squares appear at different locations on the rectangle.

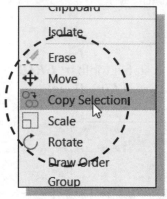

7. Inside the Drawing Area, **right-click** to bring up the **pop-up option menu**.

❖ In the center section of the pop-up menu, the set of GRIPS editing commands includes Erase, Move, Copy Selection, Scale, and Rotate.

8. In the pop-up menu, select the **Copy Selection** option.

9. In the command prompt area, the message "*Specify base point or displacement, or [Multiple]:*" is displayed. Pick the **lower right corner** as the base point. A copy of the rectangle is attached to the cursor at the base point.

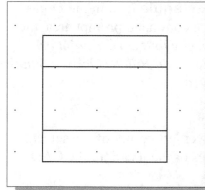

10. In the command prompt area, the message "*Specify second point of displacement, or <use first point as displacement>:*" is displayed. Enter **@0,0.75 [ENTER]**.

❖ This will position the second rectangle at the location for the vertical 30-degree angle.

11. Inside the Drawing Area, **right-click** once to bring up the option menu and choose **Enter** to exit the Copy option.

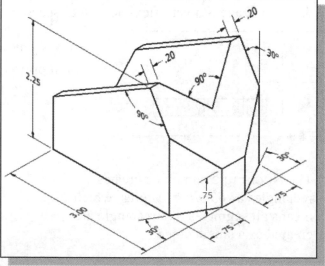

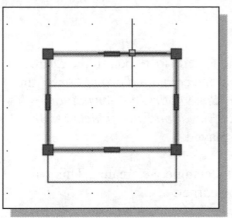

12. **Pre-select** the **copy** by picking the top horizontal line on the screen. The second rectangle, the copy we just created, is selected.

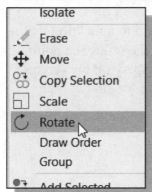

13. Inside the Drawing Area, **right-click** to bring up the pop-up option menu and select the **Rotate** option.

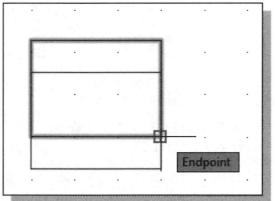

14. In the command prompt area, the message "*Specify base point:*" is displayed. Pick the **lower right corner** of the **selected rectangle** as the base point.

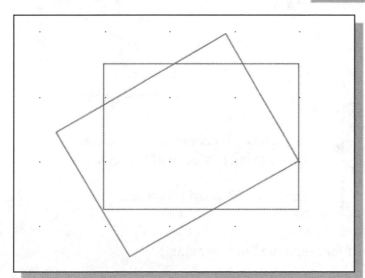

15. In the command prompt area, the message "*Specify the rotation angle or [Reference]:*" is displayed. Enter: **30** [**ENTER**].

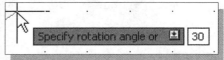

Setting up the Top View

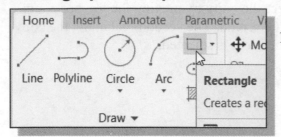

1. Select the **Rectangle** icon in the *Draw* toolbar. In the *command prompt area*, the message *"Specify first corner point or [Chamfer/Elevation/Fillet/Thickness/Width]:"* is displayed.

2. Move the cursor over the top left corner of the first rectangle we created. This will activate the *object tracking* alignment feature to the corner.

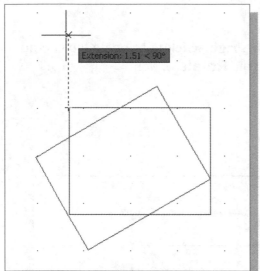

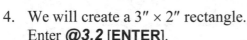

3. Move the cursor upward to a location that is about 1.5″ away from the reference point. (Read the *OTRACK* display on the screen.) Left-click once to place the first corner-point of the rectangle.

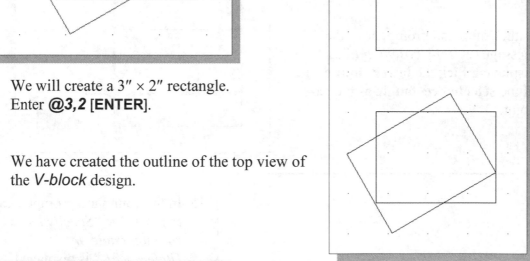

4. We will create a 3″ × 2″ rectangle. Enter **@3,2 [ENTER]**.

*   We have created the outline of the top view of the *V-block* design.

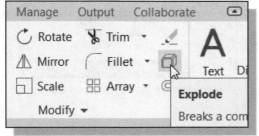

5. Pre-select the **rectangle** we just created by left-clicking any edge of the rectangle.

6. Select the **Explode** icon in the *Modify* toolbar.

*   The top rectangle now consists of four separate line segments.

Using the Offset Command

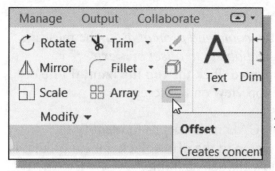

1. Click the **Offset** icon in the *Modify* toolbar. In the command prompt area, the message *"Specify offset distance or [Through]:"* is displayed.

2. In the command prompt area, enter **0.2** **[ENTER]**.

3. In the *command prompt area*, the message *"Select object to offset or <exit>:"* is displayed. Pick the **top horizontal line** of the top view on the screen.

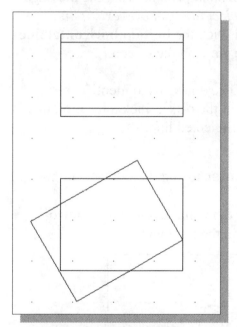

4. AutoCAD next asks us to identify the direction of the offset. Pick a location that is **below** the selected line.

5. In the command prompt area, the message *"Select object to offset or <exit>:"* is displayed. Pick the **bottom horizontal line** of the top view on the screen.

6. AutoCAD next asks us to identify the direction of the offset. Pick a location that is **above** the selected line.

7. Inside the *Drawing Area*, **right-click** and select **Enter** to end the Offset command.

8. Inside the Drawing Area, right-click to bring up the pop-up option menu and select the **Repeat Offset** option.

- Notice in the pop-up menu, none of the GRIPS editing commands are displayed; the GRIPS editing commands are displayed only if objects are pre-selected.

9. In the command prompt area, the message *"Specify offset distance or [Through]:"* is displayed. Enter **0.75** [**ENTER**].

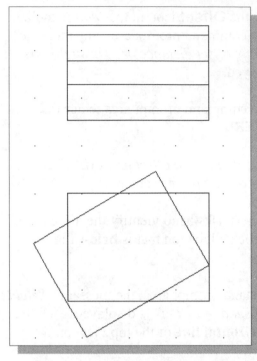

10. In the *command prompt area*, the message *"Select object to offset or <exit>:"* is displayed. Pick the **top horizontal line** of the top view on the screen.

11. AutoCAD next asks us to identify the direction of the offset. Pick a location that is **below** the selected line.

12. In the *command prompt area*, the message *"Select object to offset or <exit>:"* is displayed. Pick the **bottom horizontal line** of the top view on the screen.

13. AutoCAD next asks us to identify the direction of the offset. Pick a location that is **above** the selected line.

14. Inside the *Drawing Area*, **right-click** and choose **Enter** to end the Offset command.

- The four parallel lines will be used to construct the top v-cut feature and the two 0.75″ × 30° cut features at the base of the *V-block* in the top view.

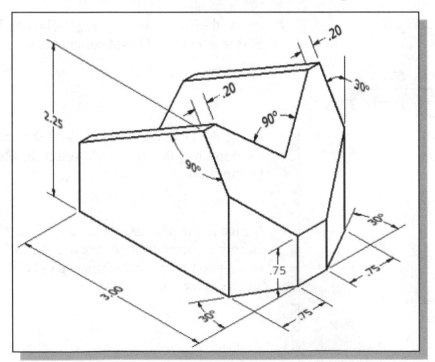

Creating Object Lines in the Front View

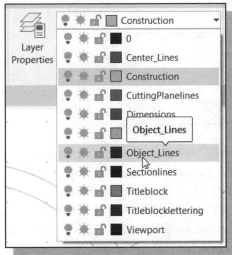

1. On the *Object Properties* toolbar, choose the **Layer Control** box with the left-mouse-button.

2. Move the cursor over the name of layer **ObjectLines**; the tool tip *"ObjectLines"* appears.

3. **Left-click once** and layer *ObjectLines* is set as the *Current Layer*.

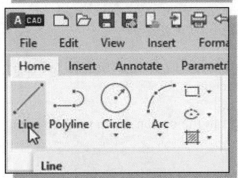

4. Select the **Line** command icon in the *Draw* toolbar. In the command prompt area, the message *"Line Specify first point:"* is displayed.

5. Pick the **lower left corner** of the bottom horizontal line in the front view as the starting point of the line segments.

6. Pick the **lower right corner** of the bottom horizontal line in the front view as the second point.

7. Select the **third** and **fourth** points as shown in the figure below.

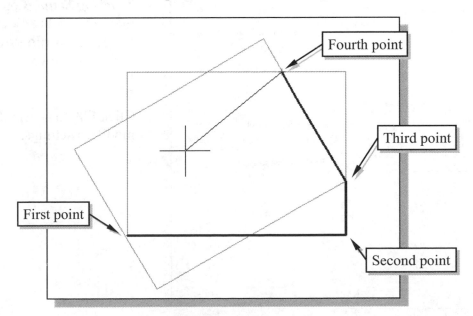

Setting the Polar Tracking Option

1. In the *Status Bar* area, turn **ON** the *Polar Tracking* option.

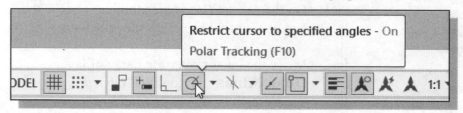

- Note that the *Polar Tracking* option is one of the *AutoCAD AutoTrack*™ features. The *AutoTrack* features include two tracking options: polar tracking and object snap tracking. When the *Polar Tracking* option is turned on, alignment markers are displayed to help us create objects at precise positions and angles. A quick way to change the settings of the *AutoTrack* feature is to use the option menu.

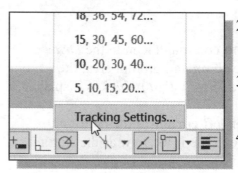

2. Move the cursor on top of the *Polar Tracking* option in the *Status Bar* area.

3. Click once with the **right-mouse-button** to bring up the option menu.

4. Select **Settings** in the option menu as shown in the figure.

5. In the *Drafting Settings* dialog box, set the *Increment angle* to **30** as shown in the figure below.

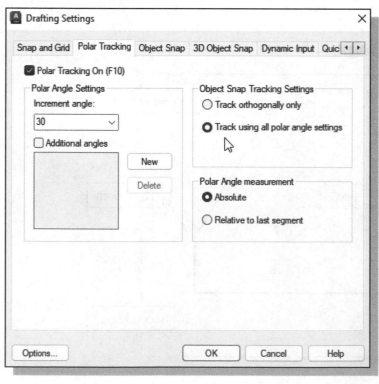

6. Under the *Object Snap Tracking Settings* option, turn **ON** the **Track using all polar angle settings** option as shown in the figure.

7. Click **OK** to accept the modified settings.

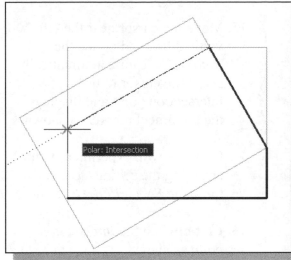

8. Move the cursor near the left vertical line as shown and notice that *AutoCAD AutoTrack* automatically snaps the cursor to the intersection point and displays the alignment marker as shown.

➢ In the following steps, we will illustrate the use of different *POLAR* settings to achieve the same result.

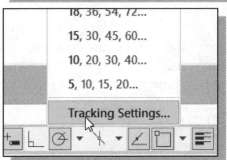

9. Click once with the right-mouse-button on the *POLAR* option in the *Status Bar* area to bring up the option menu.

10. Select **Settings** in the option menu as shown in the figure.

11. In the *Drafting Settings* dialog box, set the *Increment angle* to **90** as shown in the figure below.

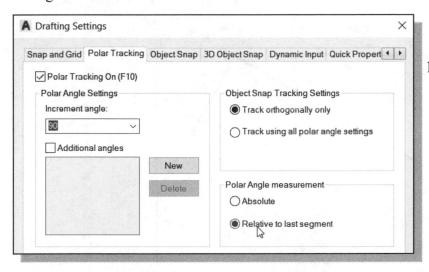

12. Under the *Object Snap Tracking Settings* option, turn *ON* the **Track orthogonally only** option as shown in the figure.

13. Under the *Object Snap Tracking Settings* option, turn *ON* the **Relative to last segment** option as shown in the figure above.

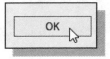

14. Click **OK** to accept the modified settings.

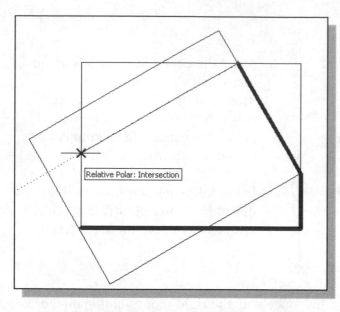

15. Move the cursor near the left vertical line and notice the *AutoTrack* feature automatically snaps the cursor to the intersection point and displays the alignment marker as shown.

➢ On your own, experiment with changing the settings to achieve the same *SNAP/POLAR* results.

16. **Left-click** at the *intersection point* as shown.

17. Click the **Close** option in the *command prompt area*. AutoCAD will create a line connecting the last point to the first point of the line sequence.

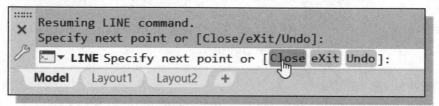

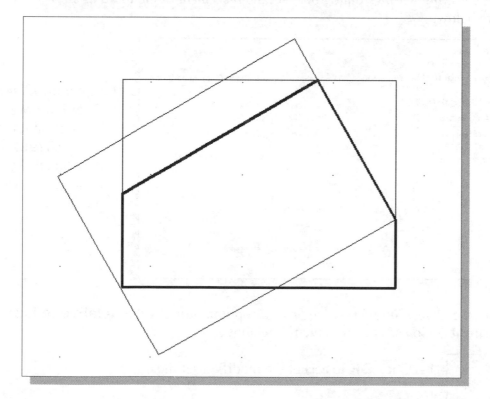

Setting up an Auxiliary View

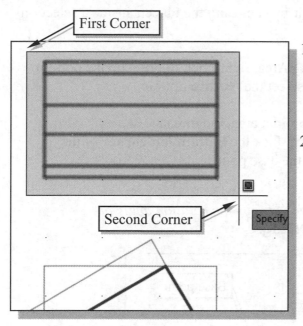

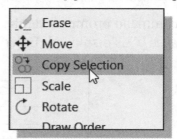

1. **Pre-select** all objects in the top view, by using two single clicks of the left mouse button, enclosing the objects inside a selection window.

2. Inside the *Drawing Area*, right-click to bring up the pop-up option menu and select the **Copy Selection** option.

3. In the command prompt area, the message "*Specify base point or displacement, or [Multiple]:*" is displayed. Pick the **lower left corner** of the top view as the base point.

4. Using the *AutoTrack* feature, place the copy of the top-view by aligning it to the inclined object line we just created. **Left-click** once to position the copy about 2″ away from the top corner of the front-view.

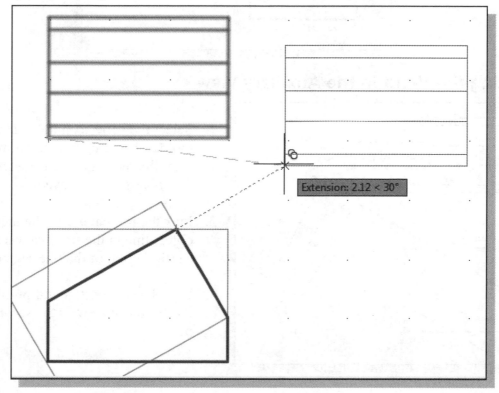

Aligning the Auxiliary View to the Front View

1. **Pre-select** all objects in the *auxiliary view* by enclosing the objects inside a selection window.

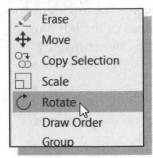

2. Inside the Drawing Area, right-click to bring up the pop-up option menu and select the **Rotate** option.

3. In the command prompt area, the message "*Specify base point:*" is displayed. Pick the **bottom left corner** of the auxiliary view as the base point.

4. In the command prompt area, the message "*Specify the rotation angle or [Reference]:*" is displayed. Enter **-60** [ENTER].

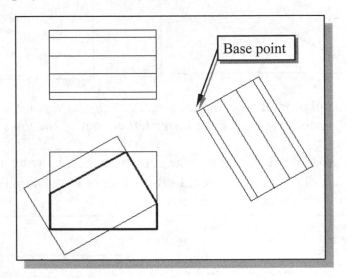

Base point

Creating the V-cut in the Auxiliary View

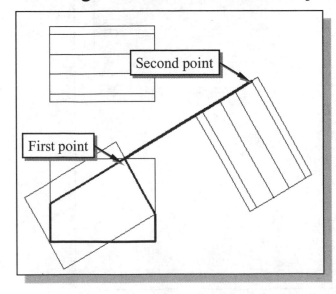

Second point

First point

1. Select the **Line** icon in the *Draw* toolbar. In the command prompt area, the message "*Line Specify first point:*" is displayed.

2. Pick the **top corner** of the inclined object line in the front view as the starting point of the line segments.

3. Pick the **second top end point** in the auxiliary view as the second point.

4. Inside the *Drawing Area*, right-click and select **Enter** to end the Line command.

5. Pre-select the line we just created.

6. Inside the Drawing Area, right-click to bring up the pop-up option menu and select the **Rotate** option.

7. In the command prompt area, the message "*Specify base point:*" is displayed. Pick the top right endpoint of the line as the base point.

8. In the command prompt area, the message "*Specify the rotation angle or [Reference]:*" is displayed. Enter **45** [**ENTER**].

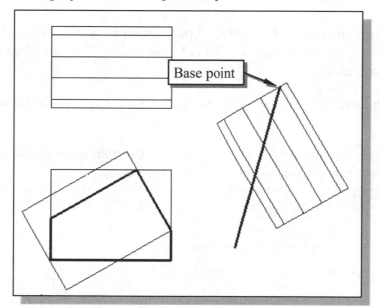

9. On your own, repeat the above steps and create the other line as shown.

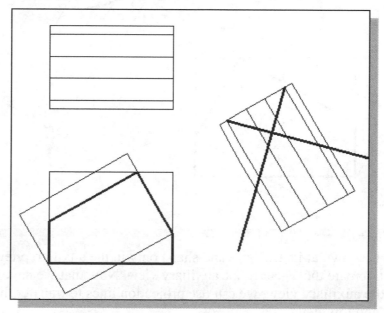

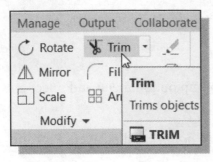

10. Select the **Trim** command icon in the *Modify* toolbar.

11. In the command prompt area, click **cuTting edges** as shown

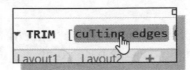

12. Pick the two inclined lines we just created in the auxiliary view as the *boundary edges*.

13. Inside the *Drawing Area*, **right-click** to proceed with the Trim command. The message "*Select object to trim or [Project/Edge/Undo]:*" is displayed in the command prompt area.

14. Pick the two lower endpoints of the two inclined lines to remove the unwanted portions.

15. Inside the Drawing Area, right-click to activate the option menu and select **Enter** with the left-mouse-button to end the Trim command.

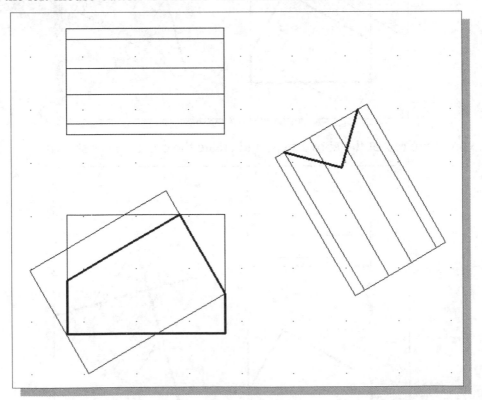

➤ The V-cut is shown at its true size and shape only in the auxiliary view. It is therefore necessary to create the V-cut in the auxiliary view. Now that we have constructed the feature in the auxiliary view, we can use projection lines to transfer the feature to the front view and top view.

Creating the V-cut in the Front View and Top View

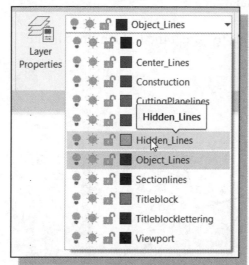

1. On the *Object Properties* toolbar, choose the *Layer Control* box with the left-mouse-button.

2. Move the cursor over the name of layer *Hidden***Lines**; the tool tip *"HiddenLines"* appears.

3. **Left-click once** and layer *HiddenLines* is set as the *Current Layer*.

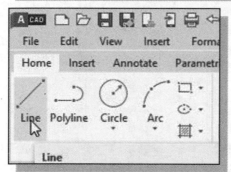

4. Select the **Line** command icon in the *Draw* toolbar. In the command prompt area, the message *"Line Specify first point:"* is displayed.

5. Pick the **vertex** of the V-cut in the *auxiliary view* as the first point of the line.

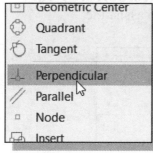

6. Inside the *Drawing Area*, hold down the **[SHIFT]** key and **right-click** once to bring up the *Object Snap* shortcut menu.

7. Select the **Perpendicular** option in the pop-up window.

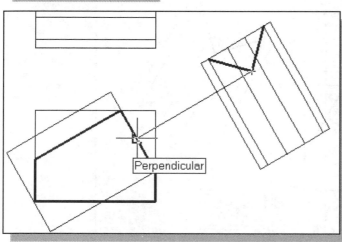

8. Move the cursor to the front view on the inclined edge and notice the perpendicular symbol appears at different locations. Select the intersection point on the inclined line as shown.

9. Inside the *Drawing Area*, right-click to activate the option menu and select **Enter** with the left-mouse-button to end the Line command.

10. On your own, use the **Trim** and **Extend** commands to adjust the hidden lines in the front view.

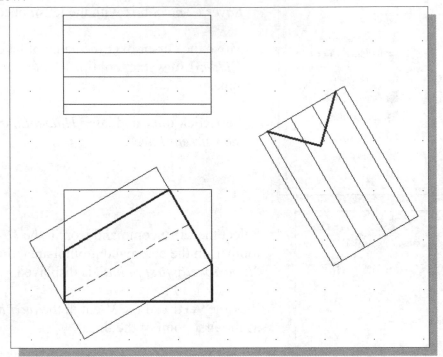

➢ On your own, first create the three construction lines and then construct the V-cut feature in the top-view. Use the Trim and Extend commands to assist the construction.

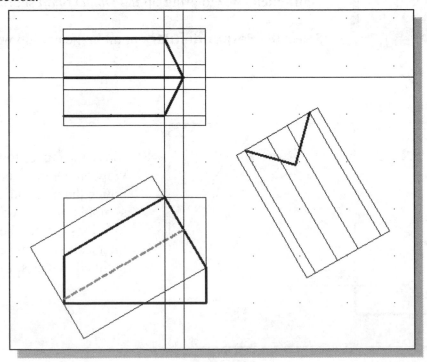

Setting the Polar Tracking Option

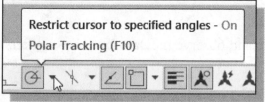

1. Move the cursor to the *Status Bar* area, over the **Polar Tracking** option button.

2. **Left-click once** on the **triangle icon** to bring up a pop-up option menu.

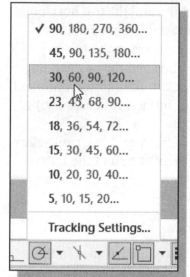

3. Select the **30, 60, 90, 120...** option by clicking once with the left-mouse-button. This is the shortcut to set the angles for the *Polar Tracking Angle Setting*.

➢ Notice the other settings that are available. We will use the absolute polar angle measurement for this example.

Completing the Top View

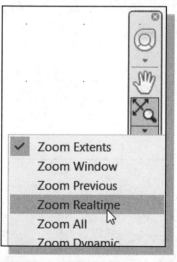

1. Click on the **Zoom Realtime** icon in the *View display* toolbar area.

2. Move the cursor near the center of the *Drawing Area*.

3. **Push and hold down the left-mouse-button**, then move upward to enlarge the current display scale factor. (Press the **[Esc]** key to exit the command.)

4. On your own, use the **Pan Realtime** option to reposition the display so that we can work on the top view of the *V-block*.

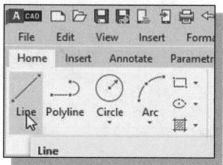

5. Select the **Line** icon in the *Draw* toolbar.

6. In the command prompt area, the message *"Line Specify first point:"* is displayed. Pick the **right endpoint** of the third horizontal line (from the top) in the top view as the starting point of the line segments.

7. Move the cursor toward the top horizontal line and observe the *AutoTracking* markers over different locations.

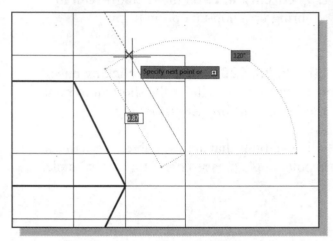

8. **Left-click** at the intersection of the polar tracking and the top horizontal line as shown. Do not select the intersection between the top horizontal line and the vertical line. (Use the Zoom Realtime command to zoom-in further, if necessary.)

9. Inside the *Drawing Area*, right-click and select **Enter** to end the Line command.

10. Repeat the above steps and create the other inclined line in the top view.

➢ On your own, complete the top view by adding all the necessary object lines in the top view. Use the Trim and Extend commands to assist the construction.

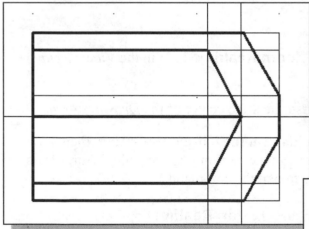

- Notice that the two 30° cut features are shown as **true size and shape** only in the top view and therefore it is necessary for us to construct the features in the top view.

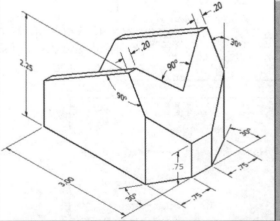

➤ On your own, create the **vertical construction line** through the corner of the 30° cut as shown.

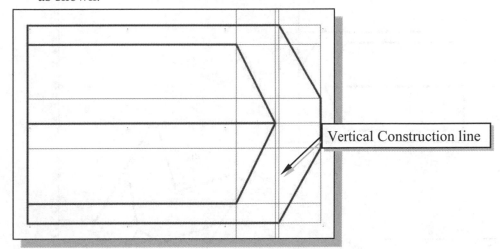

Vertical Construction line

11. Complete the front view by adding the object line along the construction line as shown.

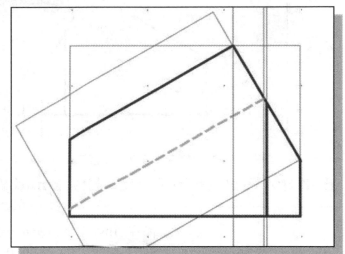

➤ On your own, complete the views by adding all the necessary object lines in the views.

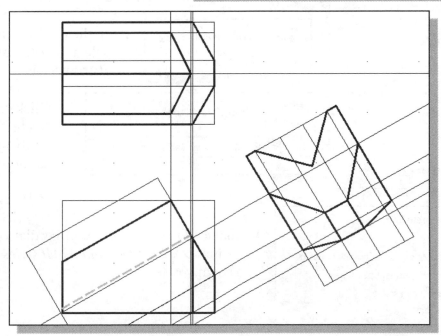

➢ Complete the drawing by adding the proper dimensions.

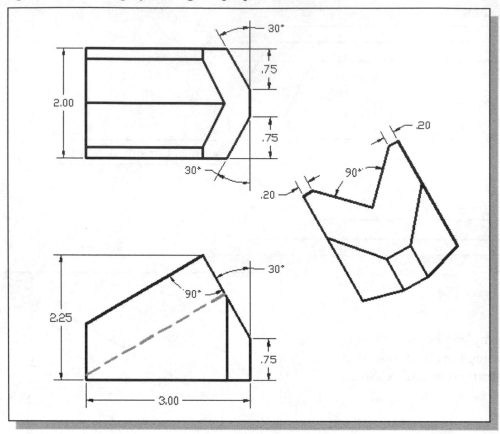

Edit the Plot Style Table - Black and White Prints

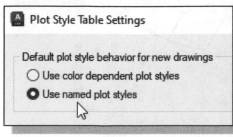

1. Inside the *Drawing Area*, right-click and select **Options** in the pop-up menu.

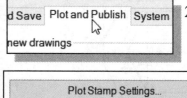

2. In the *Options* dialog box, select the **Plot and Publish** tab.

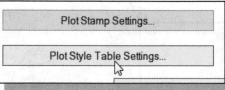

3. Click on the **Plot Style Table Settings** button as shown.

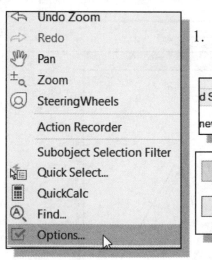

4. In the *Plot Style Table Settings* dialog box, switch *ON* the *Use named plot styles* option as shown.

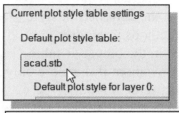

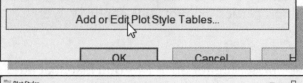

5. Choose the *acad.stb* as the default plot style table as shown.

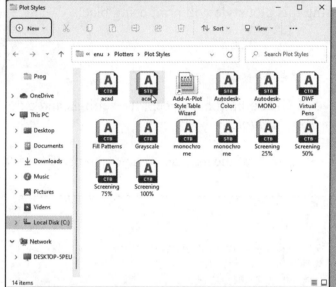

6. In the *Options* dialog box, click on the **Add or Edit Plot Style Tables** button.

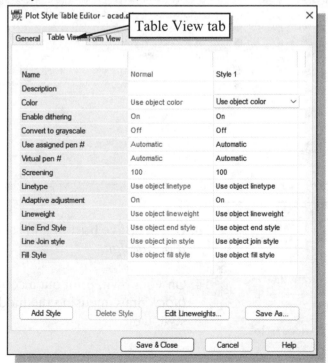

7. The *Plot Styles* folder appears on the screen.

8. Double-click the **acad.stb** icon with the left-mouse-button to open the plot style file.

9. In the *Plot Style Table Editor*, select the **Table View** tab.

Name	Normal	Style 1
Description		
Color	Use object color	Use object color
Enable dithering	On	On
Convert to grayscale	Off	Off
Use assigned pen #	Automatic	Automatic
Virtual pen #	Automatic	Automatic
Screening	100	100
Linetype	Use object linetype	Use object linetype
Adaptive adjustment	On	On
Lineweight	Use object lineweight	Use object lineweight
Line End Style	Use object end style	Use object end style
Line Join style	Use object join style	Use object join style
Fill Style	Use object fill style	Use object fill style

Plot Style Table Editor - acad.s

General   Table View   Form View

Table View tab

Add Style   Delete Style   Edit Lineweights...   Save As...

Save & Close   Cancel   Help

- The *Plot Style Table Editor* displays the plot styles that are in the current plot style table. The **Table View** and **Form View** tabs provide two methods to modify the existing plot style settings. Both tabs list all of the plot styles in the plot style table and their settings. In general, the **Table View** tab is more convenient if there are only a small number of plot styles. We can modify plot style color, screening, linetype, lineweight, and other settings. The first plot style in a named plot style table is *Normal* and represents an object's default properties (no plot style applied). We cannot modify or delete the *Normal* style.

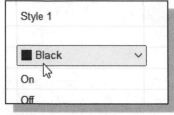

10. Change the *Color* setting for *Style 1* to **Black** so that all layers using this plot style will print using black.

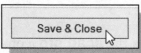

11. Pick the **Save & Close** button to accept the settings and exit the *Plot Style Table Editor*.

12. On your own, adjust and confirm the *Plot Style* of the associated layers to use **Style 1** as shown.

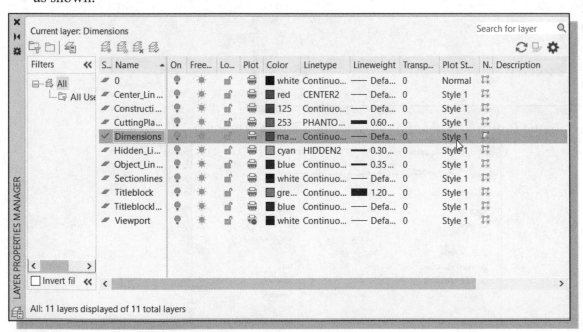

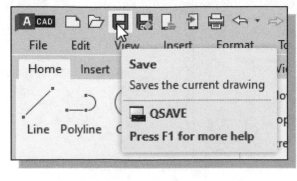

13. Use the **Save button** and save the *V-Block design*.

14. On your own, print out a copy of the *V-block* drawing using the modified plot style table.

Start a new drawing for a Metric Template File

In chapter 8, the procedure to create a template file was illustrated; in this section we will also create a metric template.

1. In the *Quick Access* toolbar, click the **New** icon.

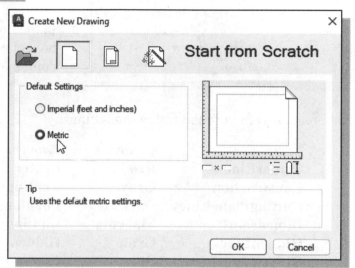

2. In the *Startup* dialog box, select the **Start from Scratch** option with a single click of the left-mouse-button.

3. In the *Default Settings* section, pick **Metric** as the drawing units.

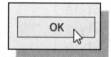

4. Click **OK** to accept the setting and start a new metric drawing.

5. On your own, set up the *grid* and *snap spacing* to **10 mm** as shown.

6. Click **OK** to accept the modified settings.

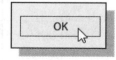

Layers Setup

1. Pick *Layer Properties Manager* in the *Object Properties* toolbar.

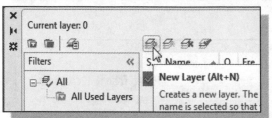

2. In the *Layer Properties Manager* dialog box, click on the **New** button (or the key combination [**Alt+N**]) to create a new layer.

3. Create **layers** with the following settings:

Layer	Color	Linetype	Lineweight	PlotStyle
CenterLines	Red	Center	Default	Style 1
Construction	Gray	Continuous	Default	Style 1
CuttingPlaneLines	Dark Gray	Phantom	0.6mm	Style 1
Dimensions	Magenta	Continuous	Default	Style 1
HiddenLines	Cyan	Hidden	0.3mm	Style 1
ObjectLines	Blue	Continuous	0.6mm	Style 1
SectionLines	White	Continuous	Default	Style 1
Title_Block	Green	Continuous	1.2mm	Style 1
TitleBlockLettering	Blue	Continuous	Default	Style 1
Viewport	White	Continuous	Default	Style 1

➢ Using the *Normal PlotStyle* enables plotting of *lineweights* defined in the specific layer. Note that the **Lineweight** settings are set for proper printing of different linetypes and some may appear thicker on screen.

4. Highlight the layer *Construction* in the list of layers.

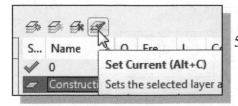

5. Click on the **Current** button to set layer *Construction* as the *Current Layer*.

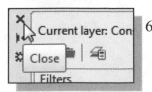

6. Click on the **Close** button to accept the settings and exit the *Layer Properties Manager* dialog box.

Set up the Metric Borders and Title Block by Copy & Paste

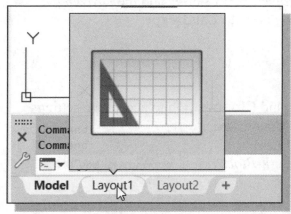

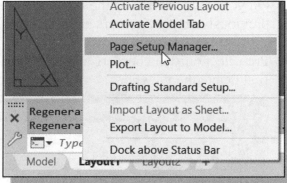

1. Click the **Layout1** icon to switch to the two-dimensional paper space.

2. Right-click once on the tab and select **Page Setup Manager**.

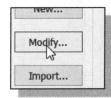

3. Choose the default layout in the *Page Setup Manager* and click **Modify**.

4. In the *Page Setup* dialog box, select a plotter/printer that is available to plot/print your design. Here we will set up the plotting for an **A-size plot** using the Microsoft **print to PDF** option. You can also use A4 paper size if your plotter supports the metric size papers.

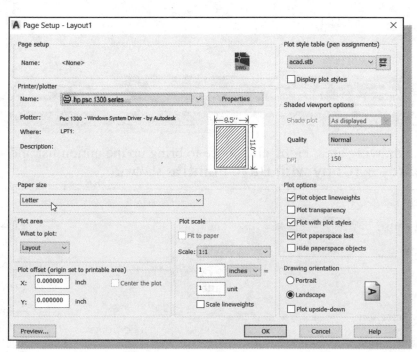

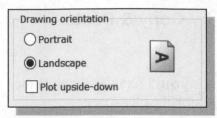

5. Set the *Paper size* to **Letter** and the *Drawing orientation* is set to **Landscape**.

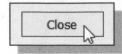

6. Click on the **OK** button and **Close** button to accept the settings and exit the *Page Setup* dialog boxes.

7. On your own, **switch** to the **Layout1** of the *V-Block drawing*. We will use the previously created borders and title block.

8. Pre-select only the borders and title block using a selection window as shown. (Hint: first delete the viewport.)

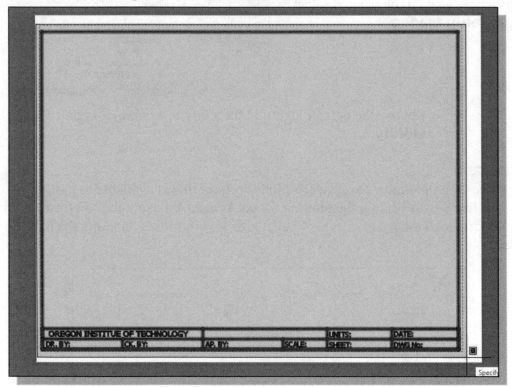

9. Inside the drawing area, **right-click** once to bring up the option list and select [**Clipboard**] → [**Copy with Base Point**] as shown.

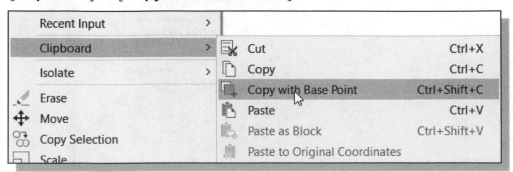

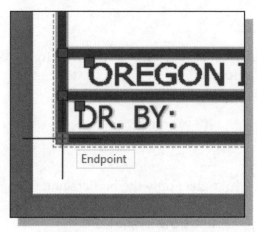

10. Zoom in and select the **lower left corner** of the borders as the base point.

11. On your own, switch back to the blank **Layout1** of the *new drawing*.

12. Inside the drawing area, right-click once to bring up the option list and select [**Clipboard**] → [**Paste**] as shown.

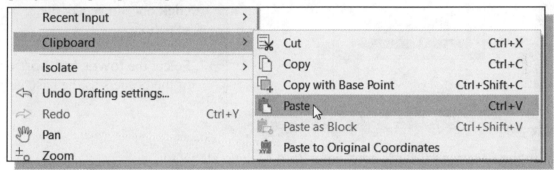

13. On your own, place the borders and the title block near the lower left corner of the page as shown. Note the borders and the title block appeared to be fairly small; this is due to the difference in units.

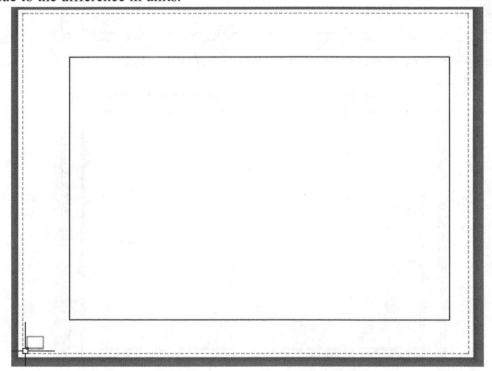

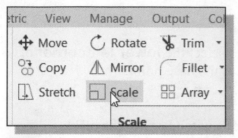

14. In the *Modify* toolbar, click the **Scale** icon to activate the command.

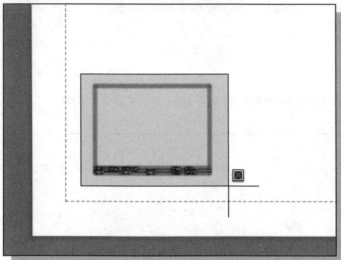

15. Select only the borders and title block using a ***selection window*** as shown.

16. Inside the drawing area, **right-click** once to accept the selection.

17. Select the **lower left corner** of the borders as the base point.

18. In the *command prompt* area, enter **25.4** as the scale factor.

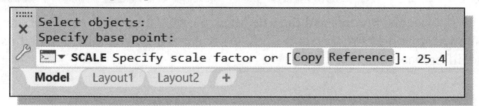

19. On your own, reposition the *borders and title block*, and add a viewport inside the borders if necessary.

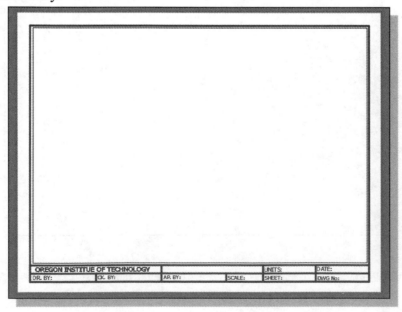

20. In the *Quick Access* toolbar, click the **Save** icon.

21. In the *Save Drawing As* dialog box, save to the default template folder (or select the folder in which you want to store the *template file)*. Enter **Acad-mm-A-H-Title** in the *File name* box.

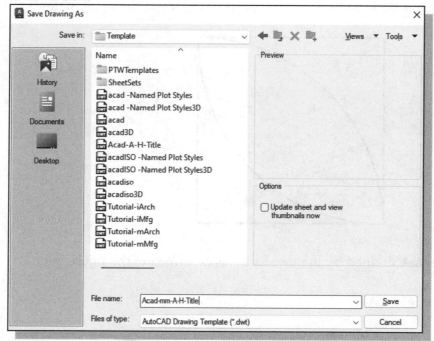

22. Pick **Save** in the *Save Drawing As* dialog box to close the dialog box.

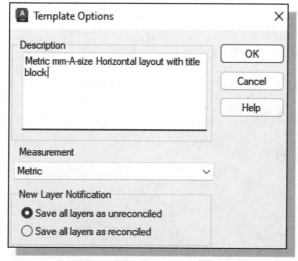

23. In the *Template Description* dialog box, enter **mm-A-size Horizontal layout with title block** in the *Description* box.

24. Pick **OK** to close the dialog box and save the template file.

➢ It is recommended that you keep a second copy of any template files on a separate disk as a backup.

Review Questions: (Time: 25 minutes)

1. What is an auxiliary view and why would it be important?

2. What is a GRIP? What are the advantages of using the GRIPS?

3. List three GRIPS editing commands you have used in the tutorial.

4. What does the *Polar Tracking* option allow us to do?

5. Find the area A defined by the two arcs, as described in the figure below.

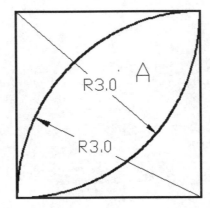

6. Find the area A defined by the three arcs, as described in the figure below.

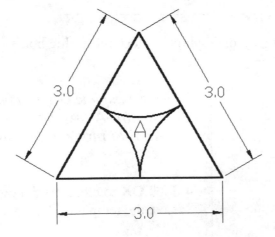

Exercises: (Time: 240 minutes)

1. Angle Base (Dimensions are in inches.)

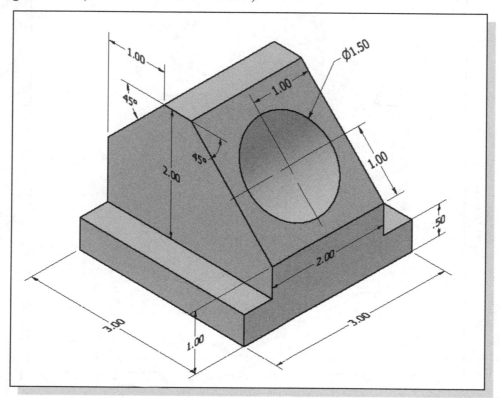

2. Indexing Guide (Dimensions are in inches.)

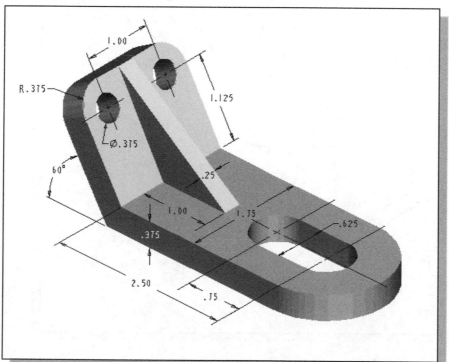

3. Spindle Base (Dimensions are in millimeters.)

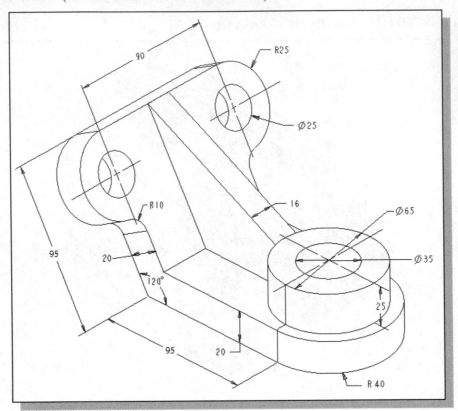

4. Transition Support (Dimensions are in inches.)

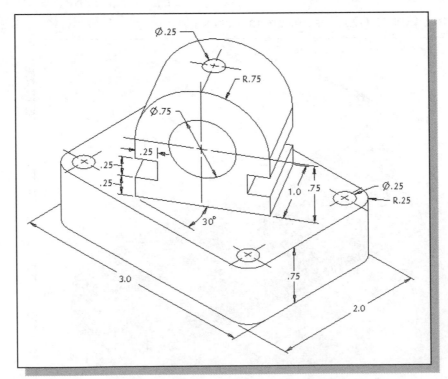

5. **Support Hanger** (Dimensions are in inches.)

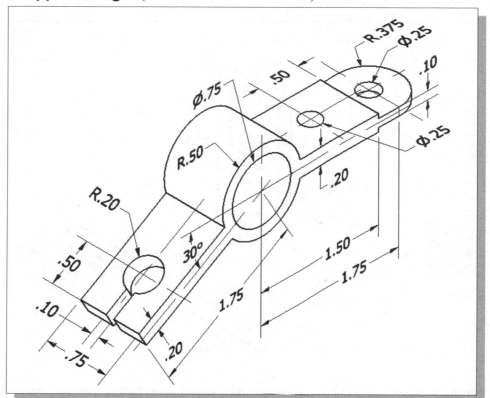

6. **Automatic Stop** (Dimensions are in inches.)

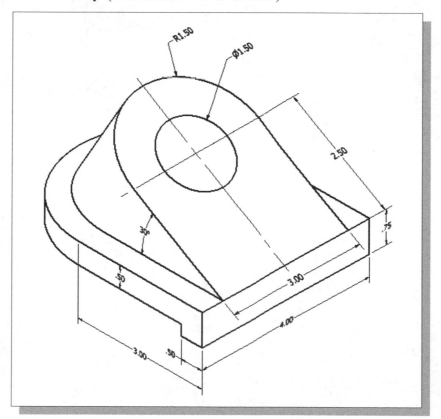

7. Angle V-Block (Four holes, Dimensions are in inches.)

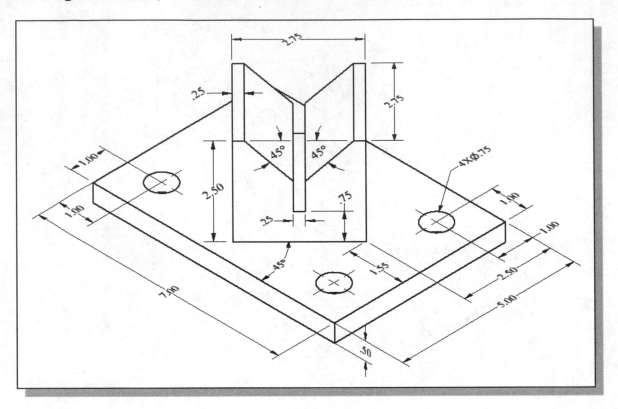

Chapter 11
Section Views

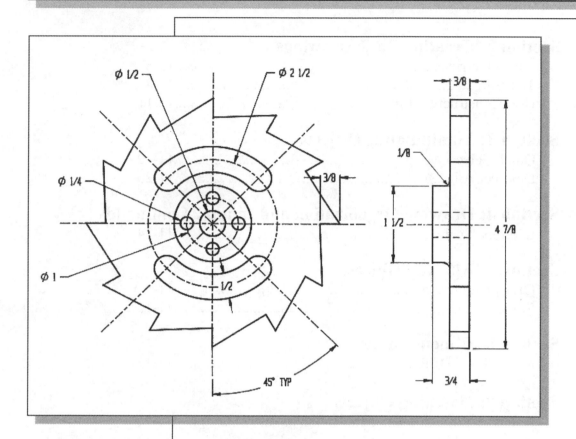

Learning Objectives

♦ **Use CAD Methods to Create Section Views**
♦ **Use the Object Snap Shortcut Options**
♦ **Change the Linetype Scale Property**
♦ **Stretch and Move Objects with Grips**
♦ **Create Cutting Plane Lines**
♦ **Use the Hatch Command**

AutoCAD Certified User Examination Objectives Coverage

This table shows the pages on which the objectives of the Certified User Examination are covered in Chapter 11.

Certified User Reference Guide

Introduction

In the previous lessons, we have explored the basic CAD methods of creating orthographic views. By carefully selecting a limited number of views, the external features of most complicated designs can be fully described. However, we are frequently confronted with the necessity of showing the interiors of parts that cannot be shown clearly by means of hidden lines. We accomplish this by passing an imaginary cutting plane through the part and creating a cutaway view of the part. This type of view is known as a **section view**. In this lesson, we will demonstrate the procedure to construct section views using **AutoCAD 2023**.

In a section view, section lines, or cross-hatch lines, are added to indicate the surfaces that are cut by the imaginary cutting plane. The type of section line used to represent a surface varies according to the type of material. AutoCAD's **Hatch** command can be used to fill a pattern inside an area. We define a boundary that consists of an object or objects that completely enclose the area. **AutoCAD 2023** comes with a solid fill and more than 50 industry-standard hatch patterns that we can use to differentiate the components of objects or represent object materials.

The Bearing Design

Starting Up AutoCAD 2023

1. Select the **AutoCAD 2023** option on the *Program* menu or select the **AutoCAD 2023** icon on the *Desktop*.

2. In the AutoCAD *Startup* dialog box, select the **Template** option.

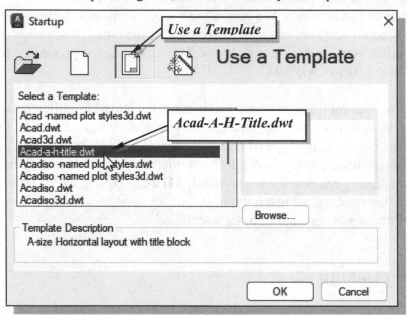

3. Select the *Acad-A-H-Title* template file from the list of template files. If the template file is not listed, click on the **Browse** button to locate and proceed to open a new drawing file.

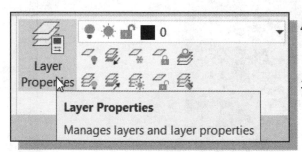

4. Pick *Layer Properties Manager* in the *Object Properties* toolbar.

5. Examine the layer property settings in the *Layer Properties Manager* dialog box. (On your own, update the *Plot Style* to **Style 1** in the template.)

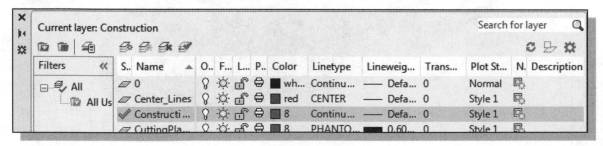

6. On your own, set layer *Construction* as the *Current Layer*, if necessary.

7. Click on the **Close** button to exit the *Layer Properties Manager* dialog box.

The Bearing Design

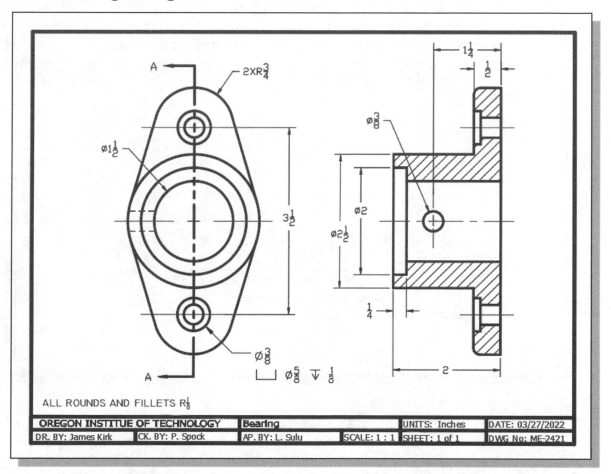

ALL ROUNDS AND FILLETS R⅛

OREGON INSTITUE OF TECHNOLOGY	Bearing		UNITS: Inches	DATE: 03/27/2022
DR. BY: James Kirk	CK. BY: P. Spock	AP. BY: L. Sulu	SCALE: 1 : 1	SHEET: 1 of 1	DWG No: ME-2421

Setting up the Principal Views

1. In the *Status Bar* area, reset the options and turn **ON** the **Dynamic Input**, **Polar Tracking**, **Object Snap**, **Object Tracking**, and **Line Weight Display** options.

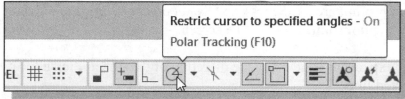

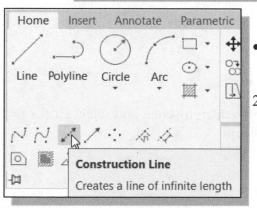

- We will first create construction lines for the front view.

2. Click the **Construction Line** icon in the *Draw* toolbar to activate the Construction Line command.

3. In the *command prompt area*, the message "*_xline Specify a point or [Hor/Ver/Ang/Bisect/Offset]:*" is displayed. On your own, create a vertical line and a horizontal line, near the left side of the screen, as shown in the figure below. These lines will be used as the references for the circular features of the design.

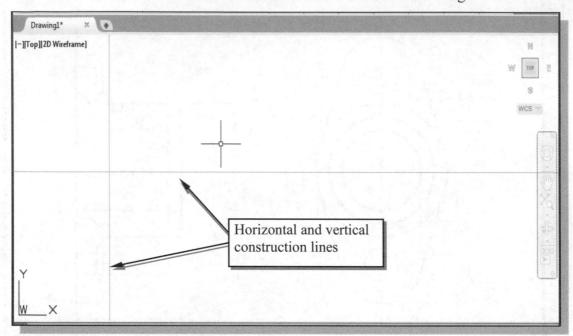

Horizontal and vertical construction lines

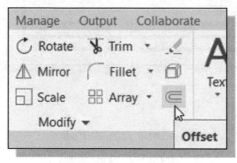

4. Click on the **Offset** icon in the *Modify* toolbar. In the command prompt area, the message "*Specify offset distance or [Through]:*" is displayed.

5. In the command prompt area, enter **1.75** **[ENTER]**.

6. In the command prompt area, the message "*Select object to offset or <exit>:*" is displayed. Pick the **horizontal line** on the screen.

7. AutoCAD next expects us to identify the direction of the offset. Pick a location that is **above** the selected line.

8. We will also create a line that is below the original horizontal line at 1.75. Pick the original **horizontal line** on the screen.

9. Pick a location that is **below** the selected line.

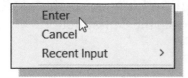

10. Inside the *Drawing Area*, **right-click** and select **Enter** to end the **Offset** command.

Creating Object Lines in the Front View

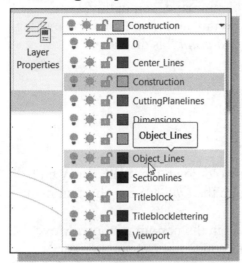

1. On the *Object Properties* toolbar, choose the ***Layer Control*** box with the left-mouse-button.

2. Move the cursor over the name of layer ***ObjectLines***; the tool tip "*ObjectLines*" appears.

3. **Left-click once** and layer *ObjectLines* is set as the *Current Layer*.

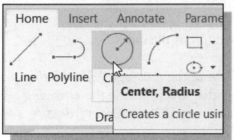

4. Click on the **Circle** icon in the *Draw* toolbar. In the command prompt area, the message "*Specify center point for circle or [3P/2P/Ttr (tan tan radius)]:*" is displayed.

5. Pick the center intersection point as the center of the circle.

6. In the command prompt area, the message "*Specify radius of circle or [Diameter]:*" is displayed. Enter ***1.25*** **[ENTER]**.

7. Repeat the **Circle** command and create the two **diameter *1.5*** circles as shown.

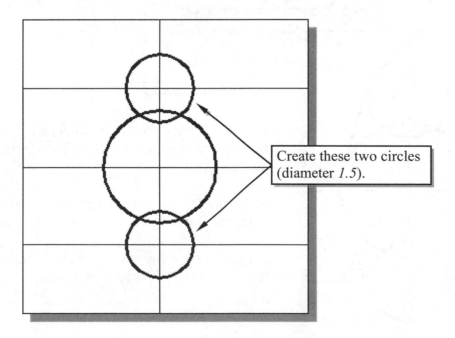

Create these two circles (diameter *1.5*).

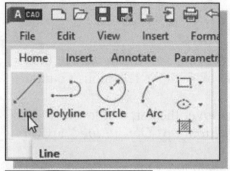

8. Select the **Line** command icon in the *Draw* toolbar. In the command prompt area, the message "*Line Specify first point:*" is displayed.

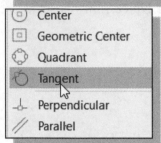

9. Inside the *Drawing Area*, hold down the **[SHIFT]** key and **right-click** once to bring up the *Object Snap* shortcut menu.

10. Select the **Tangent** option in the pop-up window. Move the cursor near the circles and notice the *Tangent* marker appears at different locations.

11. Pick the top circle by clicking on the upper right section of the circle.

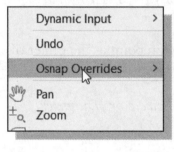

12. In the *command prompt area*, the message "*Specify the next point or [Undo]:*" is displayed. Inside the *Drawing Area*, **right-click** once and select **Snap Overrides**.

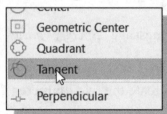

13. Select the **Tangent** option in the pop-up menu.

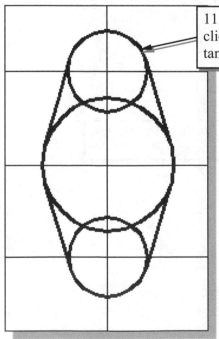

11. Pick the top circle by clicking near the expected tangency location.

14. Pick the center circle by clicking on the right side of the circle. A line tangent to the two circles appears on the screen.

15. Inside the *Drawing Area*, right-click and select **Enter** to end the Line command.

16. On your own, repeat the **Line** command and create the four tangent lines as shown.

Editing the Circles

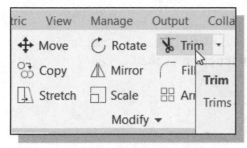

1. Select the **Trim** icon in the *Modify* toolbar. In the command prompt area, the message "*Select boundary edges... Select objects:*" is displayed.

2. Pick the four lines we just created as the *boundary edges*.

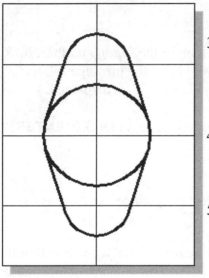

3. Inside the Drawing Area, **right-click** to proceed with the Trim command. The message "*Select object to trim or [Project/Edge/Undo]:*" is displayed in the command prompt area.

4. Trim the unwanted portions of the top and bottom circles and complete the outline of the front view as shown.

5. Inside the Drawing Area, right-click to activate the option menu and select **Enter** with the left-mouse-button to end the Trim command.

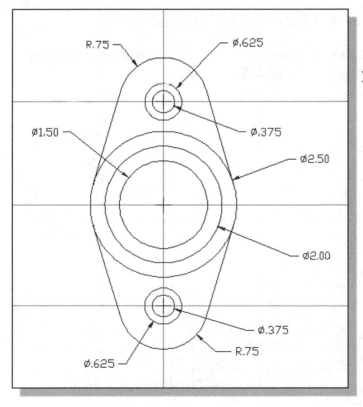

➢ On your own, create the additional circles as shown.

Setting up the Side View

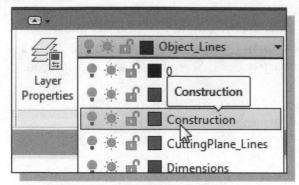

1. In the *Layer Control* box, set layer **Construction** as the *Current Layer*.

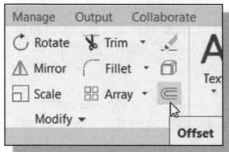

2. Select the **Offset** icon in the *Modify* toolbar. In the command prompt area, the message "*Specify offset distance or [Through]:*" is displayed.

3. In the command prompt area, enter **5.0 [ENTER]**.

4. In the *command prompt area*, the message "*Select object to offset or <exit>:*" is displayed. Pick the **vertical line** on the screen.

5. AutoCAD next expects us to identify the direction of the offset. Pick a location that is toward the right side of the selected line.

6. Inside the *Drawing Area*, right-click to end the **Offset** command.

7. On your own, repeat the **Offset** command and create two additional lines parallel to the line we just created as shown (distances of **0.5** and **2.0**).

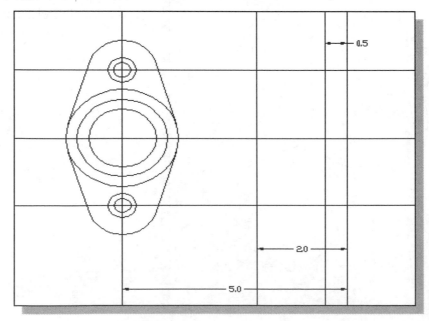

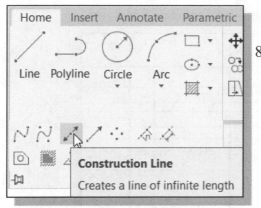

8. Select the **Construction Line** icon in the *Draw* toolbar. In the command prompt area, the message "*_xline Specify a point or [Hor/Ver/Ang/Bisect/Offset]:*" is displayed.

9. Select the **Hor** (horizontal) option in the command prompt area.

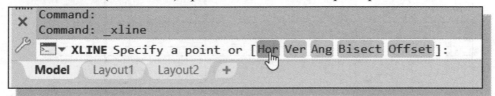

➢   The **Horizontal** option enables us to create a horizontal line by specifying one point in the Drawing Area.

10. Create **projection lines** by clicking at the intersections between the vertical line and the circles (and arcs) in the front-view.

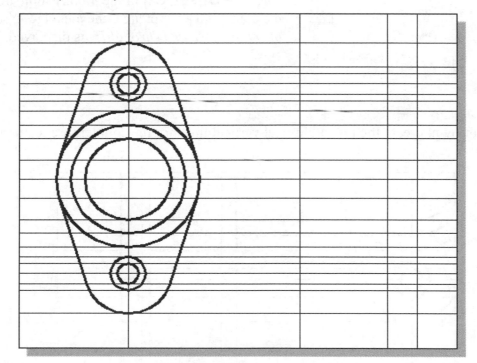

11. Also create two horizontal projection lines that pass through the two tangency-points on the ∅2.5 circle.

12. Inside the *Drawing Area*, right-click to end the **Construction Line** command.

> ➢ On your own, create object lines to show the outline of the side view.

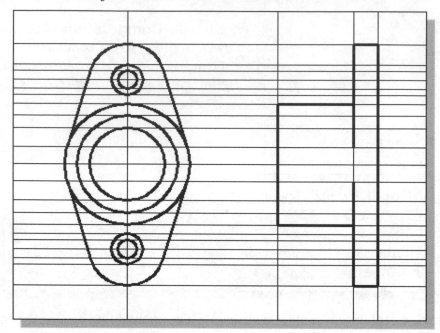

Adding Hidden Lines in the Side View

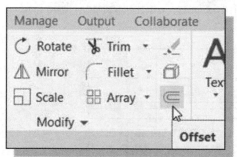

1. Click the **Offset** icon in the *Modify* toolbar. In the command prompt area, the message "*Specify offset distance or [Through]:*" is displayed.

2. On your own, create the two additional vertical lines for the counterbore features as shown.

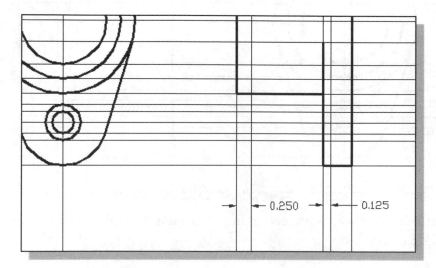

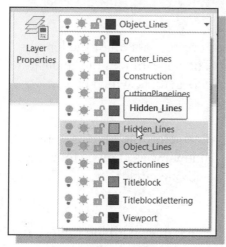

3. Set layer *HiddenLines* as the *Current Layer* in the *Layer Control* box.

4. Use the **Line** command and create the hidden lines as shown.

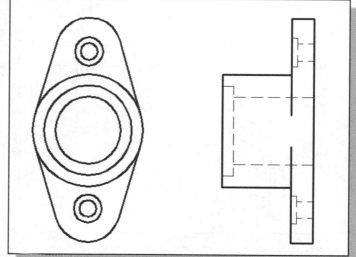

➤ On your own, complete the views by adding the side-drill, the centerlines, and the rounded corners as shown in the figure below.

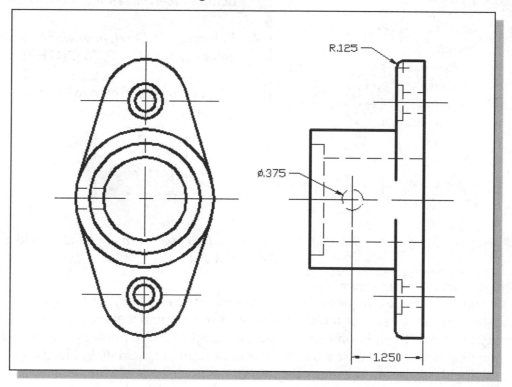

Changing the Linetype Scale Property

Looking at the side-drill feature and the centerlines we just created, not all of the lengths of the dash-dot linetypes appeared properly on the screen. The appearances of the dash-dot linetypes can be adjusted by modifying the ***Linetype Scale*** setting, which can be found under the *Object Property* option.

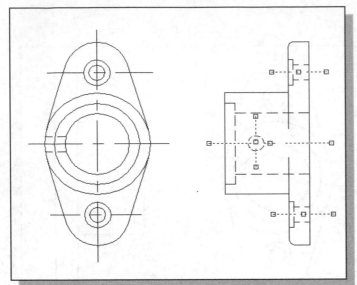

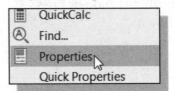

1. **Pre-select** the objects in the side view as shown (four lines and one circle).

2. Click once with the right-mouse-button to bring up the option list, and select the **Properties** option.

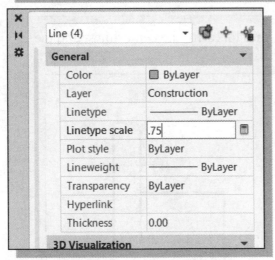

3. In the *Properties* dialog box, notice the default *Linetype scale* is 1.00.

4. Left-click on ***Linetype scale*** in the list and enter a new value: ***0.75*** [ENTER].

5. Click on the [X] button to exit the *Properties* dialog box.

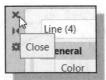

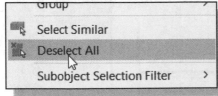

6. Inside the *Drawing Area*, **right-click** and choose **Deselect All**.

➤ The appearances of the dash-dot linetypes of the selected objects are adjusted to three-quarters the size of the other objects. Keep in mind that the dash-dot linetypes may appear differently on paper, depending on the type of printer/plotter being used. You may want to do more adjustments after examining a printed/plotted copy of the drawing. It is also more common to adjust the *Linetype Scale* for all objects of the same linetype to maintain a consistent presentation of the drawing.

Stretching and Moving Objects with Grips

We can usually *stretch* an object by moving selected grips to new locations. Some grips will not stretch the object but will move the object. This is true of grips on text objects, blocks, midpoints of lines, centers of circles, centers of ellipses, and point objects.

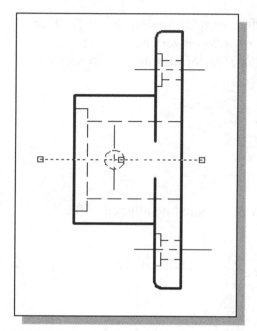

1. **Pre-select** the horizontal centerline that goes through the center of the part as shown.

2. Select the **right grip** by left-clicking once on the grip. Notice the grip is highlighted.

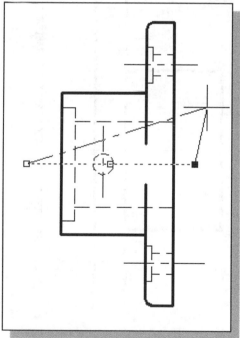

3. Move the cursor inside the Drawing Area and notice the center line is being stretched; the base point is attached to the cursor.

4. Pick a location on the screen to stretch the centerline.

5. Click on the **Undo** icon in the *Standard* toolbar area to undo the stretch we just did.

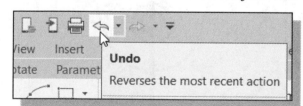

➢ On your own, experiment with moving the center grip of the centerline.

Drawing a Cutting Plane Line

Most section views require a *cutting plane line* to indicate the location on which the object is sectioned.

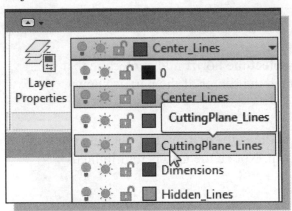

1. In the *Layer Control* box, turn **OFF** the *Construction* layer and set layer ***CuttingPlaneLines*** as the *Current Layer*.

2. Use the **Line** command and create the vertical cutting plane line aligned to the vertical centerline of the front view.

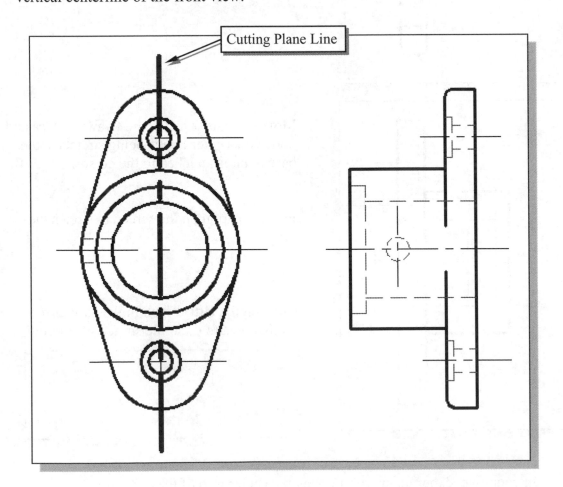

Cutting Plane Line

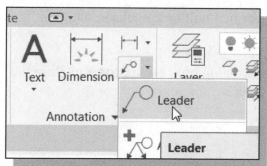

3. In the *Annotation toolbar*, select **Leader** as shown.

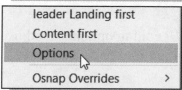

4. Inside the *Drawing Area*, **right-click** to bring up the *option menu* and select **Options** as shown.

5. Select **leader lAnding** and then **No** as shown.

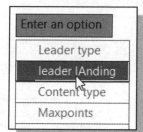 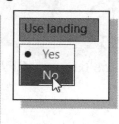

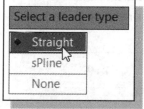

6. On your own, confirm the *leader type* is set to **Straight**.

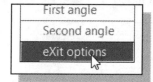

7. Select **eXit options** to end the setting options.

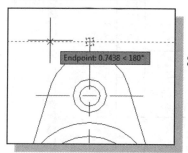

8. Use the object tracking option and set the endpoint of the leader about 0.75 to the left to set the starting point of the leader of the cutting plane line.

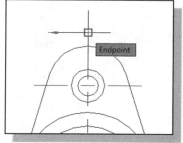

9. Pick the top endpoint of the cutting plane line to create a horizontal leader as shown.

➢ Note that AutoCAD expects us to type in the text that will be associated with the leader.

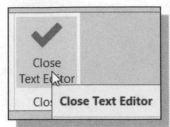

10. In the *Text Formatting* dialog box, click **Close** to close the dialog box without entering any text.

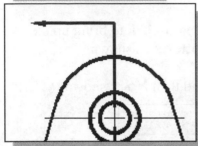

11. On your own, use the *GRIP* editing option and adjust the **arrow** as shown. (Hint: adjust the right arrow grip point.)

12. In the *Status* toolbar, turn *OFF* the *Line Weight Display* option to help the editing process.

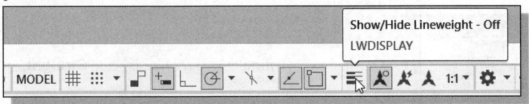

13. Repeat the **Leader** command and create the other arrow as shown.

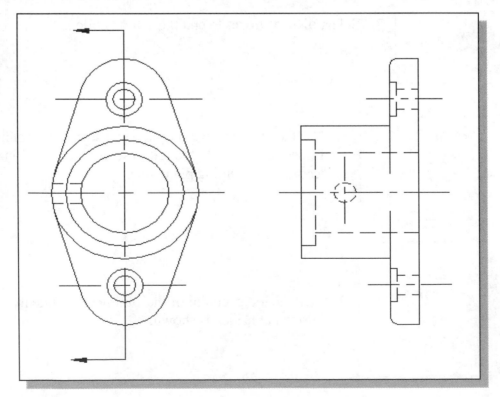

Converting the Side View into a Section View

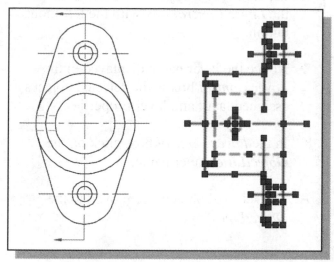

1. **Pre-select** all the objects in the side view by using a selection window.

2. Inside the *Drawing Area*, **right-click** to bring up the pop-up *option menu* and select the **Quick Select** option.

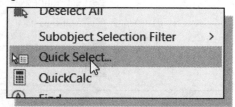

❖ The **Quick Select** option enables us to quickly select multiple objects by using various *filter* options.

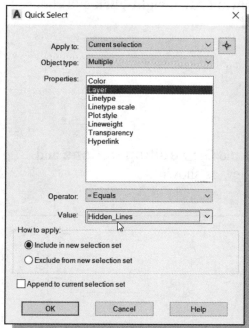

3. In the *Quick Select* dialog box, select **Layer** from the *Properties* list.

4. Set the *Value* box to **HiddenLines**.

5. In the *How to apply* section, select the **Include in new selection set** option.

6. Click on the **OK** button to accept the settings.

❖ AutoCAD will now **filter out** objects that are not on layer *HiddenLines*.

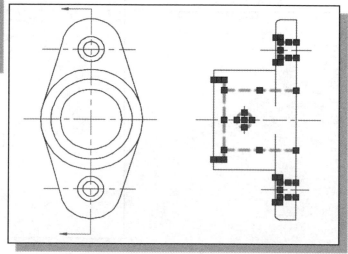

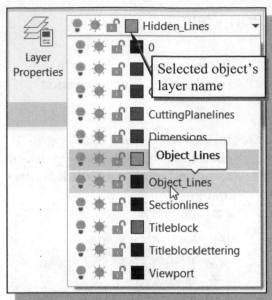

Layer Properties

Selected object's layer name

Object_Lines

7. On the *Object Properties* toolbar, choose the ***Layer Control*** box with the left-mouse-button.

❖ Notice the layer name displayed in the *Layer Control* box is the selected object's assigned layer and layer properties.

8. In the *Layer Control* box, click on the ***ObjectLines*** layer name.

➢ The selected objects are moved to the *ObjectLines* layer.

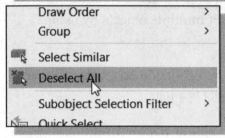

Draw Order >
Group >
Select Similar
Deselect All
Subobject Selection Filter >
Quick Select

9. Inside the Drawing Area, **right-click** and choose **Deselect All**.

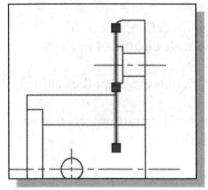

10. On your own, use the **Grip editing options** and modify the side view as shown.

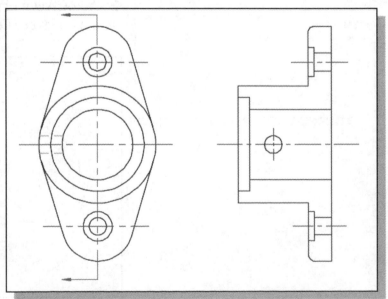

Adding Section Lines

1. In the *Layer Control* box, set layer **Section Lines** as the *Current Layer*.

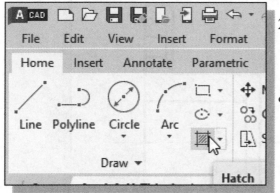

2. Select the **Hatch** icon in the *Draw* toolbar. The *Boundary Hatch* dialog box appears on the screen.

• We will use the *ANSI31* standard hatch pattern and create an associative hatch, which means the hatch is updated automatically if the boundaries are modified.

❖ We can define a boundary by **Selecting Objects** or **Picking Points**. The **Pick Points** option is usually the easier and faster way to define boundaries. We specify locations inside the region to be crosshatched and AutoCAD will automatically derive the boundary definition from the location of the specified point.

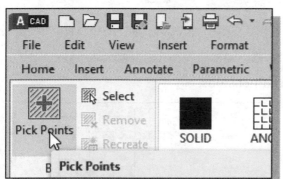

3. Click on the **Pick Points** icon.

4. In the command prompt area, the message *"Select internal point:"* is displayed.

5. **Left-click** inside the four regions as shown.

6. Inside the *Drawing Area*, **right-click** to bring up the pop-up menu and select **ENTER** to continue with the **Hatch** command.

7. Click on the **OK** button to close the *Boundary Hatch* dialog box.

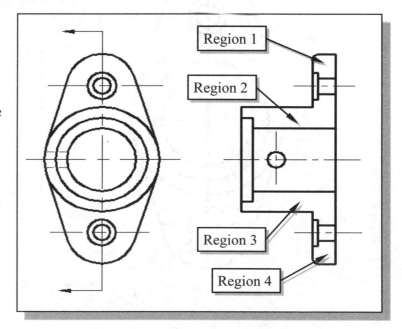

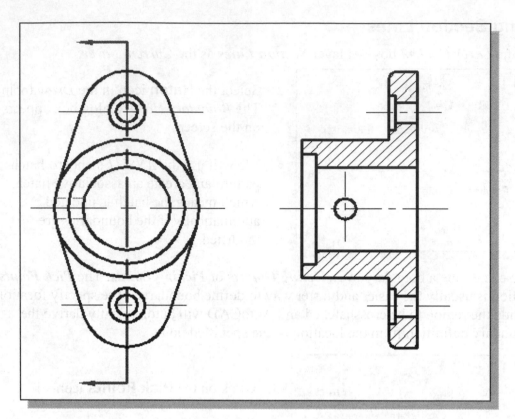

➤ Complete the drawing by adding the proper dimensions.

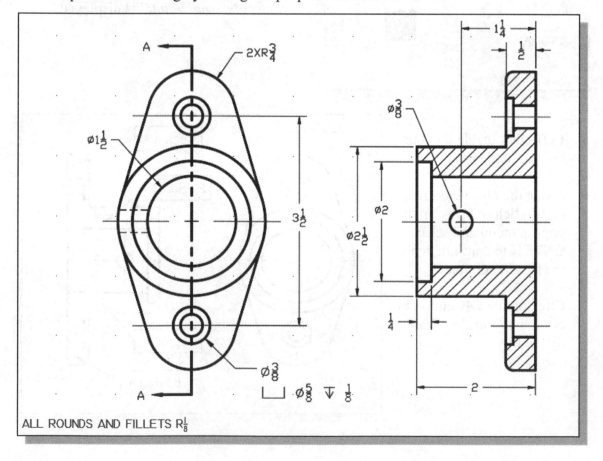

➢ On your own, create a drawing layout and print out the drawing.

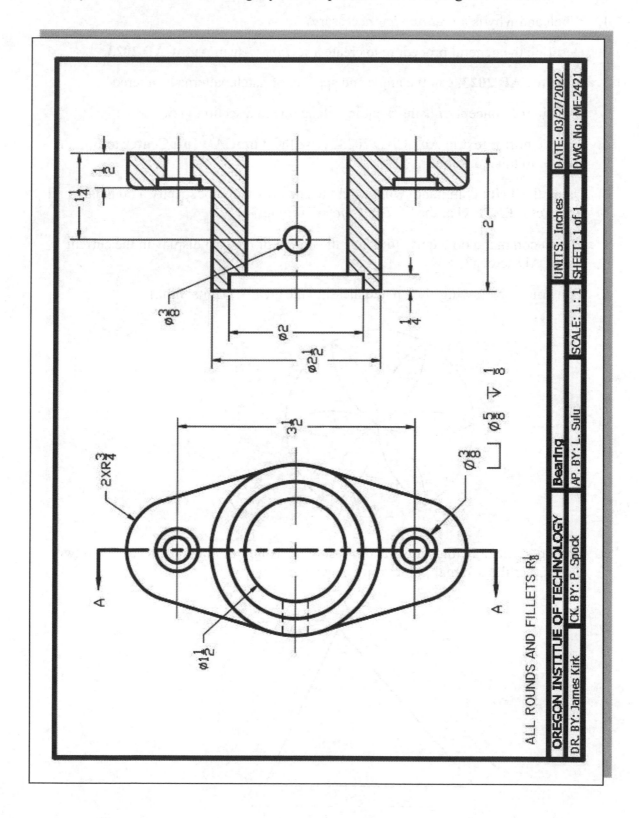

Review Questions: (Time: 30 minutes)

1. When and why is a *section view* necessary?

2. Describe the general procedure to create a *section view* in **AutoCAD 2023**.

3. In **AutoCAD 2023**, can the angle and spacing of hatch patterns be altered?

4. Explain the concept of using a cutting plane line in a section view?

5. Can we mirror text in AutoCAD 2023? (Use the **AutoCAD InfoCenter** to find the answer to this question.)

6. Using the **Trim** command, what would happen if we do not specify a boundary and just press [**ENTER**] at the "*Select Objects*" prompt?

7. Which command do we use to define the region of the grid display in the current AutoCAD screen?

8. Construct the drawing shown and measure the area A. (Units: mm.)

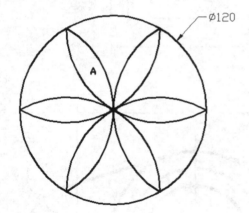

9. Construct the drawing shown and measure the length L. Show the length with two digits after the decimal point. (Units: inches.)

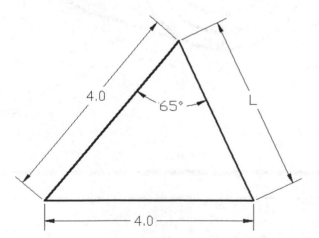

Exercises:

(All Dimensions are in inches.) (Time: 220 minutes)

1. Center Support

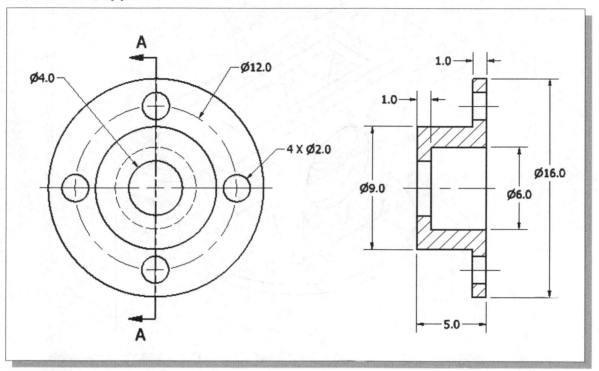

2. Ratchet Wheel

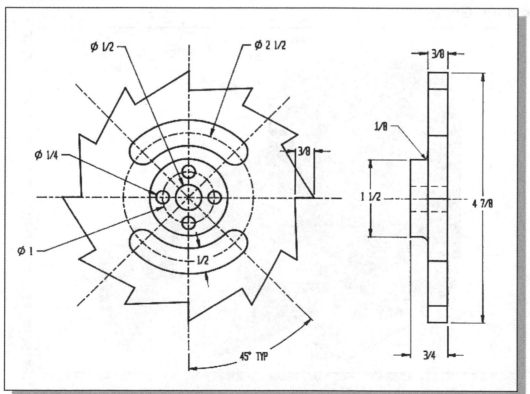

3. Mounting Bracket

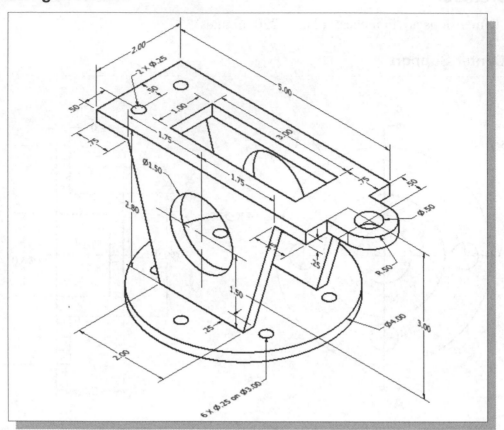

4. Position Guide

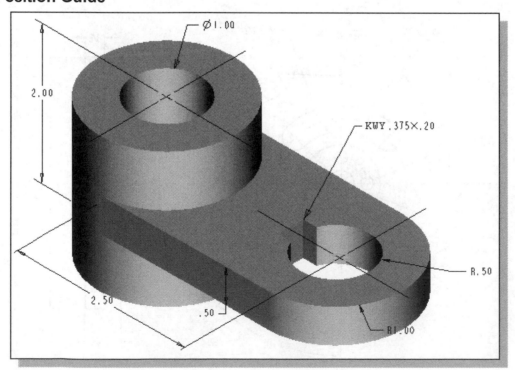

5. Yoke Support

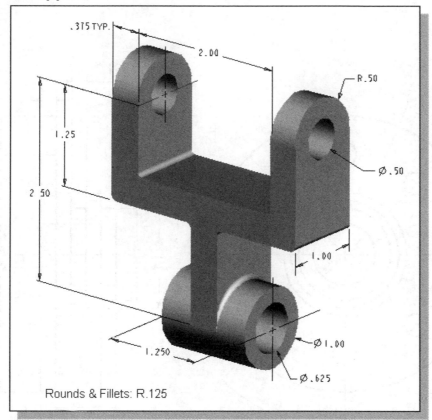

Rounds & Fillets: R.125

6. Lock Cap

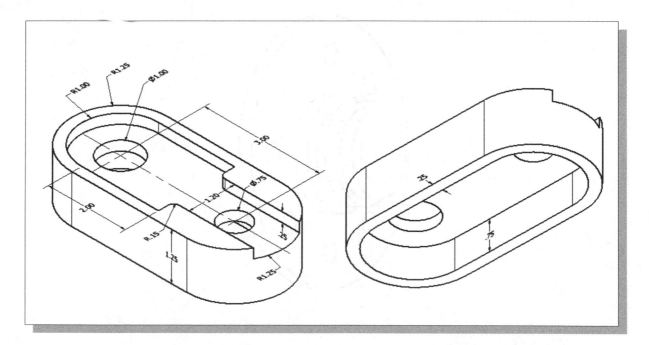

7. HUB (Create a multiview drawing with a section view of the design.)

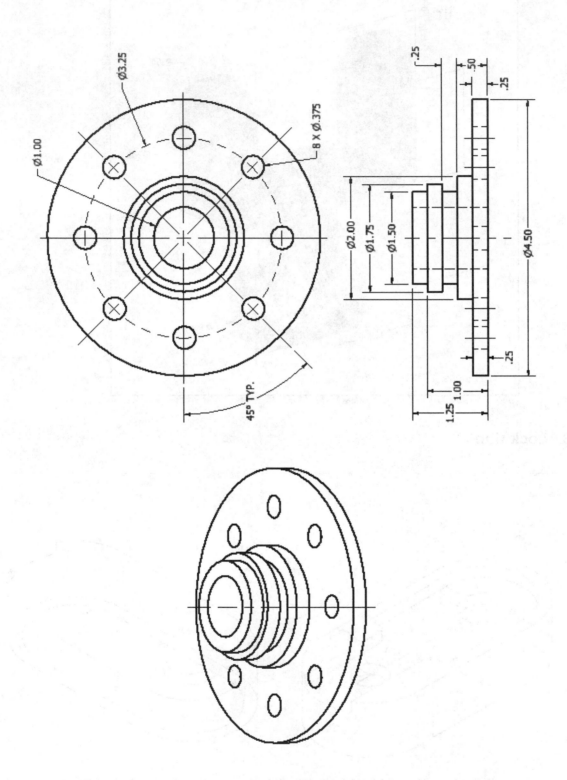

Chapter 12
Assembly Drawings and Blocks

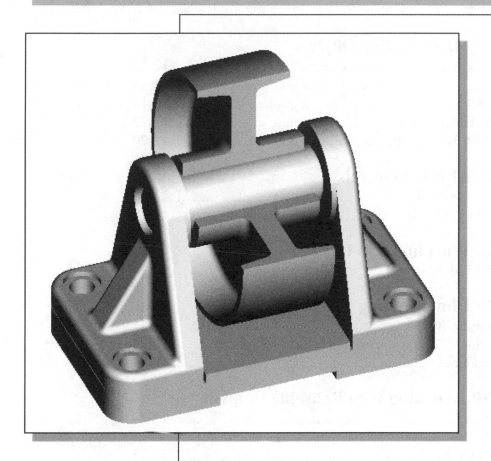

Learning Objectives

- ♦ **Create an Assembly Drawing from Part Files**
- ♦ **Using AutoCAD with the Internet**
- ♦ **Load Multiple Drawings into a Single AutoCAD Session**
- ♦ **Define a Block**
- ♦ **Create Multiple Copies using Blocks**
- ♦ **Copy and Paste with the Windows Clipboard**
- ♦ **Use the Move and Rotate Commands**

AutoCAD Certified User Examination Objectives Coverage

This table shows the pages on which the objectives of the Certified User Examination are covered in Chapter 12.

Certified User Reference Guide

Introduction

The term **assembly drawing** refers to the type of drawing in which the various parts of a design are shown in their relative positions in the finished product. Assembly drawings are used to represent the function of each part and the proper working relationships of the mating parts. Sectioning is used more extensively on assembly drawings than on detail drawings to show the relationship of various parts. Assembly drawings should not be overly detailed since precise information is provided on the detail drawings. In most cases dimensions are omitted on assembly drawings except for assembly dimensions such as important center distances, overall dimensions, and dimensions showing relationships between the parts. For the purpose of clarity, *subassembly drawings* are often made to give the information needed for the smaller units of a larger assembly. Several options are available in **AutoCAD 2023** to assist us in creating assembly drawings.

In **AutoCAD 2023**, a **block** is a collection of objects that is identified by a unique name and essentially behaves as if it is a single object. Using blocks can help us organize our design by associating the related objects into smaller units. We can insert, scale, and rotate multiple objects that belong to the same block with a single selection. We can insert the same block numerous times instead of re-creating the individual geometric objects each time. We can also import a block from a CAD file outside the current drawing. We can use blocks to build a standard library of frequently used symbols, components, or standard parts; the blocks can then be inserted into other drawings. Using blocks also helps us save disk space by storing all references to the same block as one block definition in the database. We can *explode* a block to separate its component objects, modify them, and redefine the block. **AutoCAD 2023** updates all instances of that block based on the *block definition*. Blocks can also be nested, so that one block is a part of another block. Using blocks greatly reduces repetitive work.

In **AutoCAD 2023**, we can load multiple drawings into a single AutoCAD session. This feature enables us to work with multiple drawings at the same time, and we can easily copy objects from one drawing to another by using the *Windows Clipboard*. The *Windows Clipboard* options, *copying-to* and *pasting-from,* can be used to quickly assemble objects in different files and thus increase our productivity.

AutoCAD 2023 also allows us to create a collaborative design environment where files and resources can be shared through the Internet. We can open and save AutoCAD drawings to an Internet location, insert blocks by dragging drawings from a web site, and insert hyperlinks in drawings so that others can access related documents. An Internet connection is required in order to utilize the **AutoCAD 2023** internet features.

In this lesson, we will demonstrate using the **AutoCAD 2023** Internet features to access drawings through the Internet, as well as using blocks and the *Windows Clipboard* to create a subassembly drawing. We will use the bearing part that was created in the previous lesson.

The Shaft Support Subassembly

Additional Parts

Besides the **Bearing** part, we will need three additional parts: (1) **Cap-Screw**, (2) **Collar** and (3) **Base-Plate**. Create the *Collar* and *Base-Plate* drawings as shown below; save the drawings as separate part files (*Collar*, *Base-Plate*). (Exit **AutoCAD 2023** after you create the files.)

(1) *Cap-Screw*
(We will open this drawing through the Internet.)

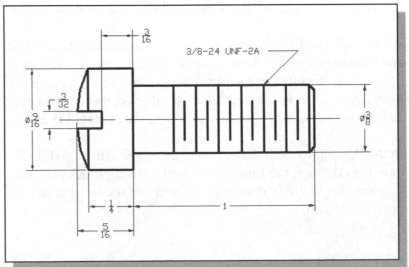

(2) *Collar*

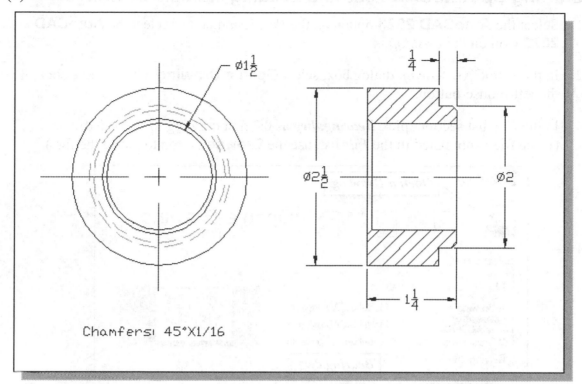

(3) *Base-Plate*

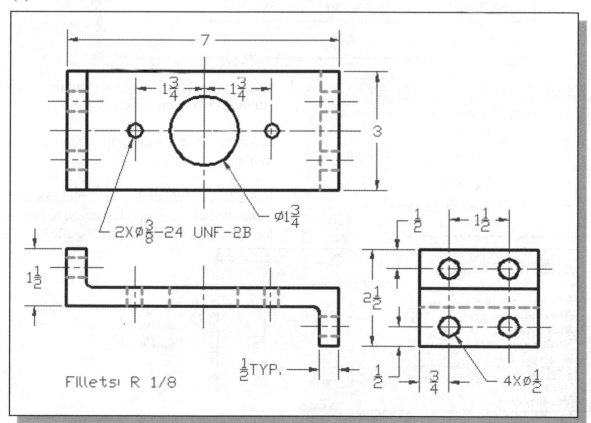

Starting Up AutoCAD 2023 and Loading Multiple Drawings

1. Select the **AutoCAD 2023** option on the *Program* menu or select the **AutoCAD 2023** icon on the *Desktop*.

2. In the AutoCAD *Startup* dialog box, select **Open a Drawing** with a single click of the left-mouse-button.

3. In the *File* list section, pick ***Bearing.dwg*** as the first drawing to be loaded.
    (If the file is not listed in the *File* list, use the **Browse** button to locate the file.)

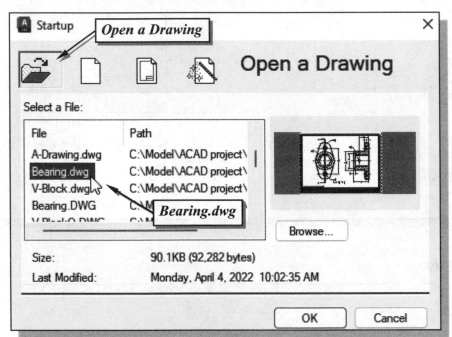

4. Click on the **OK** button to open the AutoCAD *Startup* dialog box.

5. Select the **Open** icon in the *Standard* toolbar.

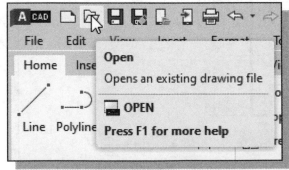

6. In the *Select File* dialog box, pick ***Base-Plate.dwg*** as the second drawing to be loaded. Click **Open** to load the file.

7. On your own, repeat the above steps and open the ***Collar.dwg*** file.

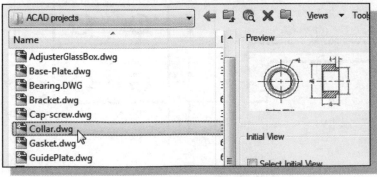

Using AutoCAD with the Internet

AutoCAD 2023 allows us to share files and resources through the Internet. Drawings can be placed and opened to an Internet location, blocks inserted by dragging drawings from a web site, and hyperlinks inserted in drawings so that others can access related documents. Note that to use the **AutoCAD 2023** Internet features an Internet connection is required.

We will illustrate the procedure to open an AutoCAD file from the Internet by *Uniform Resource Locator* (URL).

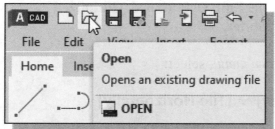

1. Click the **Open** icon in the *Standard* toolbar area as shown.

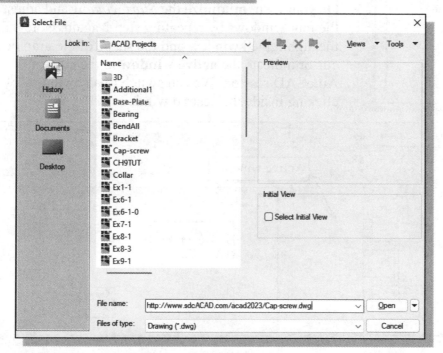

2. In the *Select File* dialog box, enter ***http://www.sdcACAD.com/acad2023/Cap-screw.dwg*** as shown.

3. Click the **Open** icon and the *Cap-screw* file is downloaded from the www.sdcACAD.com website to the local computer.

• The URL entered must be of the *Hypertext Transfer Protocol* (http://) and the complete filename must be entered including the filename extension (.dwg or .dwt).

• As an alternative, the *Cap-screw* file can also be downloaded using a web browser from the publisher's website: www.SDCpublications.com/downloads/978-1-63057-501-4

Rearrange the Displayed Windows

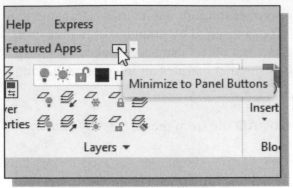

1. In the *Ribbon Toolbar* area, click on the **Minimize to Panels Buttons** once to adjust the size of the *Ribbon toolbar*.

➤ The **Minimize to Panels Buttons** can be used to adjust the size of Ribbon toolbar; clicking on the button will cycle through the **Full**, **Default** and **Minimized States.**

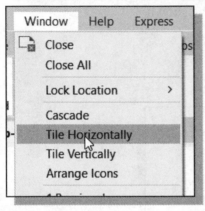

2. In the *pull-down menu*, select:

[Window] → [Tile Horizontally]

➤ On your own, **minimize** the *Start window* and adjust the four windows by repeating step 2 again. Note that the highlighted window and the shape of the graphics cursor indicate the **active window** in the current AutoCAD session. We can switch to any window by clicking inside the desired window.

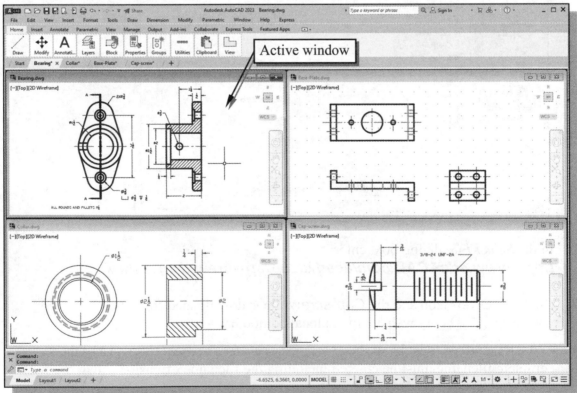

➤ On your own, adjust the display of each window by left-clicking inside each window and using the **Zoom Realtime** command.

Defining a Block

1. Set the *Cap-Screw* window as the *active window* by left-clicking inside the window.

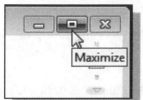

2. Click on the **Maximize** icon at the top right corner of the *Cap-Screw* window to enlarge the window.

3. In the *Layer Control* box, switch **off** the ***Dimension* layer** and leave only the *Object Lines* and *Center Lines* layers visible.

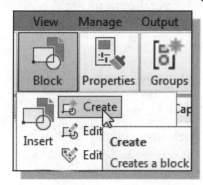

4. Pick the **Create Block** command icon in the *Block* toolbar. The *Block Definition* dialog box appears on the screen.

5. In the *Block Definition* dialog box, enter ***Cap Screw*** as the block *Name*.

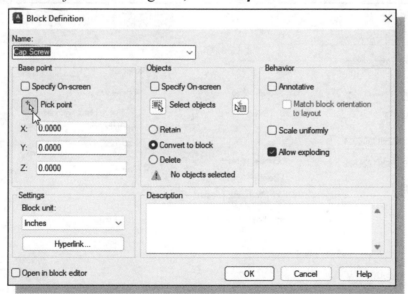

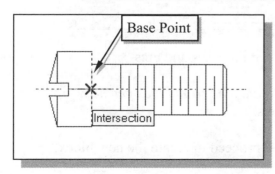

6. Click on the **Pick point** button to define a reference point of the block.

7. Pick the intersection of the centerline and the base of the *Cap-Screw* head as the base point.

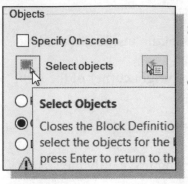

8. Click on the **Select Objects** icon to select the objects to be placed in the block.

9. Select all *object lines* and *centerlines* by using a **selection window** on the screen.

10. Inside the *Drawing Area*, **right-click** once to accept the selected objects.

- The selected objects will be included in the new block, and several options are available regarding the selected objects after the block is created. We can retain or delete the selected objects or convert them to a block instance.

    ➢ **Retain**: Keep the selected objects as regular objects in the drawing after creating the block.

    ➢ **Convert to Block**: Convert the selected objects to a block instance in the drawing after creating the block.

    ➢ **Delete**: Remove the selected objects from the drawing after creating the block.

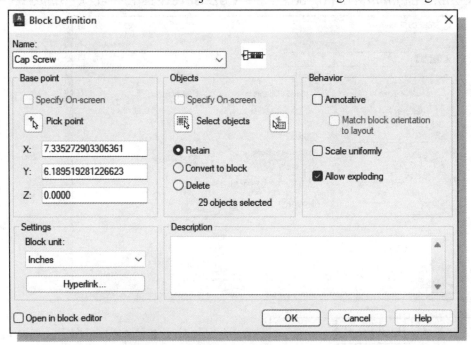

11. Pick the ***Retain*** option to keep the objects as regular lines and arcs.

    ➢ Notice in the *Preview icon* section, the *Create icon from block geometry* option displays a small icon of the selected objects.

12. Click the **OK** button to accept the settings and proceed to create the new block.

Insert and Rotate a Block

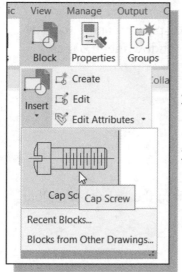

1. Pick the **Insert Block** command icon in the *Block* toolbar. The *Block Definition* dialog box appears on the screen.

2. In the pull-down list, select ***Cap screw*** as the block to be inserted.

3. Move the cursor toward the right side of the original copy of the cap screw. Left-click to place a copy of the block.

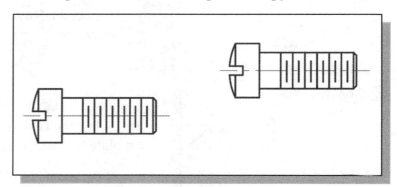

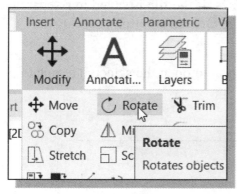

4. Pick the **Rotate** command icon in the *Modify* toolbar.

5. Pick the **cap-screw block** on the right.

6. **Right-mouse click** once to accept the selection.

7. Choose the same ***Base Point*** we used to define the block.

8. On your own, use the mouse and **Rotate** the selected block ***90*** degrees as shown.

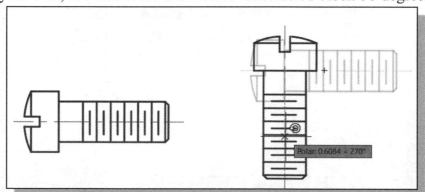

➢ In **AutoCAD 2023**, a **block** is a collection of objects that is identified by a unique name and essentially behaves as if it is a single object.

Starting the Assembly Drawing

1. Switch back to the tiled-windows display by using the **Tile Horizontally** option through the window menu bar as shown.

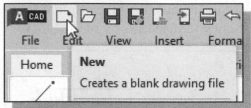

2. Select the **New** icon in the *Standard* toolbar area.

3. In the AutoCAD 2023 *Create New Drawing* dialog box, select the **Use a Template** option to open the template list as shown.

4. Select the ***Acad-A-H-Title*** template file from the list of template files. If the file is saved in a separate folder, click on the **Browse** button to locate the file.

5. Click the **OK** button to open the selected template file.

6. On your own, resize the new window as shown.

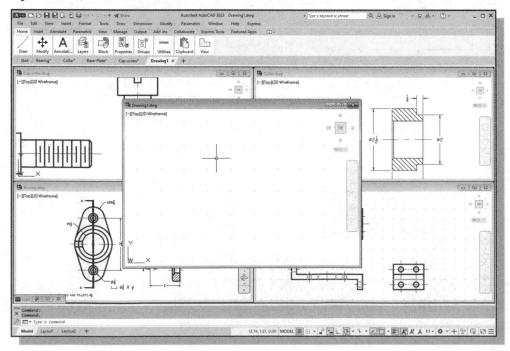

Copying and Pasting with the Windows Clipboard

1. Set the **Base-Plate** window as the *current window* by left-clicking inside the window.

2. In the *Layer Control* box, switch off all layers except the *ObjectLines, HiddenLines*, and *CenterLines* layers.

3. Select the **front view** of the *Base-Plate* by enclosing the front-view using a selection window.

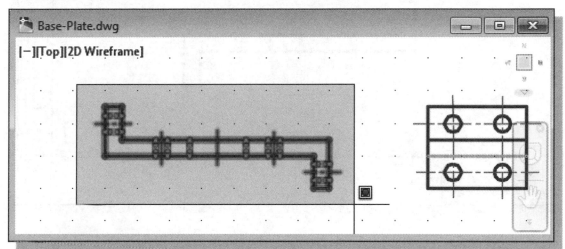

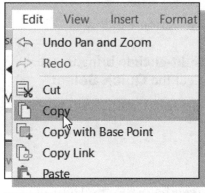

4. Use the **Copy to Clipboard** option in the *Edit* pull-down menu.

5. Set the **New Drawing** window as the *current window* by left-clicking inside the window.

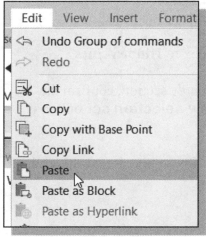

6. Select the **Paste** icon in the *Edit pull-down* menu as shown.

7. Position the front view of the *Base-Plate* near the center of the *Drawing Area* as shown on the next page.

❖   Note that we are using the *Windows Clipboard* options; the selected items are copied into the new drawing and all of the layer settings are retained.

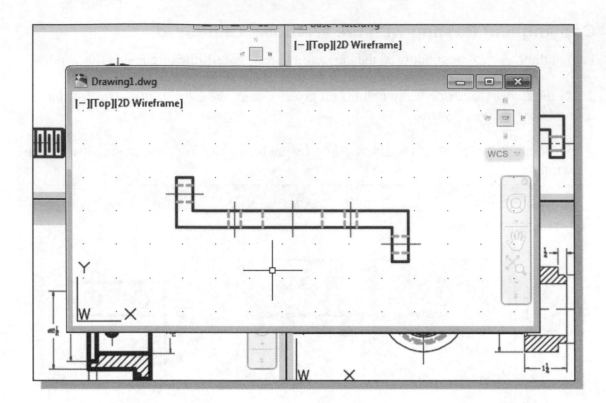

Converting the View into a Section View

1. **Pre-select** all the objects in the new drawing by using a selection window.

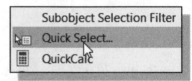

2. Inside the *Drawing Area*, **right-click** to bring up the pop-up option menu and select the **Quick Select** option.

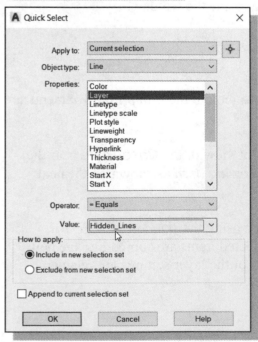

3. In the *Quick Select* dialog box, select **Layer** from the *Properties* list.

4. Set the *Value* box to **HiddenLines**.

5. In the *How to apply* section, confirm the **Include in new selection set** option is selected.

6. Click on the **OK** button to accept the settings.

7. AutoCAD will now **filter out** objects that are not on layer *HiddenLines*.

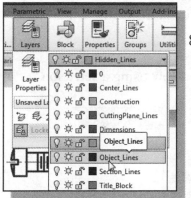

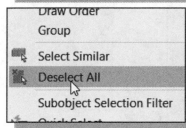

8. In the *Layer Control* tab, click on the **ObjectLines** layer name to move the selected objects to the *ObjectLines* layer.

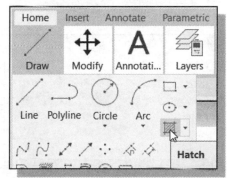

9. Inside the *Drawing Area*, **right-click** and select **Deselect All**.

10. In the *Layer Control* box, set layer **SectionLines** as the *Current Layer*.

11. Select the **Hatch** icon in the *Draw* toolbar. The *Boundary Hatch* dialog box appears on the screen.

12. On your own, create the hatch-pattern as shown.

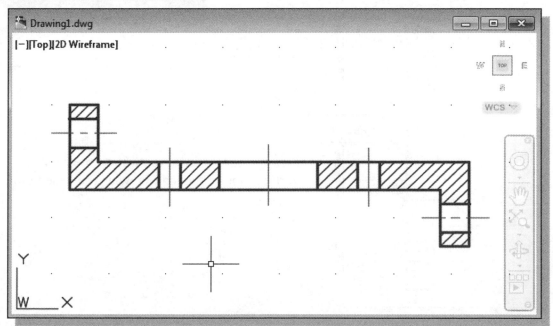

Adding the Bearing to the Assembly

1. Set the **Bearing** window as the *current window* by a left-click inside the window.

2. In the *Layer Control* box, switch **off** the *Dimension* layer.

3. Select the **side view** of the *Bearing* by enclosing the side view using a selection window.

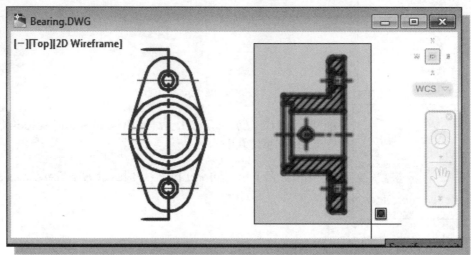

4. Inside the *Drawing Area*, **right-click** and select the **Copy with Base Point** option under *Clipboard* as shown.

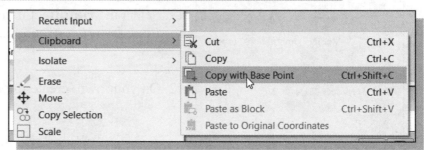

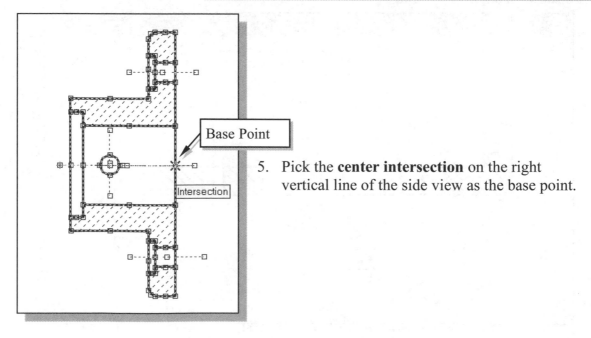

5. Pick the **center intersection** on the right vertical line of the side view as the base point.

6. Set the new **Drawing1** window as the *current window* by left-mouse-clicking on the associated icon of the window.

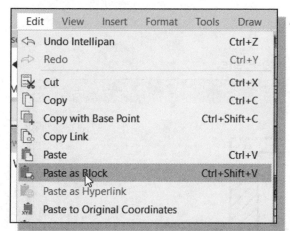

7. In the pull-down menu, select

[Edit] → [Paste as a Block]

8. Align the side view of the *Bearing* to the top center-intersection of the *Base-Plate* as shown.

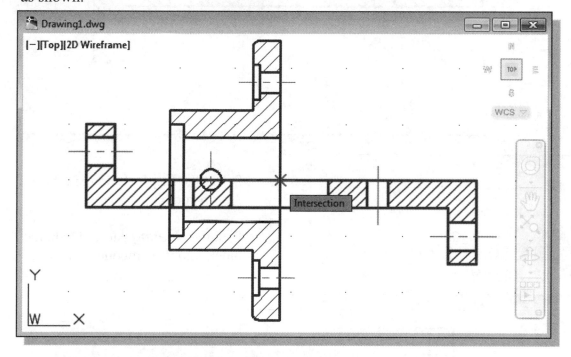

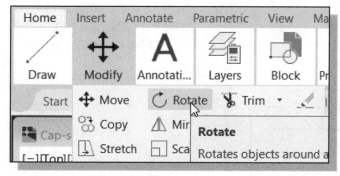

9. Click on the **Rotate** icon in the *Modify* toolbar.

10. Pick the pasted view of the *Bearing* we just placed into the assembly drawing. Notice the entire view is treated as a block object.

11. Inside the *Drawing Area*, **right-click** to accept the selection and proceed with the Rotate command.

12. Pick the base point as the rotation reference point.

13. Rotate the *Bearing* part to the top of the *Base-Plate*.

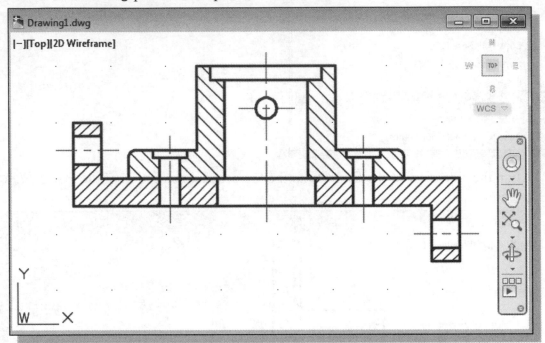

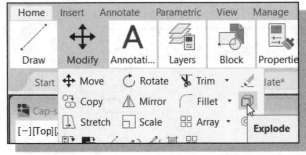

14. Select the **Explode** icon in the *Modify* toolbar.

15. Pick the *Bearing* part to break the block into its component objects.

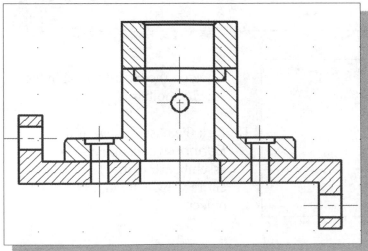

➢ On your own, copy and paste the collar on top of the bearing part as shown.

16. Use the **Explode**, **Trim**, **Delete** and **Properties** commands and modify the assembly as shown.

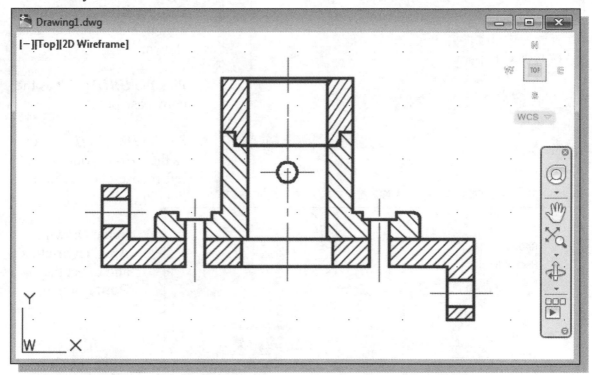

Adding the Cap-Screws to the Assembly

1. Set the *Cap-Screw* window as the *current window* by a left-mouse-click inside the window.

2. Pre-select the vertical *Cap-Screw*. Since all objects belong to a block we can quickly select the block.

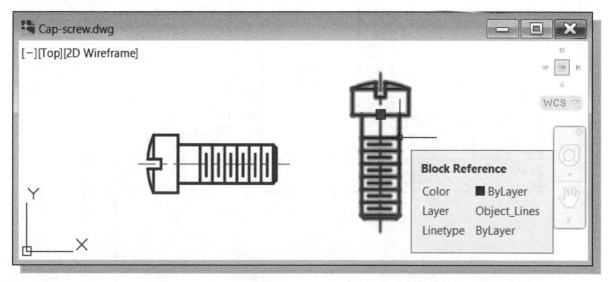

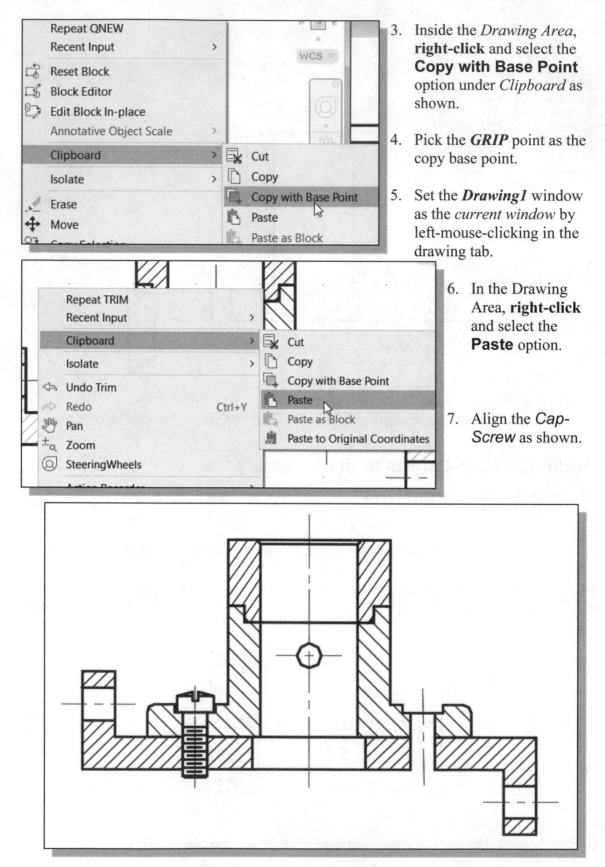

3. Inside the *Drawing Area*, **right-click** and select the **Copy with Base Point** option under *Clipboard* as shown.

4. Pick the ***GRIP*** point as the copy base point.

5. Set the ***Drawing1*** window as the *current window* by left-mouse-clicking in the drawing tab.

6. In the Drawing Area, **right-click** and select the **Paste** option.

7. Align the *Cap-Screw* as shown.

➢ On your own, repeat the **Paste** command and place the other *Cap-Screw* in place.

Creating Callouts with the Multileader Command

1. In the pull-down menus, select:
   [Format] → [Multileader Style]

2. Click the **New** icon to start a new multileader style.

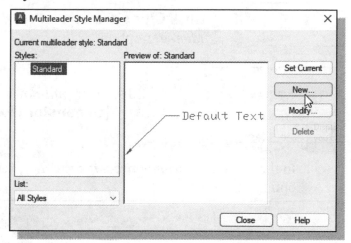

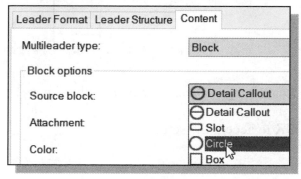

3. Enter **_Balloon_Callout_** as the _New style name_ as shown.

4. Click **Continue** to proceed with the new style setup.

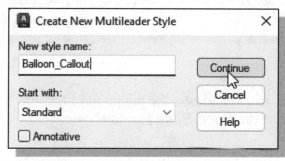

5. Click the **Content** tab and select **Block** in the _Multileader type_ option as shown.

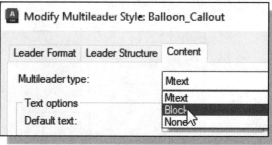

6. Select **Circle** in the _Source block_ option as shown.

7. Click **OK** to accept the settings.

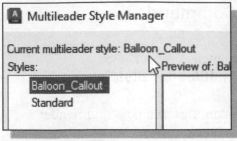

8. Confirm the **Balloon_Callout** style is the current active style.

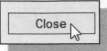

9. Click **Close** to accept the settings and exit the *Multileader Style Manager* option.

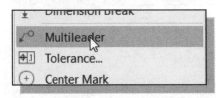

10. In the *pull-down menus*, select:
    [Dimension] → [Multileader]

11. In the *command prompt area*, *if necessary*, select the **leader arrowhead first** option as shown.

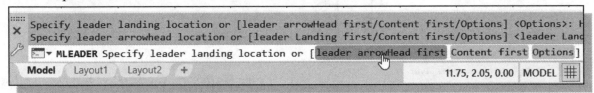

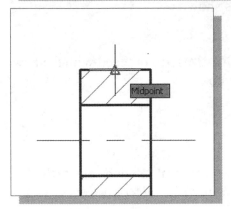

12. Place the arrowhead near the midpoint of the left top edge of the *Base-Plate* part as shown.

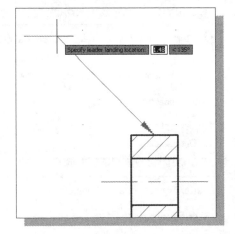

13. Select a location that is toward the left side of the **Base Plate** part as shown in the figure.

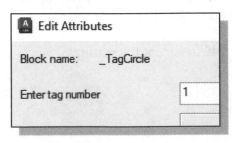

14. Enter **1** as the *tag number* as shown in the figure.

15. Click **OK** to accept the setting.

16. On your own, repeat the above process and create the four callouts as shown in the figure below.

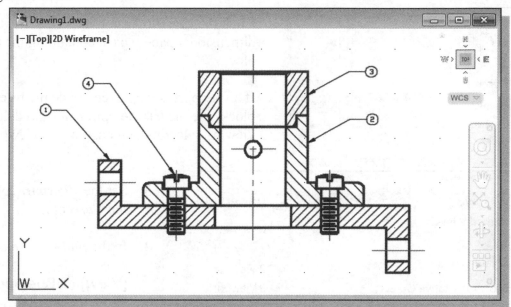

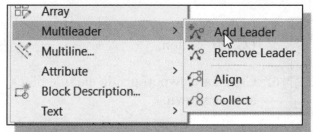

17. In the *pull-down menus*, select:

[Modify] → [Object] → [Multileader]→[Add Leader]

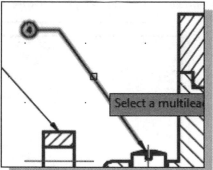

18. Select the leader attached to the *Cap-Screw* part as shown.

19. Place another leader pointing to the other *Cap-Screw* as shown.

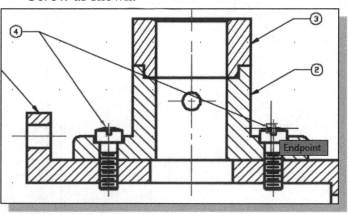

➤ For the *Shaft Support* design, the second leader is not necessary as only one type of *Cap-Screw* is used in the design.

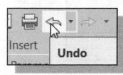

20. Click the **Undo** button to undo the last step.

Creating a Viewport in the A-size Layout

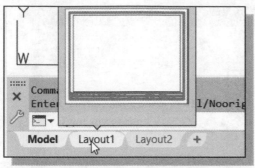

1. Click the **Layout1** tab to switch to the two-dimensional paper space containing the title block.

2. If a viewport is displayed inside the title block, use the **Erase** command and delete the view by selecting any edge of the viewport.

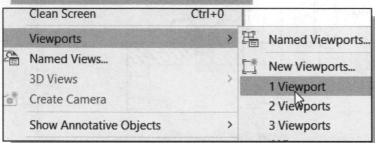

3. Set the *Viewport* layer as the *Current Layer*.

4. In the *pull-down menus*, select:
   [View] → [Viewports] → [1 Viewport]

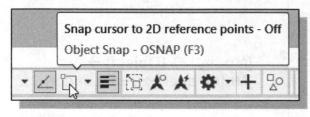

5. In the *Status Bar* area, turn *OFF* the *OSNAP* option.

6. Create a viewport inside the title block area as shown.

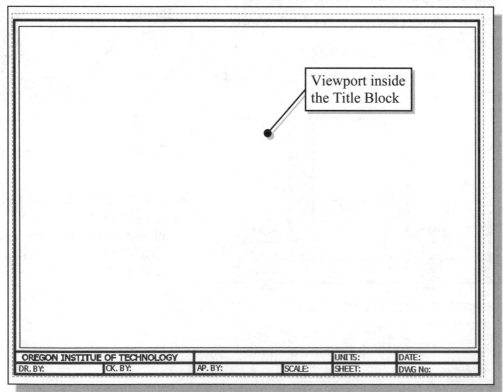

Viewport inside the Title Block

Viewport Properties

1. Pre-select the **viewport** by left-clicking once on any edge of the viewport.

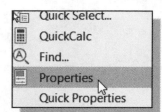

2. Inside the Drawing Area, **right-click** and select the **Properties** option.

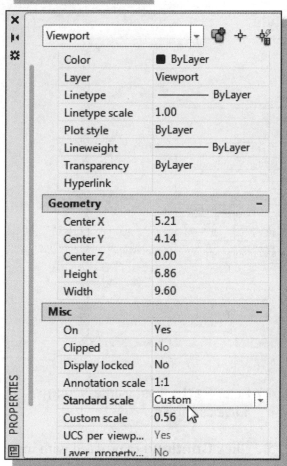

3. In the *Properties* dialog box, scroll down to the bottom of the list. Notice the current scale is set to **Custom, 0.5602.** (The number on your screen might be different.)

4. **Left-click** the *Standard scale* box and notice an arrowhead appears.

5. Click on the arrowhead button and a list of standard scales is displayed. Use the scroll bar to look at the list.

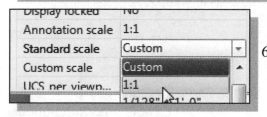

6. Select **1:1** in the *Standard scale* list. This will set the plotting scale factor to half scale.

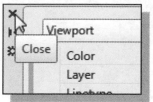

7. Click on the **[X]** button to exit the *Properties* dialog box.

8. On your own, turn **OFF** the display of the *Viewport* layer.

Adding a Parts List to the Assembly Drawing

1. Set the ***Titleblock_lettering*** layer as the *Current Layer*.

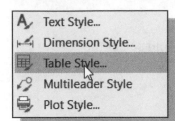

2. In the pull-down menus, select:

    [Format] → [Table Style]

3. Click the **New** icon to start a new **Table Style**.

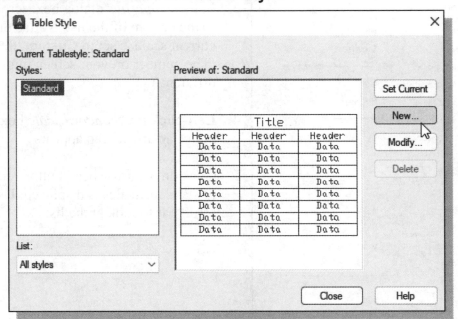

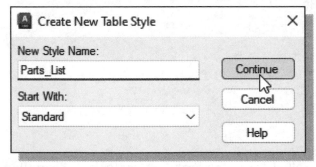

4. Enter **Parts_List** as the *New style name* as shown.

5. Click **Continue** to proceed with the new style setup.

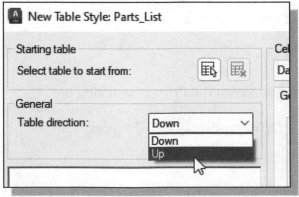

6. Set the *Table direction* to **Up** by selecting in the *Table direction* option list as shown.

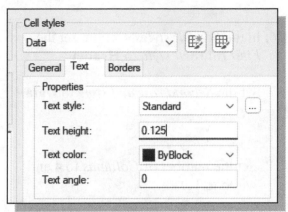

7. Select the **Text** tab in the *Cell Styles* section as shown.

8. Set the *Text height* to **0.125** as shown.

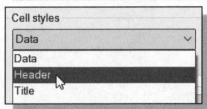

9. Select the **Header** tab in the *Cell Styles* section as shown.

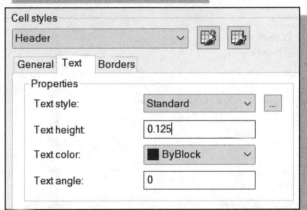

10. Select the **Text** tab in the *Cell Styles* section as shown.

11. Set the *Text height* to **0.125** as shown.

12. Click **OK** to accept the settings.

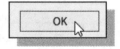

13. Click **Close** to accept the settings and exit the New Table Style option.

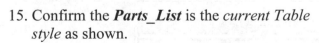

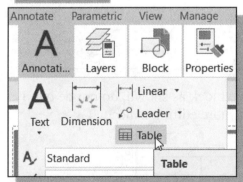

14. Click on the **Table** icon in the *Annotation* toolbar.

15. Confirm the ***Parts_List*** is the *current Table style* as shown.

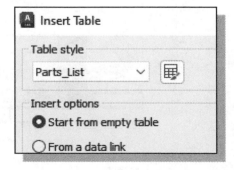

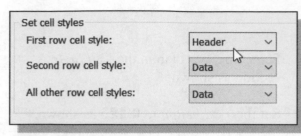

16. In the *Set cell styles* section, set the *First row cell style* to **Header**.

17. Set the *Second row cell style* to **Data**.

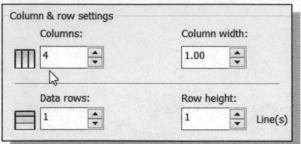

18. Set the number of *Columns* to **4** and width to **1.0**.

19. Click the **OK** button to accept the settings as shown.

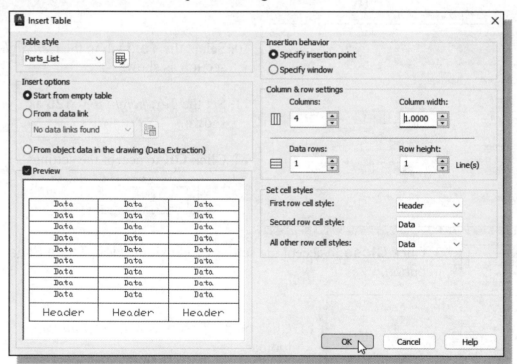

20. Place the table near the lower left corner of the title block as shown; do not be overly concerned about the actual position of the table as more editing is needed.

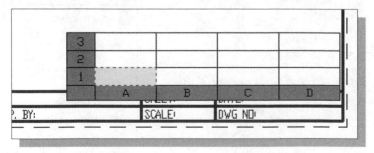

21. Enter "**No.**", "**DESCRIPTION**", "**REQ**" and "**MATL**" as the four headers of the parts list table as shown. (Click **Close** to end the text input option.)

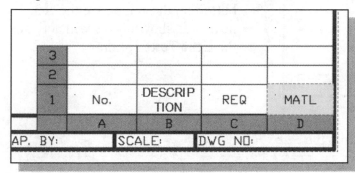

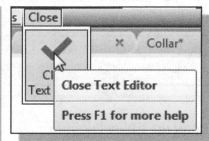

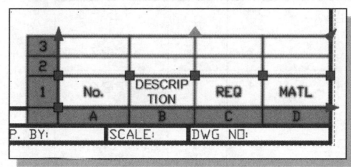

22. Select the *Parts_List* table by clicking on any one of the edges.

➤ Notice the different control GRIPS that are available to resize and reposition the table.

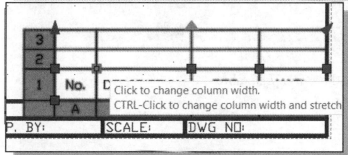

23. Drag the second square GRIP point toward the left to adjust the width of the second column.

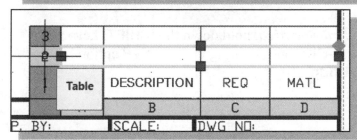

24. Click on the second row number to enter the **Edit Table** command.

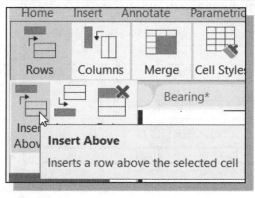

25. In the row tab, click on the first icon to show the different Row option. Click the first option to **Insert Row above**.

26. On your own, click the second option and **Insert Row below**.

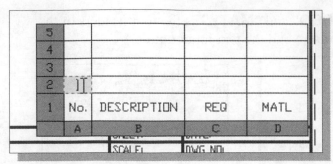

27. Double-click with the left-mouse-button on the cell above the **No.** header as shown. This will activate the **Edit Text** option.

28. Enter the following information into the *Parts_List*.

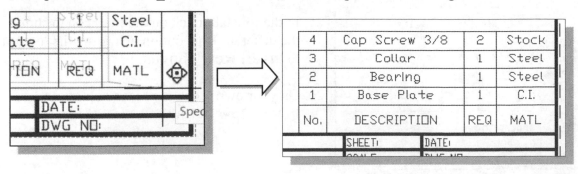

4	Cap Screw 3/8	2	Stock
3	Collar	1	Steel
2	Bearing	1	Steel
1	Base Plate	1	C.I.
No.	DESCRIPTION	REQ	MATL

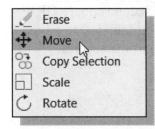

29. Pre-select the *Parts_List* table.

30. Inside the *Drawing Area*, **right-mouse-click** and select the **Move** option.

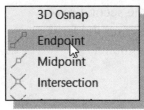

31. Inside the *Drawing Area*, hold down the **[SHIFT]** key and **right-mouse-click** once to bring up the *SNAP* option list. Select the **Endpoint** option.

32. Reposition the *Parts_List* table as shown and complete the drawing.

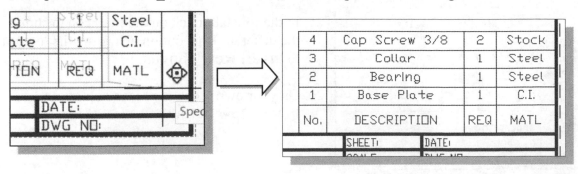

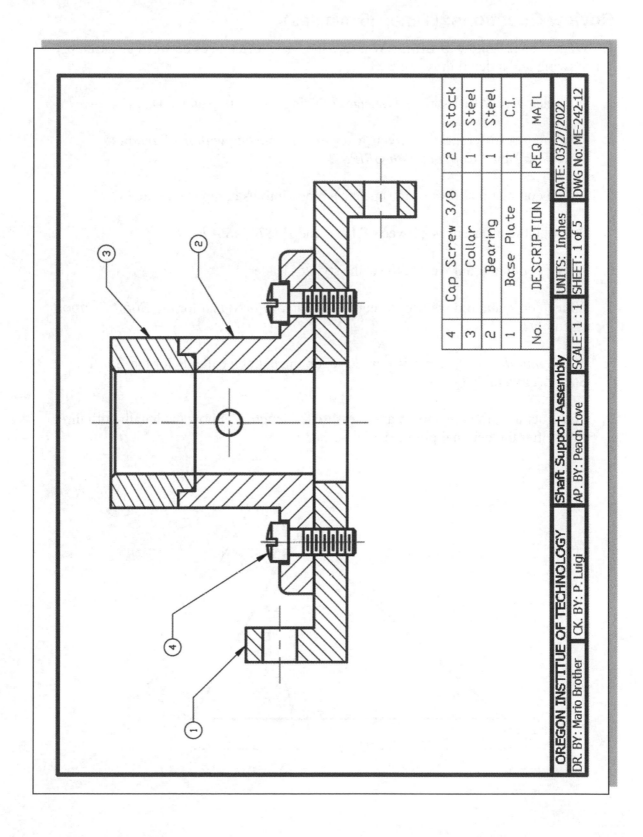

No.	DESCRIPTION	REQ	MATL
4	Cap Screw 3/8	2	Stock
3	Collar	1	Steel
2	Bearing	1	Steel
1	Base Plate	1	C.I.

UNITS: Inches   DATE: 03/27/2022

SCALE: 1 : 1   SHEET: 1 of 5   DWG No: ME-242-12

Shaft Support Assembly

OREGON INSTITUE OF TECHNOLOGY

DR. BY: Mario Brother   CK. BY: P. Luigi   AP. BY: Peach Love

Review Questions: (Time: 25 minutes)

1. What is an *assembly drawing*? What are the basic differences between an assembly drawing and a detail drawing?

2. What is a *block*? List some advantages of using blocks in AutoCAD.

3. *What are the differences between "Copying and pasting with the Windows Clipboard" and "Copying with GRIPS"?*

4. Which command allows us to separate a block into its component objects?

5. Describe the differences between *PASTE* and *PASTE AS A BLOCK*.

6. Which command did we use to set the style of leaders?

7. Which command did we use to set up the style of *Parts_List* for the **Shaft Support** assembly?

8. Which *multileader type* did we use to set up the balloon callouts for the **Shaft Support** assembly?

9. Construct the drawing shown and measure the length **L**. Show the length with three digits after the decimal point. (Units: inches.)

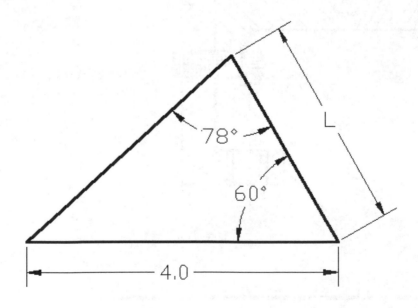

Exercises: (Time: 240 minutes)

1. Wheel Assembly (Create a set of detail and assembly drawings. All dimensions are in mm.)

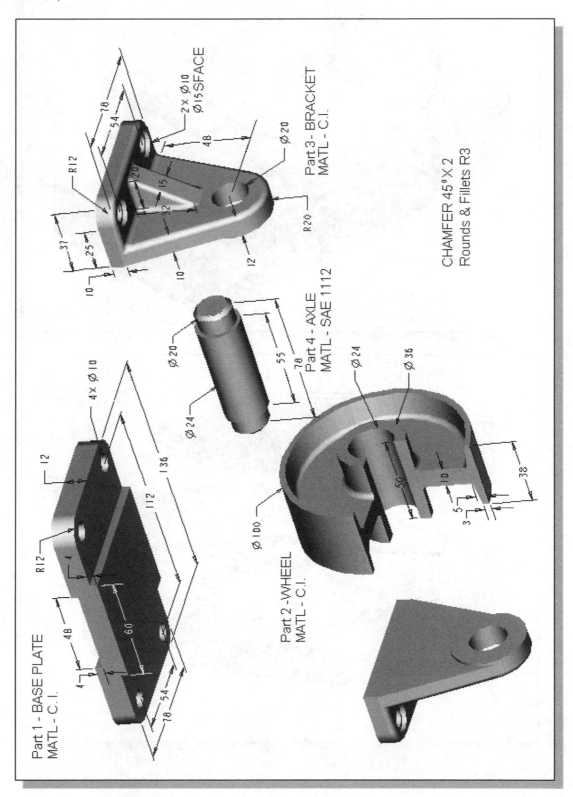

2. Leveling Assembly (Create a set of detail and assembly drawings. All dimensions are in mm.)

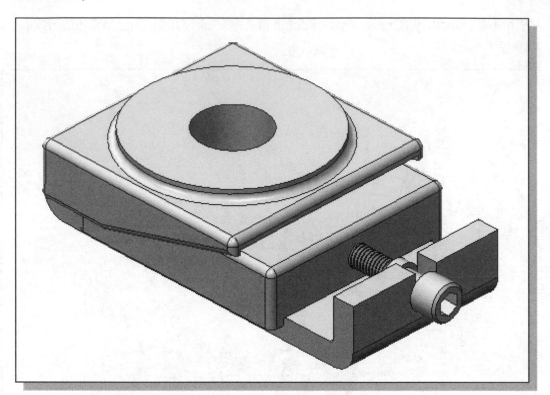

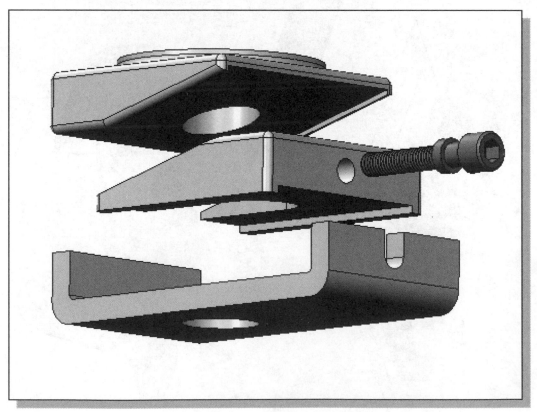

(a) **Base Plate**

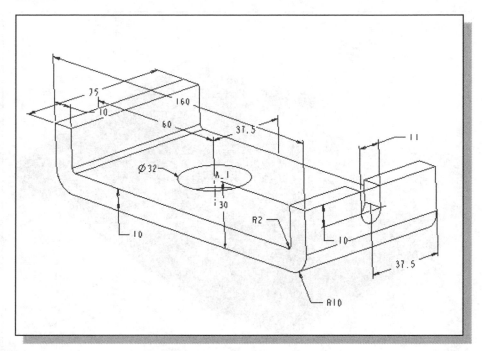

(b) **Sliding Block** (Rounds & Fillets: R3)

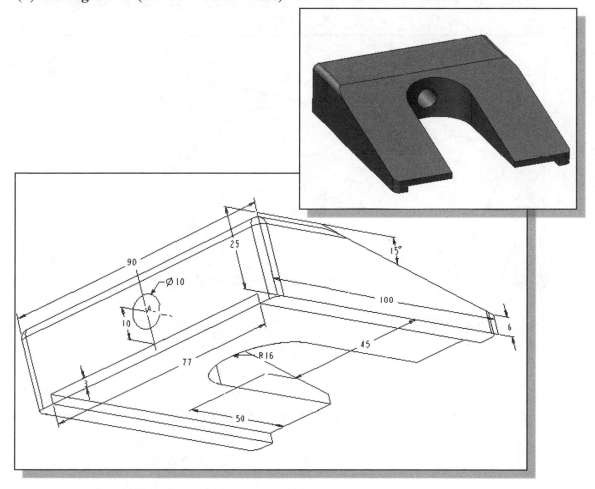

(c) **Lifting Block** (Rounds & Fillets: R3)

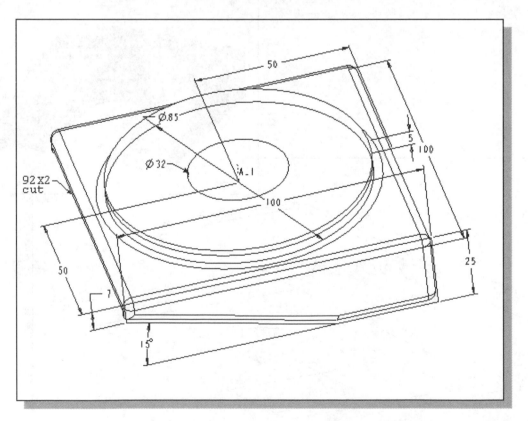

(d) **Adjusting Screw** (M10 X 1.5)

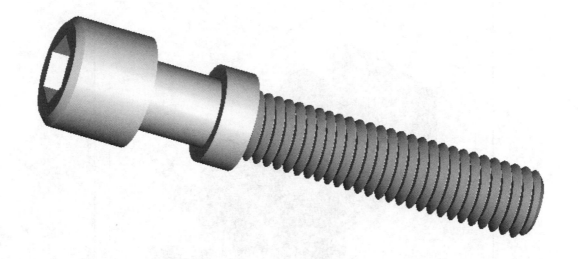

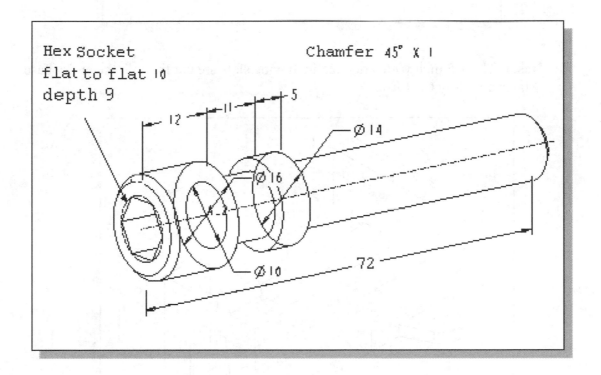

Hex Socket
flat to flat 10
depth 9

Chamfer 45° X 1

12   11   5

Ø 14

Ø 16

Ø 10

72

3. Vise Assembly (Create a set of detail and assembly drawings. All dimensions are in inches.)

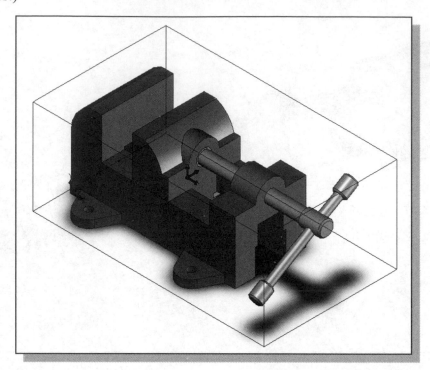

(a) **Base:** The 1.5 inch wide and 1.25 inch wide slots are cut through the entire base. Material: Gray Cast Iron.

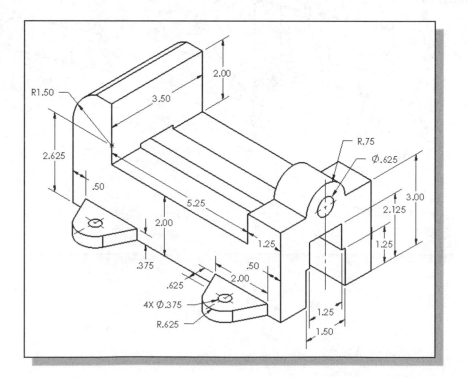

(b) **Jaw:** The shoulder of the jaw rests on the flat surface of the base and the jaw opening is set to 1.5 inches. Material: Gray Cast Iron.

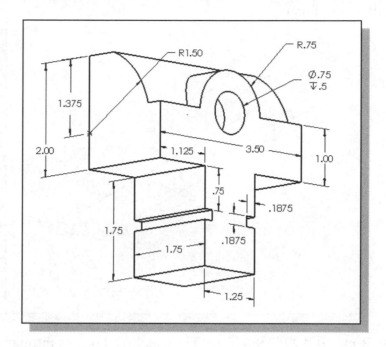

(c) **Key:** 0.1875 inch H x 0.3 inch W x 1.75 inch L. The keys fit into the slots on the jaw with the edge faces flush as shown in the sub-assembly to the right. Material: Alloy Steel.

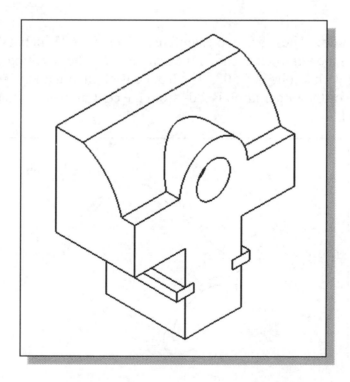

(d) **Screw:** There is one chamfered edge (0.0625 inch x 45°). The flat ∅ 0.75" edge of the screw is flush with the corresponding recessed ∅ 0.75 face on the jaw. Material: Alloy Steel.

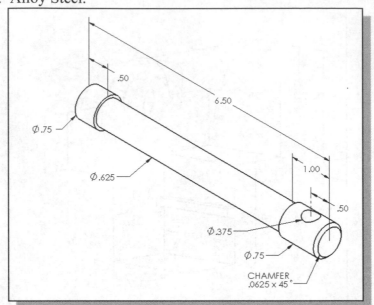

(e) **Handle Rod:** ∅ 0.375" x 5.0" L. The handle rod passes through the hole in the screw and is rotated to an angle of 30° with the horizontal as shown in the assembly view. The flat ∅ 0.375" edges of the handle rod are flush with the corresponding recessed ∅ 0.735 faces on the handle knobs. Material: Alloy Steel.

(f) **Handle Knob:** There are two chamfered edges (0.0625 inch x 45°). The handle knobs are attached to each end of the handle rod. The resulting overall length of the handle with knobs is 5.50". The handle is aligned with the screw so that the outer edge of the upper knob is 2.0" from the central axis of the screw. Material: Alloy Steel.

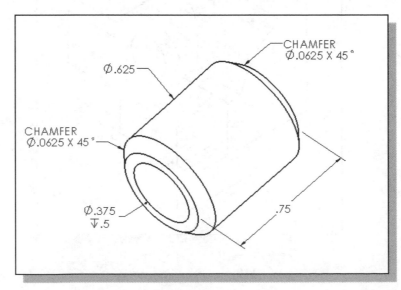

INDEX